Orphan

THE QUEST TO SAVE CHILDREN
WITH RARE GENETIC DISORDERS

OTHER TITLES BY THE AUTHOR

The Strongest Boy in the World: How Genetic Information Is Reshaping Our Lives (2006), Cold Spring Harbor Laboratory Press, Cold Spring Harbor, NY

Is It in Your Genes? The Influence of Genes on Common Disorders and Diseases That Affect You and Your Family (2004), Cold Spring Harbor Laboratory Press, Cold Spring Harbor, NY

Abraham Lincoln's DNA and Other Adventures in Genetics (2000), Cold Spring Harbor Laboratory Press, Cold Spring Harbor, NY

The Surgical Solution: A History of Involuntary Sterilization in the United States (1991), The Johns Hopkins University Press, Baltimore

To Do No Harm: A Journey Through Medical School (1987), Auburn House Publishing Company, Dover, MA

Genetics, Law, and Social Policy (1977), Harvard University Press, Cambridge, MA

Orphan

THE QUEST TO SAVE CHILDREN
WITH RARE GENETIC DISORDERS

PHILIP R. REILLY, MD, JD

CSH PRESS
COLD SPRING HARBOR LABORATORY PRESS
Cold Spring Harbor, New York • www.cshlpress.org

Orphan
THE QUEST TO SAVE CHILDREN
WITH RARE GENETIC DISORDERS

Publisher and Acquisition Editor	John Inglis
Director of Editorial Development	Jan Argentine
Project Manager	Inez Sialiano
Director of Publication Services	Linda Sussman
Production Editor	Kathleen Bubbeo
Production Manager	Denise Weiss
Cover Designer	Mike Albano

Library of Congress Cataloging-in-Publication Data

Reilly, Philip, 1947-
Orphan: the quest to save children with rare genetic disorders/Philip R. Reilly.
pages cm
Summary: "Orphan is about the struggle to save the lives of children who, because of an unlucky roll of the genetic dice, are born with any one of several thousand rare genetic disorders. Many are burdened with diseases that carry mysterious names, some of which you will read about for the first time in this book, along with compelling stories about the physicians, scientists, and parents who have taken them on. Orphan is more than a book about disease and research-it gives voice to thousands of people who, all too often, have endured terrible illnesses, bravely faced arduous clinical trials, and (sometimes) known victories, almost always in silence. It recounts extraordinary breakthroughs and hopes for the future. Many of the disorders that will end our lives are in some part genetically influenced. We really are all orphans, and this book is for all of us"-- Provided by publisher.
Includes bibliographical references and index.
ISBN 978-1-62182-137-3 (hardback)
1. Genetic disorders in children. 2. Genetic disorders in children--Research. 3. Medical genetics. I. Title.

RJ47.3.R45 2015
618.92'0042--dc23

2015009762

10 9 8 7 6 5 4 3

For a complete catalog of all Cold Spring Harbor Laboratory Press publications, visit our website at www.cshlpress.org.

To my wife, Nancy, and to my
children, and to all
parents who rise each day to help
children born with rare genetic disorders

Contents

Preface

THIS BOOK IS ABOUT THE STRUGGLE TO SAVE the lives of children who, because of a roll of the genetic dice, are born with any one of more than several thousand rare genetic disorders.

I have spent much of my life in medicine trying to help people who are afflicted with some of these disorders. In many cases, people are burdened with diseases carrying mysterious names that you will read for the first time in this book. Often physicians have no therapies to offer. Care is custodial, and the main goal is to relieve suffering. Although each genetic disorder is rare—often so rare as to be invisible—collectively these disorders touch hundreds of thousands of people . . . and of course their families.

Over the last three decades I have come to know many families burdened with such disorders. Over and over again, I have met mothers and fathers who I am certain are among the great, unsung *heroes* of our world. Each day they wake to face challenges that most of us cannot even imagine. You will meet a few of these amazing parents in this book.

I think, for example, of the parents of daughters with Rett syndrome. Apparently healthy for the first year or so of life, by the age of 6 or 7 many of these girls have lost all speech, cannot walk, have seizures (sometimes on a daily basis), and have abnormal breathing patterns that force some parents into a state of unrelenting fear that their child might die of respiratory arrest. When a child develops Rett syndrome, the family is changed profoundly and forever. So severe is the disorder and so needy is the child that parents often alter the very structure of their homes to provide daily care.

Children born with severe dystrophic epidermolysis bullosa (rDEB) do not have the proper form of a protein that holds the outer layer of the skin to its inner layer. The slightest trauma causes the skin to slough off. Sometimes, strangers mistakenly think that these children have been scalded with hot water. Parents often spend 2 hours each day bandaging the children in a desperate effort to control chronic skin lesions. During childhood, despite the best medical care, some of these children lose the use of their hands and feet as chronic injuries cause the digits to fuse. The more severely affected live with unremitting pain. Then, sometime in midlife, many succumb to an unusually aggressive cancer of the skin.

Here is one more example. Until recently, children born with one of the many severe lysosomal storage disorders (diseases in which different genetic errors prevent cells from clearing certain metabolites, causing a buildup of toxic material that impairs many organ systems) were doomed to a slow and painful death. Best known among these may be Tay–Sachs disease, a disorder that manifests at about age 2, stopping a healthy child in his or her tracks, and leading inexorably to blindness, neurological decline, and death by age 4 or 5. There are about 50 rare genetic disorders like Tay–Sachs disease. Fortunately (as I discuss in Chapter 6) there are now therapies for *some* of them (but, alas, not yet for children with Tay–Sachs disease).

This book recounts the century-long effort of small groups of physicians and scientists to take on some of these genetic diseases. As you will see, in many cases just a few physician–scientists—whom I also regard as *heroes*—have made an immense contribution to blazing a path toward new therapies.

Clinical genetics occupies a tiny niche in the vast medical ecosystem. So, it seems right that I briefly recount how I came to spend much of my professional life in it. Although the son of a surgeon, I did not become interested in science and medicine until after college. I first became aware of striking advances in genetics in about 1972, and despite my inadequate training, had the good fortune to find a mentor who took me under her wing as I redirected my career. In the fall of 1973, I started graduate work in human genetics in the laboratory of a remarkable physician–scientist named Margery Shaw at the University of Texas Graduate School of Biomedical Sciences in Houston.

After 2 years studying with Margery Shaw and her colleagues, I decided to become a clinical geneticist. After earning my medical degree at Yale in 1981, I undertook a residency in internal medicine at Boston City Hospital. My real training in clinical genetics began in 1985 when I joined the Eunice Kennedy Shriver Center for Mental Retardation as a member (and later director) of the primary care group that cared for about 800 adults with severe neurological disorders who resided at the Walter E. Fernald State School, a large state-run institution near Boston.

The physicians at Fernald knew the cause of the disability in only about one-quarter of the population for which we cared. But, even then, my medical colleagues and I suspected that many of our patients were burdened with as-yet-undiagnosed genetic disorders that had harmed brain development. During the course of my time there (1985–2000), new procedures (such as improved cytogenetic testing, DNA testing, and MRI scanning) enabled us to diagnose an increasing, but still relatively small, number of patients. For example, we found that several adults had Fragile X syndrome and that several others had severe intellectual disability due to deletions or duplications in one of their chromosomes. But diagnostic success never led to therapeutic benefit.

As the years passed I became increasingly frustrated with how little we could offer most patients. Eventually, I decided that I would like to help catalyze efforts to develop new drugs to help children who were burdened with rare genetic disorders. In medicine, we often call such diseases "orphan disorders." This is largely because (until recently) the progress in biomedical research since World War II had largely bypassed them. When hundreds of thousands of people are dying each year from heart disease and cancer, it is hard to convince the National Institutes of Health or the pharmaceutical industry to allocate many millions of dollars to better understand and develop new drugs for diseases that afflict only a few thousand (or even fewer) people. The term "orphan disease" became firmly set in our lexicon in the mid-1980s after the federal government enacted legislation that offered special tax benefits and market exclusivity to companies that developed drugs for rare disorders. In the United States, to gain the benefits of that law, the drug under development must afflict less than 200,000 persons in the nation. This includes many cancers and almost all single-gene disorders.

In the summer of 2008, I had the good fortune to start work at a venture capital firm in Boston known as Third Rock Ventures. From 2008 to 2015, TRV created, built, and launched about 35 innovative biotech companies. One of the areas in which it has done much work is in starting companies to develop therapies for rare genetic disorders. Of course, my comments about any of these companies do not represent the views of anyone else.

Although I miss the practice of medicine, I view my current work as a determined response to a world in which there is still little to offer children and adults affected with most of these terrible disorders. I often have the privilege of interacting with parents of affected children, some of whom have started foundations that are dedicated to taking on the immense challenge of developing new therapies. Today, such groups are quintessential partners in the struggle to develop new therapies for rare genetic disorders.

Despite the immense discord among peoples on this planet, we are united by our shared membership in the human gene pool. When I think about those who suffer and die from rare genetic disorders, I recall the great English poet, John Dunne, who wrote, "No man is an island, entire of itself; every man is a piece of the continent, a part of the main. If a clod be washed away by the sea, Europe is the less, as well as if a promontory were, as well as if a manor of thy friend's or of thine own were: any man's death diminishes me, because I am involved in mankind, and therefore never send to know for whom the bell tolls; it tolls for thee."

Acknowledgments

DURING THE COURSE OF WRITING THIS BOOK, I met many adults with genetic disorders and interacted with many families whose lives are devoted to helping children born with rare genetic disorders. I thank each of them for allowing me into their homes and sharing the intimate, painful, but always inspiring, stories of their lives. I cannot name all of them, but I will mention a few who helped me achieve a deeper understanding of some of the genetic disorders about which you will read: David and Lynn Paolella (phenylketonuria), Helen Sarpong (sickle cell anemia), Tim Hibbard and his parents (dystrophic epidermolysis bullosa), Mary Kaye Richter and her son (X-linked hypohidrotic ectodermal dysplasia), Andrew and Amanda Stricos (Huntington disease), Nick Johnson and his wife Sue (Friedreich's ataxia), Monica Coenraads (Rett syndrome), and Matt Wilsey (whose daughter has an exceedingly rare condition called NGLY1 deficiency).

I cannot possibly thank all the physicians and scientists (some now deceased) who in some way—often unknown to them—contributed directly or indirectly to this book. Margery Shaw was my first and best mentor in clinical genetics. I first came to know Art Beaudet when I was a medical student; over the decades I have learned much from him about genetic medicine. I had the great pleasure to know (and occasionally work with) Victor McKusick for many years. For three decades, Harvey Levy has taught me about inborn errors of metabolism. Mike Kaback taught me much about population-based genetic testing programs. During my years at the Shriver Center, Ed Kolodny taught me about neurogenetic disorders, as did Raymond Adams. I thank Marty Steinberg for adding to my knowledge of sickle cell anemia. Dr. Arnulf Koeppen helped me to understand the pathophysiology of Friedrich's ataxia.

I owe a special thanks to Mark Levin, Bob Tepper, and Kevin Starr, the founders of Third Rock Ventures. Among the many innovative biotech companies that TRV has launched there are several that were created to develop new therapies for rare genetic disorders. It has been a privilege to be a part of that process. Over the years at TRV, I have worked closely with several brilliant young people in the early stages of helping to develop new companies to treat rare genetic diseases. I think especially of Abe Bassan, Neil Kumar, James McLaughlin, and Walter Kowtoniuk.

I owe thanks to Mark de Souza and Jim Fordyce for involving me early in their idea to develop a novel treatment for dystrophic epidermolysis bullosa. Thanks, also, to Neil Kirby and Kathleen Kirby for adding to my understanding of drug development. I thank Dr. Ken Huttner for teaching me about an orphan disorder called X-linked hypohidrotic ectodermal dysplasia (XLHED). Henri Termeer (the former CEO of Genzyme) has inspired me on many occasions.

One of the most rewarding periods of my life was the year and one-half I spent working with Nick Leschly and Mitch Finer to help build a gene therapy company called bluebird bio. Nick, who has served as CEO of the company since its inception, and Mitch, a brilliant gene scientist who until recently served as the chief scientific officer, have led the effort to build a company that is helping to make gene therapy a reality.

My wife, Nancy, spent many hours listening to me talk about this project. She faithfully read evolving drafts and identified many passages in which I failed to communicate technical information as clearly as I should. She also unfailingly encouraged me when doubts about the project slithered through my mind.

Introduction

I SPENT SEVERAL YEARS AS A PRACTICING PHYSICIAN overseeing the care of hundreds of mentally retarded (that is how we described them in the 1980s and 1990s) adult patients who resided at the Walter E. Fernald State School, a large, century-old, state-operated institution in Waltham, Massachusetts. For the most part, these folks were cut off from the world we inhabit. Most of them had almost no use of language, and many had severe behavioral problems. Virtually all of them required round-the-clock custodial care. One building at Fernald was home to patients who were profoundly mentally retarded and deaf and blind. They were among the last group of individuals born (in the early 1960s) with severe congenital rubella, just before rubella vaccine became a routine part of health care. The rubella virus had devastated them while they grew in their mothers' wombs. Scattered among the different housing units were about 60 older patients with Down syndrome, people with untreated phenylketonuria, men afflicted with Fragile X syndrome, and persons with rare genetic disorders with unfamiliar names such as Rubenstein–Taybi syndrome and Cornelia de Lange syndrome. But the largest diagnostic group by far (~75%) included those for whom the cause of their disability was *unknown*.

Despite the best efforts of the staff, the living conditions at the school were grim. Although it happened 25 years ago, I still recall the sense of despair I felt one hot summer afternoon when I elected to check on a patient in one of the buildings. In a large room filled only with plastic furniture that reeked of urine, I found the woman alone, rocking herself, and softly uttering sounds that carried no message for me. When I approached, she turned away. Although I was the medical director of the program, it was beyond my power to make her life better in any way. I remember reflecting over the following weeks about therapeutic futility. It was that summer that I began a journey that, among other things, led me to write this book.

Rare genetic disorders—a group of disorders that we often call "orphans"—are, collectively, quite common. About four million babies are born each year in the United States. About 120,000 (3%) will be diagnosed with a disorder that is caused (or heavily influenced) by a mutation in a single gene. Some babies will be diagnosed within 2 or 3 days. In the United States virtually all babies are screened at birth for (depending on the state) about 25 to 50 severe single gene disorders. Screening will identify about 5000 affected babies in time to start treatment. But there are more than 1000 well-described genetic disorders and many more that are poorly described. We do not yet have the means to screen for most of them. Many genetic disorders manifest in mid-childhood (some muscular dystrophies), others in late childhood (Friedreich's ataxia, some forms of retinitis pigmentosa), and still others in adulthood (Huntington disease). Many thousands of children afflicted with ultrarare disorders will receive a correct diagnosis only after their parents spend a year or more on a diagnostic odyssey that is frustrating and expensive. A few will go undiagnosed for decades.

Rare genetic disorders touch almost every family. If there is no child with a genetically driven disability in your extended circle, you are among the lucky minority. About one-half of all pregnancies end (usually very early) due to chromosomal or genetic abnormalities, about one in a 1000 children is diagnosed at birth with spina bifida, about one in 300 infants is born with a heart defect, and about one in 100 children will be diagnosed with autism spectrum disorder. About 3% of children have significant intellectual disability. There are many thousands of people in the United States with phenylketonuria, muscular dystrophy, cystic fibrosis, β-thalassemia, sickle cell anemia, hemophilia, Marfan syndrome, Huntington disease, and an ever-growing list of newly described genetic disorders. One hundred years ago the wards in children's hospitals were full of patients suffering from pneumonia and related infectious diseases. Today, a sizable percentage of the in-patient population carries a primary diagnosis of a genetic disease.

This book recounts some of the important milestones in the fight to help children with rare genetic diseases. In most cases I tread softly over the more technical aspects of the accomplishment, instead trying to tell stories that stress the human element. The chapters are links in a chain, a coherent story of the heroic efforts by parents, physicians, and scientists to help children born with mysterious and, frequently, untreatable genetic disorders. I do not use the word "heroes" lightly, but heroes do stride through the pages of this book. Many stories are painful, but some do have happy endings. I am pleased to say that I think that there will be many more happy endings in the future. I hope the stories in this book lead you to share my optimism.

Above all else, I wrote this book to convey a message of *hope*. The history of efforts to help children with genetic disorders provides irrefutable evidence that disorders that scientists and caregivers once viewed as beyond their therapeutic reach can be overcome. Sometimes, the therapeutic answer seems profoundly simple. As you will see in Chapter 1, the development of a special diet to treat children with phenylketonuria (PKU) by greatly reducing the amount of phenylalanine they consume has converted a severe form of mental retardation into a manageable, albeit chronic, disorder compatible with living normally. Today, rigorous control of diet (to avoid elements in foods that most of us eat safely each day) comprises the essential therapy for several genetic diseases.

In Chapter 3 I focus on blood. Until World War II most of the hundreds of thousands of children in the world who were born with β-thalassemia (in which the cells fail to produce enough hemoglobin to transport oxygen to the body's organs) died in childhood of profound anemia. The advent of blood banks and transfusion medicine in the early 1950s opened a new approach to treatment that converted (for those children who had access to such care) a fatal disorder into a chronic, but manageable, disease. Until not so long ago, children born with classical hemophilia (those whose cells could not make enough of the clotting protein called Factor VIII) rarely survived to adulthood. The purification of cryoprecipitate (which carries Factor VIII protein in great concentration) in the 1960s revolutionized the care of children with hemophilia.

Beginning four decades ago, a few intrepid physicians began to advocate for bone marrow transplants to treat certain single-gene disorders. This heroic intervention succeeded in some diseases, but failed in others. In the 1980s, a few scientists, especially at the National Institutes of Health, pushed the boundaries of biochemistry to create the new field of enzyme replacement therapy (ERT). The industrialization of the production of some crucial enzymes opened up a new era of therapy and helped to accelerate the growth of the biotech industry. Under the steady hand of CEO Henri Termeer, a company called Genzyme developed several ERT therapies that greatly improved the lives of children and adults with some of the lysosomal storage disorders. Today, as I discuss in the second half of the book, the burgeoning biotech industry is embracing a growing array of new technologies, including gene therapy, exon skipping, gene editing, and the delivery of structural proteins to take on the challenges posed by rare genetic disorders.

It is extremely difficult to meet the scientific, clinical, and regulatory challenges of drug development, but each year more groups determine that they will take on those challenges. Today, there are new partners in the mix of groups that develop novel drugs. Parents of children with rare genetic disorders have emerged as influential players at the biotech table. Foundations created by parents who refuse to surrender to rare disorders develop patient registries, support natural history studies, contribute to discussions about what constitutes approvable clinical end points, raise money to fund research, lobby regulatory agencies to be more innovative in how they oversee clinical trials, and push hard to have Congress earmark more research dollars for their diseases. As anyone who has had close contact with families that include children with severe genetic disorders knows, there are countless mothers who rise each day determined to move mountains.

When I was a medical student, the median life expectancy of a child with cystic fibrosis was about 12 years; today an affected child can expect to live into his 50s. Even a decade ago, most experts could not conceive of a means to attack the fundamental cause of cystic fibrosis. In 2012, after nearly two decades of effort, Vertex, a biotech company in Boston, won approval of a small-molecule drug to partially correct the molecular defect in one form of the disease. This has opened a door that may lead to the development of similar drugs for other molecular forms of cystic fibrosis. Further, it has established a new paradigm for drug development in general. Vertex's triumph required great vision, access to immense resources, and incredible tenacity. Victory came in no small part because the Cystic Fibrosis Foundation poured tens of millions of dollars into the effort. Unfortunately, in many cases the road to approval of a new drug will be long. However, some new technologies, especially gene therapy and gene editing, offer the hope that as our experience with them grows, the development pathway will be considerably shorter. I discuss them in two of the later chapters.

No tale about efforts to treat rare genetic disorders should sidestep the impact of the ever-growing power of diagnostic technologies, which I cover in Chapter 4. For decades, our ability to assess risk for genetic disorders has exceeded our abilities to treat affected individuals. Since the early 1970s pregnant women have had the option to undergo testing to determine if the fetus is burdened with a chromosomal disorder. Also since the 1970s members of certain groups have had the option to determine if they carry a mutation that (if they marry another carrier) puts them at risk for bearing a child with a severe genetic disorder such as sickle cell anemia, Tay–Sachs disease, and β-thalassemia. For a few disorders, abortion and avoidance of entering into at-risk marriages became an important option. Today, we are on the verge of being able to offer individuals *vastly* greater amounts of genetic risk information in regard to their reproductive planning. But for most genes we still have a great deal to learn, so large-scale DNA testing will for some years pose difficult questions of how to interpret some test results.

In several of the later chapters I discuss emerging technologies that allow us to hope that the dreams they have for their children will be realized. Our ever-deepening knowledge of human biology is expanding the boundaries of therapeutic possibility.

The final chapter confronts the fact that *we are all orphans*. Each one of us carries many mutations across many genes that will (interacting with environmental forces) affect our overall health, how we will age, and when we will die. Do we want to confront this information? In that chapter, I also confront the difficult fact that the cost of developing new therapies for a small number of patients demands that the price of these drugs be extraordinarily high. As medical therapy becomes more personalized (as is rapidly becoming the case with cancer therapies), society will have to determine what, if any, restraints it should place on pricing. Yet, if limits are imposed, drug discovery may cease in certain areas. Indeed, the more important problem may be how to find ways to encourage drug development to save the lives of children with ultrarare disorders, of which there are thousands.

The title *Orphan* is meant to convey two messages. The first is that there are a vast number of genetic disorders that burden *only a few* people, but for which we must find a way to harness our biomedical research engine to develop cures. The second is to emphasize what we all know—that nothing is more precious than our children, and we must not let a bad ticket in life's genetic lottery leave them and their families desperate and alone.

When I reflect on the range of my experiences as a physician—from toiling in the 1980s in the back wards of a state institution housing hundreds of severely disabled people to the last 7 years during which I have helped to create several new biotech companies that intend to translate cutting-edge science into breakthrough therapies—I feel a great sense of privilege. One of the main reasons is that I have met so many people whose stories have touched me. I retell some of those stories here. They will touch your lives.

1

Diet

DURING THE LATE 1980S, I WAS THE PHYSICIAN in charge of a primary care program for about 800 adults who were living at the Fernald State School in Waltham, Massachusetts. Built in the late 19th century at a time when society felt the best option for mentally retarded (as they were then called) children was to care for them in largely bucolic, institutional settings for life, Fernald—a lovely campus of red brick buildings shaded by stately oak trees—recalled a medical era long forgotten in the gleaming modern hospitals of Cambridge and Boston.

Among my patients were two sisters in their early 30s. Carol was born in the fall of 1958; Nancy was born in 1960. Both were tall, slender, attractive women with naturally blond hair and light-blue eyes. From a distance one might believe that they were twins. Both women were also profoundly mentally retarded, with IQ scores below 50. Neither one could speak or care for herself. Both were extremely anxious in the presence of everyone (including me), except for their elderly parents, who visited them occasionally on Sunday afternoons, and the poorly paid Haitian caregivers who dressed and fed them each day. Carol and Nancy are among the last few people born in the United States who bear the full burden of a rare single-gene disorder called phenylketonuria (PKU).

Starting in 1962, the states began to test all newborns for PKU. This was because a small group of researchers had shown that by immediately placing infants born with PKU on a special low-protein diet containing only trace amounts of the amino acid phenylalanine, they could avert the severe intellectual disability that would otherwise enshroud them. Because of this unprecedented intervention, in just a few years PKU virtually disappeared from the differential diagnosis of severe developmental delay in young children. I recall hosting pediatric residents from Boston Children's Hospital who worked in its PKU clinic (even when they are adhering to their brain-saving diet, the blood chemistry of these children needs to be carefully monitored) so they could meet Carol and Nancy to better grasp just how profoundly devastating PKU is if untreated. The conquest of PKU is one of the great achievements in our quest to treat hundreds of rare, mysterious, genetic disorders, most of which do not receive the attention that will be necessary if we are to conquer them. But, before I recount the fascinating story of PKU, a brief explanation of inborn errors of metabolism is in order.

The most common genetic disorder in the world is probably a disorder of the red blood cells that, long before it carried the scientific name of glucose-6-phosphate dehydrogenase (G6PD) deficiency, was known as *favism.* That term derives from the Latin name given to the fava or broad bean plant, an ancient crop cultivated throughout southern Europe, the Middle East, and parts of Africa. Historians of medicine have long hypothesized that the admonition by Pythagoras, one of the first great physicians, to avoid eating the fava bean derived from observations that some men became acutely ill if they did. In its severe form, the disease causes sudden, massive destruction of red blood cells. The severely affected person develops abdominal pain, yellow eyes and skin (jaundice), and profound weakness. Usually, over time, he recovers (so long as he avoids the bean) as the body makes new red blood cells. However, the disease has a wide range of severity, and it can be fatal. Why men? Today we know that the gene responsible for G6PD deficiency sits on the X chromosome. Like other X-linked disorders, it is far more common in men because they do not have a second, protective X chromosome with a normal copy of the gene, whereas women do.

It is plausible that ancient people learned that some men should avoid the bean simply because so many were at risk for the severe consequences of eating it. In many parts of the world, around one in 10 persons carries a variation in the gene that results in a defective G6PD enzyme that, given the right environmental exposure, triggers the sudden destruction of red cells. In the world today, literally millions of men carry one of hundreds of different mutations that predispose to favism.

How can deleterious mutations become so common among humans? The answer, as you might have guessed, is that the mutations that cause G6PD deficiency also confer a benefit. For more than a century, physicians have realized that favism is a disease of the tropics, which means it is found mostly where malaria is endemic. In the late 1960s and 1970s scientists showed that mutations in the *G6PD* gene conferred some resistance to infection from the malaria parasite.

Our understanding of the molecular basis for G6PD deficiency is closely connected with the long-running battle against malaria. Beginning in the late 1920s, scientists who were studying an antimalarial drug called pamaquine noticed that a few days after they ingested it, a few of the prison "volunteers" developed dark urine, turned yellow, and had a rapid fall in the hematocrit. Over the next two decades, physicians realized that sensitivity to the drug was familial, and that some persons from some populations were more at risk than those from others. About 1948 it became clear that persons sensitive to drugs like pamaquine and its newer version primaquine also were sensitive to eating fava beans. By then scientists knew enough biochemistry to infer that both the food and the drug sensitivity could be caused by a defect in the pathway by which cells produce a compound called glutathione, which is essential to maintaining the integrity of the red cell membrane. In 1958 a team at the National Institutes of Health (NIH) led by Dr. Paul Marks, who would go on to become the longest serving director of Memorial Sloan Kettering Cancer Institute in New York, published a paper delineating the precise nature of the biochemical deficiency.

About 1865 Gregor Mendel, a Moravian monk and autodidact, deduced from his study of sweet pea plants that there must be discrete (but completely unknown) particles, inherited without change through generations, that programmed the elements of physical existence. But his work, published in an obscure journal, was barely noticed. It was not until 1900, when three European scientists independently rediscovered the laws of Mendelian inheritance—the rules that explain how single-gene disorders are inherited as dominant or recessive conditions—that an intellectual scaffold permitted organized study of genetic disorders. But it was the flimsiest of scaffolds, and with a few exceptions many decades would elapse before scientific progress permitted the first meaningful therapy for these rare disorders. Even today, the immense scientific advance in understanding human genetic diseases that was catalyzed by the success of the Human Genome Project—the decade-long effort to decipher human DNA—must still be regarded as no more than an early success in the long journey to develop treatments for genetic disorders.

Black Diapers: Archibald Garrod and Inborn Errors of Metabolism

No one more deserves the title of the *founder* of human biochemical genetics than Archibald Garrod, the son of a professor of medicine, who won a *first* at

Oxford, and then earned his degree in Medicine at St. Bartholomew's Hospital in London in 1886. Although his mentors singled him out early as a future professor of medicine, many other able young physicians were in line ahead of him, so in 1892 Garrod moved from St. Bartholomew's to the Great Ormond Street Hospital for Sick Children. There, during the period 1898–1902, building on his astute observations of children with unusual syndromes, he developed the theory that these disorders were caused by "inborn errors of metabolism," a phrase in wide use today.

To appreciate the immensity of Garrod's contribution, we must grasp the theory of disease that was prevalent in 1890. In essence, physicians, enthralled with the discoveries made by Pasteur, Koch, and other progenitors of the germ theory, had come to believe that nearly all disease (excepting alcoholism and trauma) was caused by the action of some invisible external agent or an imbalance of several in the body. This was not so very different from the ancient Greek view that health and disease were determined by various "humours." The idea that the absence of a particular chemical in a cell could lead to a discrete, well-defined disorder was largely out of reach. Organic chemistry was the province of the dye industry and the understanding of biochemistry was growing mainly out of studies of fermentation.

Garrod's name is forever linked with a rare single-gene disorder called alkaptonuria, which had been known to physicians for many years. If one is aware of the disorder, the diagnosis of alkaptonuria is extraordinarily simple. Mothers bring their infants to doctors because they are alarmed to see diapers stained with urine as "black as ink." It was also recognized that although these children were relatively healthy, that in adulthood they would often mysteriously develop severe back pain and incapacitating arthritis. In the 1890s, the few academic physicians who studied the matter believed that there must be some species of *bacteria* living in the gut of these patients that interfered with the metabolism of tyrosine, leading in turn to the excretion of a metabolite called homogentisic acid, a pigment that caused the black urine. Garrod, who earlier in his career (along with Frederick Hopkins who would in 1929 win a Nobel Prize for proving the existence of vitamins) had studied the excretion by the kidneys of several organic compounds, was not so sure.

One day, after seeing a child with alkaptonuria, he made the inspired guess that the disease arose as a consequence of a malfunction of some chemical in the body that led to a failure to properly metabolize tyrosine. As luck would have it, the child's mother was pregnant. When that baby was born, Garrod asked the nurses to examine each of her diapers. Fifty-seven hours later, a nurse brought him the first black diaper. By combing through hospital records Garrod discovered two facts about alkaptonuria that fit perfectly with the newly discovered laws of inheritance: (1) the disease often appeared in two or more siblings, but *never* in the parents, and (2) the disorder was often seen in the offspring of marriages between first cousins. Thus, he became one of the first physicians to infer that a disorder was due to the inheritance of recessive mutations in the same gene from each parent.

It is said that after this discovery, Garrod became obsessed with the study of human urine. Over the next few years he established that three more rare disorders—cystinuria, albinism, and pentosuria—were also recessively inherited, and that they could be best explained by the malfunction of some chemical in the body that, when normal, properly metabolized some other chemical. In two of the three—cystinuria and pentosuria—patients excrete unusually large amounts of a metabolite in their urine. In classic cystinuria, patients lack a proper version of a protein that transports several amino acids. The disease is so named because patients are at high risk for having cystine kidney stones. Pentosuria is a clinically benign condition in which patients who lack the normal form of a certain enzyme intended to metabolize a sugar (L-xylulose) to an alcohol, instead excrete a large amount of that sugar in their urine. It is usually discovered by accident.

Garrod was awarded his post at the prestigious St. Bartholomew's Hospital just 1 year after publishing the first of his two papers in human biochemical genetics. From that moment his career was meteoric; just 7 years later he was made a Fellow of the Royal Society of Medicine. In 1908 he delivered the famous Croonian Lectures to the Royal College of Medicine under the title, "Inborn Errors of Metabolism." Although Garrod's career led him on many different diagnostic odysseys, he always taught his students "to have a love for the unusual and for rarities in medicine, because they so frequently lead to discoveries." Garrod's lectures, published first in *The Lancet* and shortly thereafter as a book of the same title, constitute a true paradigm shift in conceptions of disease causation. Today, a first edition of his book is a much sought after prize among those eccentric folks (myself included) who collect rare medical books. Just a few days before writing these lines I attended an antiquarian book fair in Boston where I was tempted by a pristine copy, but could not justify the $2250 price tag!

During the next few decades, many basic scientists, especially Beadle and Tatum in the United States (who shared a Nobel Prize for their work on mapping enzymatic errors) pursued ideas that Garrod had discussed in his lectures, and established the "one gene–one enzyme theory." They used organisms such as *Drosophila,* corn, and yeast to prove that each protein in the body was the product of a single gene that coded for it. But the first great advance in *treating* rare human inborn errors of metabolism began because of the persistence of a heartbroken, but dedicated, *mother* and the curiosity of a young physician-scientist.

PKU

In the early 1930s in Norway, Borgny Egeland was struggling to understand why both her children had failed to acquire language, had never learned to walk, and were now profoundly retarded. She went from physician to physician telling the same story: Both children had seemed healthy at first, but had fallen far behind their developmental milestones by their second birthdays. In addition, both had a constant musty odor in their urine and about their bodies, both were of lighter complexion than their parents, and both had a seizure disorder. The first three physicians that she approached did not become interested in trying to decipher such an obviously rare disorder, but the fourth, Asbjørn Følling, did.

As luck would have it, before going into medicine Følling had studied chemistry and had been a professor of nutrition. After he found that available tests to look for abnormal amounts of carbohydrates or proteins in the urine of Borgny Egeland's children gave normal results, he tested their samples with a compound called ferric chloride, which was known to change the color of urine in the presence of ketones. When added to the urine of these two children, the liquid turned dark green—an unexpected finding. This provided a hint that the children had a disorder involving proteins. Følling now confronted a question that it seemed he had been training throughout his professional life to answer: What was the abnormal substance and how was it linked with the children's profound disability?

Drawing on the help of several other chemists, but mostly working alone, Dr. Følling was able to show that the urine in these children contained a vast excess of phenylpyruvic acid. After repeating the test several times, he wisely decided to screen the urine of other mentally retarded children. Among 430 such children, he found eight that also had very high levels of phenylpyruvic acid. He then studied their families and was able to show (much as had Garrod with alkaptonuria) that a recessively inherited gene caused this mysterious condition, making it one of the first orphan diseases to be deciphered. Based on current knowledge of biochemistry (which had vastly increased over the three decades since Garrod's work), Følling guessed that the children's cells could not perform some chemical step in the conversion of phenylalanine to tyrosine. This would explain why they were light skinned; tyrosine is needed to make melanin, the key skin pigment.

Følling next sought a means to quantify the amount of the chemical in the urine so that he could screen widely for the disorder and understand variation in its concentrations. He asked bacteriologists at the university where he worked if there was a strain of bacteria that could be harnessed for this purpose. From one he learned about a strain of *Proteus vulgaris* that could not break down phenylalanine, the very amino acid that Følling thought was elevated in this new disease and the growth of which would be inhibited in the presence of high levels of that amino acid. By using a bacterial inhibition assay (a semiquantitative way to determine if the levels of phenylalanine were so high as to block bacterial growth), he developed the first version of a simple test to screen individuals for this disorder. Although *P. vulgaris* could grow in blood from healthy children, it could not grow in blood from the children that Borgny Egeland had brought to him.

In the 1930s there was no treatment for phenylketonuria, or PKU as the disease is now commonly called, but a positive test in an affected child did at least alert the parents to the one in four risk in each pregnancy of having another affected child. Even more important, however, was that Følling's great discovery suggested a treatment. If excess blood levels of phenylalanine or one of its metabolites caused the mental retardation by crossing the blood–brain barrier and poisoning brain cells, then perhaps by sharply reducing dietary intake of phenylalanine from birth it might be possible to prevent or ameliorate the mental retardation. Unfortunately, the cataclysm of the events leading up to and through World War II caused most biomedical research that was not related to the war effort to halt, including this one.

Shortly after the war ended, Horst Bickel, a German pediatrician who was one of the first physicians in the world to dedicate his career to developing treatments for orphan inborn errors of metabolism, made the next big leap in the quest to cure PKU. From 1949 to 1955, Bickel worked at the Children's Hospital in Birmingham, England, where he directed one of the world's first pediatric metabolic units (the forerunner of today's genetic clinics). In that era when pediatricians evaluated children with relatively normal features who had severe developmental delay, they routinely used Følling's urine test to screen for PKU. Usually the disease was not diagnosed in affected children until about age 2, at a stage when physicians thought it was too late for dietary therapy to help an already badly damaged brain. In 1953 Bickel examined a 2-year-old girl with newly diagnosed PKU. He proposed to the child's mother that even though it would probably not correct her mental retardation, that a low-phenylalanine diet might prevent her from deteriorating further and might improve her behavior. It was well known that children with untreated PKU often had severe behavioral problems. The mother agreed to try.

Bickel was an expert biochemist. In addition to instructing the mother on what phenylalanine-rich foods (including all meats and dairy products) that her daughter should not eat, he developed a slurry of amino acids that could be given as a drink and that would ensure that she would have adequate amounts of protein while ingesting very little phenylalanine. The mother adhered rigorously to the dietary therapy and reported regularly to Bickel on the child's progress. By 9 months she was convinced that her daughter's interest in the world had increased and that her behavior had improved.

Next came the crucial step. Bickel asked the mother to stop the special diet and feed the child as she had done in the past. Now on a normal (but for her very high) phenylalanine diet, within 2 days the child took an obvious turn for the worse. A few days later Bickel placed the child back on the low-phenylalanine diet, and in a matter of days, her behavior obviously improved. He was able to correlate changes in behavior with blood levels of phenylalanine. Although the experiment involved a single patient, because the biochemistry of the disease was so clear (often, as we shall see, the case with orphan disorders) Bickel thought it highly likely that other PKU children would benefit as well. But in 1954 there was not yet newborn screening of infants for PKU, so even if a commercially available diet could be developed, it could not be used early enough to avert mental retardation, only at most to improve the ease of managing the child's behavior.

Nevertheless, by 1956 experts were optimistic that the devastation wrought by PKU could be prevented. To do so, two challenges would have to be overcome: (1) the development of safe and effective, commercially available low-phenylalanine foods, and (2) the creation of a low-cost accurate mass screening test that could identify babies with PKU shortly after they were born. Without early identification to reduce exposure of the brain to toxic levels of phenylalanine, the diet would likely be of only minimal value. Both tasks were difficult. Although it was possible to develop a safe and nutritious low-phenylalanine infant formula, nutrition experts worried it would not be palatable and that it would be difficult to convince a food company to make it because the product would be consumed by only about 400 newborns a year in the United States. As for the screening test, it posed a financial and logistical nightmare. For such a program to work, about 12,000 tests would have to be performed to identify just *one* child with PKU. Would the cost of screening a population justify the expense? Would society even accept such a screening program? By 1956 researchers had already shown that, when used too early, the urinary test developed by Følling resulted in too many false negative and false positive test results to use as a tool to screen all babies (Følling's test was much more accurate at about 8 weeks of life when levels of phenylpyruvic acid became high in the urine, but by then the brain was *already* badly harmed). If deployed too early, the test threatened to do more harm than good—by causing babies who received an erroneous diagnosis of PKU to be put on a low-phenylalanine diet that was *not* good for them. Also, there was little reason to deploy a screening program until doctors could be sure that they could quickly place infants who had been correctly diagnosed with PKU on a special low-phenylalanine diet.

In 1949 a scientist named Louis Woolf developed the first crude low-phenylalanine supplement, which he created by hydrolyzing (breaking down) proteins. But no good formula was easily available until the late 1950s. Fortunately, about 1957, Mead Johnson, an Indiana-based infant food company that was founded in 1900 by a man who was desperately concerned about his son's poor health, agreed to create Lofenalac, the first widely available baby formula for infants and children with PKU. The product was based on the discovery that by passing casein (a major component of cheeses) hydrolysates over charcoal filters, food scientists could remove most phenylalanine from the milk protein. The resulting material could be used as the base element in an infant formula and for making low-phenylalanine foods. Lofenalac is essentially a low-protein, high-caloric powder that is supplemented with certain amino acids, vitamins, and minerals. Researchers were able to show that when young PKU children were put on a Lofenalac diet, their blood levels of phenylalanine fell into a range that was close to normal. This suggested that, started early enough, the diet might permit the brain to develop normally. With the launch of Lofenalac in 1958, the *medical food* industry was born.

The next hero in the history of developing a treatment for PKU is Dr. Robert Guthrie. Born in 1916 in Marionville, Missouri, Guthrie grew up in Minnesota where he developed a lifelong passion for sailing on Lake Minnetonka. He earned his MD at the University of Minnesota in 1942 and he stayed on to earn a PhD in bacteriology in 1946. Guthrie spent the next 12 years working at the Roswell Park Institute for Cancer Research in Buffalo, New York. A devoted father of six children, Guthrie's professional life changed when his second son was born with mental retardation. It took an even more dramatic turn in 1957 when his niece was diagnosed with PKU.

Guthrie became active in local groups interested in helping children with disabilities and he became friends with Dr. Robert Warner, a pediatrician at the Children's Hospital of Buffalo, who was trying to promote research on the causes of mental retardation. Warner persuaded Guthrie to move from the cancer institute to the hospital and to redirect his research efforts to understanding rare genetic disorders in children. Deeply affected by his niece's condition and knowing the limits of the Følling urine test, Guthrie set out to develop a test that could diagnose PKU at birth—possibly in time to avert or greatly reduce the then inevitable mental retardation.

Drawing on his training in bacteriology, Guthrie reasoned that he could use a strain of bacteria that could only grow in the presence of high levels of phenylalanine as a method to identify children with excess levels. In a remarkably short time, he developed what became known as an automated bacterial inhibition assay test. Simply put, a strain of the bacteria *Bacillus subtilis* was grown on gel agar plates that were infused with a chemical called β-2-thienylalanine, which competed with phenylalanine and prevented the bacteria from growing. However, if a sufficiently large amount of phenylalanine was added to the medium, the bacteria would grow, overcoming the inhibiting factor. A technician glancing at the gel would see a whitish growth halo of the bacteria.

Once he had tinkered with the system so it would respond to the levels of phenylalanine typically found in the blood of a patient with PKU, Guthrie, who knew that the relevant metabolites were stable in dried blood, had the transformative idea of developing a universal PKU screening test. In the newborn unit a nurse could quickly prick the heel of a newborn and spread a drop of blood on a filter paper that could then be sent to a central laboratory. There a technician could punch out a tiny circle of dried blood and place it on a particular spot on the gel. A spot in which bacteria grew would indicate the sample came from an infant with very high levels of phenylalanine. Technicians could scan hundreds of wells quickly and easily pick out the positive tests.

In 1960 Guthrie tried out his new method by obtaining blood samples from scores of mentally retarded residents living in the nearby Newark,

New York state school. In blinded testing, he correctly identified all the patients who were known to have PKU and also found four other patients who had never before been diagnosed! Knowing that the medical establishment and state departments of public health would demand more proof than his current research offered, Guthrie quickly obtained funding to conduct a large pilot study of PKU screening. In little more than a year, hospitals in 29 states collected more than 400,000 samples from newborns and mailed them to a laboratory he set up to manage the huge sample flow. To compare the new blood test with the Følling urine test, Guthrie asked the hospitals to give each mother a urine collecting kit and to send him a sample taken when the baby was 3 weeks old. Among the 400,000 babies who underwent the heel stick, Guthrie found 275 who tested positive. Repeat testing confirmed that 37 of them had phenylketonuria; the rest had only transient elevations that he could not fully explain. He also found that four of the 37 truly affected babies did not test positive on the *urine* test. It produced too many false negatives. Thus, he proved that the blood test he had developed was superior.

Most of the participating hospitals continued to screen. The leaders of the National Association for Retarded Children quickly began lobbying for state-based funding for screening programs, and in 1964 the federal Children's Bureau urged universal screening. In an unprecedented series of events, state legislatures, often after hearing lectures by Guthrie—who had become an indefatigable champion of mass newborn screening—began to enact mandatory testing laws. By 1968 virtually every child in the United States (as well as children in the United Kingdom and other European nations) was being tested for PKU. To ensure accuracy, each baby who tested positive was retested. If testing confirmed high levels of phenylalanine, the baby was immediately put on a special diet. At newly created specialty clinics, moms of the affected kids were educated as to why they should not breast-feed and why they must feed their baby a special low-phenylalanine formula and that their babies would need to be on a rigorously controlled diet for life. If they adhered to the diet, there was a very good chance that their infants would escape the severe mental retardation to which two tiny mutations in their genomes had predestined them.

Shortly before Asbjørn Følling began his effort in Norway to understand the cause of mental retardation in a Borgny Egeland's little girl, another mother was confronting the same sorrow in Nanjing, China. In 1917 Pearl Comfort Sydenstricker, the daughter of American missionaries, who had grown up in China and attended college in the United States, returned to China, and married John Lossing Buck, an agricultural economist. In 1921 the woman, now Pearl S. Buck, gave birth to her only biological daughter, Carol. A few months later the family moved to Nanjing, where Pearl taught English literature at the University of Nanking. Like all mothers, she was bursting with happiness about her little daughter. She later recalled her pride as "people spoke of her unusual beauty and of the intelligence in her deep blue eyes."

But as the months passed, and Carol did not develop normally, Pearl became increasingly fearful. In 1925 she asked a child psychologist who was visiting the university to examine Carol. When he told her that he thought her daughter had a serious medical problem, Pearl, already growing estranged from her husband, took Carol to the United States. A series of expert evaluations, concluding with a visit to the Mayo Clinic in Rochester, Minnesota, confirmed her worst fears. Doctors there advised her that although they could not tell her the cause, they were sure that Carol's intellectual deficits were permanent and (in keeping with the times) that the best course was to have her placed in a residential institution that cared for persons with severe mental retardation. Years later Pearl Buck recalled, "I don't know of any blow in all my life that was as rending. It was as if my very flesh were torn. It was beyond belief, and yet I knew I had to believe it, and to shape my life around the fact."

Pearl and Carol returned to China, and the mother threw herself into the effort of trying to educate her daughter. Over the next 5 years she saw some small successes, but it was clear that Carol would never be able to care for herself. In 1929 Pearl traveled with Carol to the United States and placed her in the Training School in Vineland, New Jersey, one of the most forward-looking residential institutions of the time. She returned to China, but it was already clear that her marriage was going to end. With her husband unwilling to shoulder the expenses at Vineland (preferring to move Carol to a less expensive facility), Pearl soon returned to the United States, moving to Perkasie, Pennsylvania to be close to her.

In 1931 Pearl was a near-penniless, single mother responsible for Carol's expensive care. Having dreamed from childhood of being a writer, she approached the Presbyterian Mission Board in New York, asking for support. The board offered her $500 to write a children's story about missionaries, and one of the board members was so moved by her story that she personally loaned her $2000. Pearl returned to China and wrote the story she had contracted to produce (called "The Young Revolutionist"), but only after she wrote a book that had long been taking shape in her mind, *The Good Earth*. When it was published in the United States, this tale of peasant life in China stayed on the best-seller list for 2 years, earning Pearl more than $1,000,000, some of which she used to endow lifetime care for Carol at Vineland. The debut novel won the Pulitzer Prize for fiction in 1932. It was the first of many awards that Pearl would win over 40 years for her many publications. Just 6 years after Pearl won the Pulitzer, in 1938 she became the first woman to win the Nobel Prize for Literature.

Though millions of people read *The Good Earth* over the ensuing decades, virtually no one realized that the mentally retarded baby daughter cared for so fiercely by the protagonists, the farmers Wang Lung and his wife, O-lan, was modeled on Carol. As one of Pearl Buck's biographers wrote, "The nameless child, who serves throughout the novel as a symbol of humanity's essential helplessness, is Pearl's anguished, barely disguised memorial to Carol."

Until the 1960s, families usually shrouded the existence of a mentally retarded child in secrecy. For 25 years Pearl Buck (who adopted many children with both her first and second husband) refused to mention Carol. The topic was off the table for the many journalists who interviewed her. Then in 1950, now 58 years old, Pearl S. Buck, Nobel Laureate, finally coming to terms with her shame and sorrow, wrote an article for *The Ladies Home Journal* called, "The Child Who Never Grew." Her frank disclosure about her daughter's retardation and her regrets about how she dealt with it, elicited such a tsunami of response from readers that an expanded version was published in book form a few months later. Thus, Pearl Buck became one of the first formidable voices to speak out about the needs of people with disabilities.

Twelve years later, when Eunice Kennedy Shriver, an older sister of President John F. Kennedy, first wrote about her sister Rosemary (who was mentally retarded and living in a private institution) in *The Saturday Evening Post,* she cited Buck's earlier writing as having given her the courage to speak out in support of people with disabilities. The wife of French President Charles de Gaulle, who had a disabled child, also credited Pearl Buck for giving her the courage to go public with her sorrow. Sometime in the early 1960s, doctors at Vineland told Pearl that Carol's problems were due to phenylketonuria. They had used the test based on Asbjørn Følling's research to make the diagnosis. The beautiful deep blue eyes that Pearl had so admired in infant Carol were caused by the genetic disorder, which among other things causes a lack of dark pigments in the eyes.

From 1985 to 2000, I worked at the Eunice Kennedy Shriver Center for Mental Retardation in Waltham, Massachusetts. For five of those years I was in charge of the medical team that cared for about 800 institutionalized mentally retarded persons at the Walter E. Fernald State School (the Shriver

Center was located on its campus). Among my patients were the two young women with PKU who I mentioned earlier. They were tall, thin, blue-eyed, blond women who from across the room appeared normal. But, when I performed their annual physical examinations, they emitted frightened, whining noises. They could not speak; they seemed never at ease, and all their movements were awkward. They had lived at Fernald for almost their whole lives. I met their parents once. I still remember the father's tears. About his wife and himself, he said, "We live with unending sorrow."

Although I did not know Eunice Kennedy Shriver well, I did meet with her on a number of occasions. What I remember best was the way she greeted the persons who lived at Fernald. When she visited with a group, she would look each person squarely in the eyes, firmly shake their hands, and speak to each just as she would to potential donors at a political fundraiser. She refused to acknowledge that they were "different." I also remember a Board of Directors meeting in the early 1990s when Eunice nominated her brother, Senator Ted Kennedy, to be president of her foundation. Her eyes twinkled as she said, "He really deserves to be president of something." As I write these words, I feel a tiny connection to that young woman in Nanjing, China who gave birth to a child with PKU in 1920, a birth that drove her in 1931 to write a book to earn money to support that child, a book that made her rich and earned a Nobel Prize. Twenty years later she would write another book that would help lift the curtain hiding orphan genetic disorders.

The advent of newborn screening for PKU came at an auspicious moment; it strongly reinforced policies being developed by President Kennedy, who had a sister with mental retardation and who, along with his sister, Eunice, had decided to support advances in understanding this mysterious collection of disorders, making it a health priority. In 1962 President Kennedy created a national award to recognize pioneering work in the field. Fittingly, the first recipient was Dr. Asbjørn Følling. According to his son (also a physician), when Dr. Følling, who did not like to travel, received the notice of the award and an invitation to accept it in Washington, D.C., he did not even reply, apparently not understanding the prestige associated with being selected. It took a second call from the White House to get him on the plane in Norway! Also fittingly, Dr. Guthrie won the same award in 1977.

When mass screening for PKU was launched in the mid-1960s there were still many important unanswered questions. Were all children with elevated blood phenylalanine destined for mental retardation? How low must dietary therapy drive blood levels to prevent disability? Could phenylalanine levels be driven too low, resulting in another form of mental retardation? At some point in life—perhaps after the brain completed its development in adolescence—could the patients go off the diet? Today, thanks to much research, we have good answers to these questions. To maximize brain development all children with persistently elevated levels of blood phenylalanine must be placed and rigorously kept on a low-phenylalanine diet at least through adolescence. Even when dietary compliance is nearly perfect, the IQ scores of well-treated patients run 5–10 points lower than those of their unaffected brothers and sisters. If adults with PKU reject their diet, they experience noticeable behavioral and cognitive problems, albeit not nearly as severe as the impact of the disease on untreated infants. It is not necessary or wise to put infants on a diet with no phenylalanine; after all, it is an amino acid that the cells need to make most proteins. The task is to determine the lowest safe level.

Forty years of successful PKU screening has created a new medical challenge. All over the western world there are young women with treated PKU who wish to marry and have children. If they become pregnant and do not adhere to the special diet they are highly likely (nearly 90%) to bear children with severe mental retardation, heart defects, and other problems. This terrible outcome results from the fact that the high levels of phenylalanine and its metabolites cross the placenta and have toxic effects on the developing fetus. It is an ongoing challenge to clinical geneticists to keep track of these women and make sure they get the very special prenatal care they need. Canada created a national registry to track these women, a course of action that the privacy-sensitive United States probably will not tolerate.

Today there are about 40,000 persons with PKU living normal lives—except for adhering to special diets—in the United States and Europe. Each year about 400 babies born in the United States are diagnosed with PKU. Not surprisingly, as the decades have passed and infants have outgrown special formulas, the challenge of keeping children, teenagers, and adults on diets that are not too palatable has become ever more important.

After discovering the cause of PKU and developing a highly accurate newborn screening test, the next great task in conquering this orphan disease was to develop a highly nutritious, very low protein (the source of most phenylalanine), reasonably priced formula for infants. As I have mentioned, in 1958 Mead Johnson launched the first baby formula for infants with phenylketonuria, Lofenalac, which has long dominated that niche market in the United States. Parents purchase the formula as sets of packets that one mixes with water and a measured amount of evaporated milk just before feeding the infant. Hungry babies seem not to mind the taste, which almost all children and adults find slightly unpleasant—a sort of chalky, slightly bitter, yet rather bland taste. During the 1980s and 1990s most PKU children continued to derive most of their calories from Lofenalac, despite the fact that as they grew older, it became more difficult to persuade them to consume it.

The advent of dietary therapy for PKU initiated a sustained debate about how to regulate such products and how to inform patients about risks in other products. The Food and Drug Administration (FDA) recognized a different regulatory framework was needed for approving the creation of an infant formula that was low in protein (and very low in phenylalanine) than for a new blood pressure drug. In 1973 when Abbott Pharmaceuticals launched Ensure, the first widely used lactose-free nutritional supplement for persons of all ages with a wide variety of clinical problems, the need for a new regulatory framework became even more urgent.

The FDA created a new class of substances that it called "medical foods." In essence, a medical food is an orally consumed product that has been formulated and processed to serve particular patients with a defined metabolic problem for which the patient is under a physician's care. The formal definition of a medical food was incorporated into law in 1988 with the passage of the "Orphan Drug Act" (which we will discuss later).

One of the first things that parents who discover that they have a child with PKU do is to become experts in nutrition. Fortunately, there are some foods—mostly vegetables—that PKU children can eat, but the list of forbidden foods is formidable. Meat, fish, and milk products top the list. Imagine managing the life of a child who cannot eat pizza, ice cream, or hamburgers! It sounds simple enough to say that to avert mental retardation, you merely have to make sure your child adheres to a special diet, but what if there are few palatable foods to provide?

During the 1980s and 1990s several small companies, recognizing the growing market for PKU foods, began to market low-phenylalanine foods (breads, frozen dinners, desserts) directly to families. These products are more expensive than regular foods, which led many states to enact laws that mandated that (depending on family income) PKU families could receive several thousand dollars a year to offset some of the extra costs. But the real problem was not the cost; it was the taste. Even the best of this first generation of nonformula PKU foods often had an unappealing texture and a slightly bitter taste. I have tasted some that I could just barely make myself swallow. Imagine battling with a 7 year old to consume such products two or three times a day!

Not surprisingly, the parents of two children with PKU finally decided to do something about this problem. In 1992 David and Lynn Paolella were a young, successful, happily married couple. David was a rising star in architecture in the Boston area and Lynn, a jeweler, was eagerly awaiting the birth of their second child. When they learned on his third day of life that their newborn son had PKU, they, like any parents, delved deeply into the disease. Over the years they adhered closely to the prescribed diet and worked closely with the staff at the PKU clinic at Boston Children's Hospital to monitor their son's blood phenylalanine levels. But as their son reached school age and began to rebel against the formula and the poorly tasting commercially available foods, a rebellion with which—given the taste—they were strongly sympathetic, they took a dramatic leap.

Lynn, who had given up her career to ensure that her children (by this time they had three children—two with PKU) adhered to the strictest possible diet (consider the challenge posed by sleepovers and birthday parties), began to devote her time to creating new recipes for low-phenylalanine meals. By trial and error, Lynn, who told me that a lot of her early efforts were mostly enjoyed by the family dog, gradually developed meals that tasted far better than did prepackaged commercially available PKU products. Painfully aware of how dissatisfied the families were with what was available in the marketplace, in 2000, David ended his career as an architect, and he and Lynn set out on the potentially quixotic task of building a company that would produce a line of meals for all children and adults with PKU. I met David a few years ago when Dr. Harvey Levy, a professor at Harvard Medical School and one of the nation's leading experts on PKU, suggested he would be a great person to teach me about how treating PKU affects everyday family life.

Today, Lynn and David's company, Cambrooke Therapeutics, offers a wide range of low-phenylalanine prepared foods that can be purchased online. Over the last few years they have been pioneering the creation of a wide variety of protein substitute drinks (PKU formulas) that rely on a new method for processing cheese (appropriately developed at the University of Wisconsin) to yield a more palatable and functional protein source. Cambrooke's medical foods constitute a major advance in taste along with improved nutrition. David and Lynn have been tireless advocates for PKU families and have taken tremendous financial risks to do everything they can to help improve the life experience and outcomes for PKU patients and their families. From Borgny Egeland to Lynn and David Paolella, the history of PKU has many heroes.

Since the early 1960s, the standard of care for treating infants born with PKU has been the prompt and unrelenting use of a low-phenylalanine diet coupled with periodic blood evaluations to make sure that blood phenylalanine levels are at a safe level. The results have been astounding. About 40,000 persons in the United States and Europe, ranging in age from infancy to 50 are—thanks to Følling's curiosity, Guthrie's test, and Bickel's diet—living regular lives. Children born with PKU are schooled in regular classrooms; teenagers compete in high school sports and edit school newspapers; affected young adults excel in college and build solid careers. They look normal, act normal, and are normal, except for one aspect of their lives—their diet.

When universal newborn screening for PKU was introduced, there was still much to learn about both the disease and the novel dietary therapy. One of the big unanswered questions was whether or not patients would have to be maintained on the diet for life. It now appears that to achieve and maintain the best possible cognitive function, persons with PKU do indeed have to be forever on the diet. Over the last four decades there have been many studies of the cognitive development of persons with PKU. Studies of affected adults show that if they go off diet for even a few weeks, they experience harmful behavioral changes and a discernible decline in cognitive skills. For obvious ethical reasons, no one is going to conduct an experiment in which they pull the patients off the diet for a lengthy period of time to see how much damage ensues.

Despite the risks of doing so, many persons with PKU, especially teenagers and young adults, frequently cheat on their diets. Imagine a 12-year-old girl at a birthday party who cannot have any cake or ice cream or a 16-year-old boy hanging out with his friends on a Saturday night who cannot—ever—join them for burgers or a pizza. Just like we do, they love the tastes of these foods, but they are forbidden. No one really knows how quickly brain damage accrues when a young adult goes off the diet. But the best evidence says that noticeable behavioral changes such as easy irritability happen within a few days.

The need to stay on diet is reinforced by the fact that among persons with the most severe form of PKU (as with most diseases there is a range of severity that is only partially understood), after strict adherence for months on end, blood levels of phenylalanine are still quite a bit higher than they are in unaffected persons. Thus, although the PKU diet has long and justly been trumpeted as a great victory in the fight against an orphan genetic disorder, it falls well short of a cure. Diet has changed PKU from a devastating form of mental retardation to a chronic disorder that can be kept at bay only if the patient commits to living in a special environment. This raises the question: Can we develop an even better therapy, perhaps one that frees the person with PKU from his dietary chains? In the case of PKU, as well as for other inborn errors of metabolism for which dietary therapy is an option, there is an obvious strategy to try.

Like many other enzymes, phenylalanine hydroxylase, the enzyme that is defective in PKU, depends on a cofactor (a small molecule like a vitamin) to help it work. If the PKU is attributed to a genetic mutation that greatly lessens, but does not eliminate, production of that enzyme, leaving it partially functional, it might be possible to squeeze more action out of it by providing a lot more of the cofactor. Over the years, scientists have studied the relevant PKU cofactor—tetrahydrobiopterin (BH_4)—in an effort to understand this question. Fortunately, BH_4 can be safely given to humans. In 1999 Japanese researchers undertook studies in which they gave unusually high doses of BH_4 to persons with PKU. They showed that in nine of 37 PKU subjects, treatment with BH_4 lowered blood phenylalanine levels by 30% in just 8 hours. When the group was treated daily for several weeks, blood levels dropped 30% or more in 17 out of 37 (46%) of patients. This impressive drop in nearly one-half of the subjects prompted major interest in developing a new therapy to treat PKU. It catalyzed one of the most rapidly successful drug development programs in the history of the pharmaceutical industry.

In the same year the Japanese reported their findings, BioMarin, a young biotech company devoted to orphan genetic disorders, raised $67 million in its IPO (initial public offering). BioMarin had been founded in 1997 to develop a new drug called Aldurazyme to treat an orphan genetic disease called Hurler syndrome. Its then medical leader, a physician named Emil Kakkis, is an expert in a class of childhood-onset genetic diseases called lysosomal storage disorders. In 2004, the year Aldurazyme won FDA approval, BioMarin geared up a new program to develop a pharmaceutical version of BH4 to improve the treatment of PKU.

The company goal was to develop a compound called sapropterin hydrochloride and show that it sharply lowered blood phenylalanine levels in PKU patients. Unlike most drug development projects, BioMarin did not have to spend many millions searching for a novel chemical compound nor did it face the daunting challenge of manufacturing a new drug with acceptable purity in large quantities. Although BioMarin would have to run safety studies, prior experience virtually guaranteed the compound could be safely administered to humans. The company really faced a single, albeit difficult, hurdle—proving that the drug was efficacious.

In a negotiation that is almost without precedent in the annals of the drug industry, the FDA agreed with the company in advance that an approval "end point" for the study would be if the drug could be shown to lower blood phenylalanine levels by 30% in a specified fraction of patients who were maintained on their regular diet. The FDA did not demand that the company show any neurological or other benefit to the patient. Both parties essentially agreed to accept the premise that if the blood levels of phenylalanine achieved with diet could be further lowered by giving a safe cofactor, then the patient might benefit in two ways: maintain consistently lower phenylalanine levels and suffer less damage if he or she deviated from a rigid adherence to the diet.

In just less than 3 years, BioMarin, working with academic experts, completed several clinical trials of its experimental drug. First it showed that among a large group of PKU patients aged 8 to 48 that sapropterin at a dose of 10 mg/kg a day for just 8 days resulted in a drop of blood phenylalanine levels of 30% or more in about one in five persons. As suspected, some PKU patients benefited much more from cofactor therapy than did others (with the variation being largely a consequence of the specific nature of the many different possible mutations in the gene). In the next trial, which focused on the responders in the first trial, the scientists randomly divided 88 patients into two groups: one to receive sapropterin and the other to receive a placebo. At the end of 6 weeks of therapy the treated group showed a dramatic decrease in blood levels, whereas levels in the placebo group rose slightly. The third and final study focused on 90 PKU kids aged 4 to 12 who were on diet and ran blood levels above a certain value. The children were all given sapropterin at 20 mg/kg each day for 8 days. At the end, 56% of the children had experienced at least a 30% reduction in their blood levels from that they had maintained on dietary restriction.

Although it was concerned that the company had not linked lowered blood levels to clinical benefit and required that BioMarin conduct a number of follow-up studies, the FDA approved the drug (marketed as Kuvan) for use in December of 2007. The drug was launched at a price that, depending on the weight of the patient, would cost between about $50,000 and $150,000 a year. In 2011, just 3 years after product launch, BioMarin reported net product revenues of $117 million from the sale of Kuvan. From these numbers, it appears that about 15,000 persons in the western world with PKU are now regularly taking Kuvan. It will, however, be many years before we will know for certain how much Kuvan is helping. If patients wind up taking it largely to let them deviate from diet and if that means the net change in median blood phenylalanine levels is not lowered, then in the long run patients taking Kuvan may have no better clinical status than patients who just stick to their diets. Still, Kuvan does promise to help those patients with PKU who are most responsive to cofactor therapy; further it will make eating a more pleasant experience for the rest of them.

Today, a little more than a century since Archibald Garrod's great insight about the cause of alkaptonuria and 50 years since Robert Guthrie launched the test that became the foundation for universal newborn screening, there is no question that dietary therapy for PKU is one of the great achievements in the struggle to help children with inborn errors of metabolism. But, it would be wrong to believe the battle against PKU is over. Even with the innovative work of David and Lynn Paolella, lifelong adherence to an expensive and relatively unpalatable diet is an imperfect therapy. Kuvan constitutes an important adjunctive therapy, but it only helps about 30% of PKU patients. Work on two innovative therapies offers some hope for the future.

The first involves harnessing an enzyme made by some bacteria called phenylalanine ammonia lyase (PAL), which breaks phenylalanine down into two harmless metabolites, as a drug to treat patients with PKU. Scientists have been working on this project for nearly a decade; they have found that although the enzyme dramatically reduces blood levels of phenylalanine, when given repeatedly it evokes a potentially harmful immune response. Efforts are under way to block or reduce that response by "masking" the drug with a technique called "pegylation" (creating a compound affectionately called PEG-PAL). The second and more daring approach (discussed more fully in Chapter 7) is gene therapy. In essence, the idea is to use certain viruses that are known *not* to cause human disease, as vectors to carry the DNA sequence coding for the normal phenylalanine hydroxylase gene and to use those vectors to transduce (invade) liver cells. Once in those cells, the vector will use the cell's normal machinery to make the normal version of the enzyme. Theoretically, a single appropriate dose of such a viral vector could control PKU for several years (depending on how frequently liver cells are replaced through natural turnover). Recent advances using adeno-associated viral (AAV) vectors offer promise, but there is much work to be done. I am optimistic; I believe that children with PKU who are today following a very strict low-phenyalanine diet will in 10 years have less demanding and more efficacious treatment options.

Phenylketonuria is the best known of several rare genetic disorders that can be partially ameliorated by rigorous adherence to a diet tailored to circumvent the problems posed by a defective enzyme. These include, for example, several forms of tyrosinemia (a disorder of tyrosine metabolism), galactosemia (a disorder of carbohydrate metabolism), and a very rare disorder called maple syrup urine disease (in which a failure to metabolize the branched chain amino acids is associated with a sweet odor in the urine). In each case the histories—from characterizing the disease to developing an ameliorative dietary intervention—are congruent.

One of the most underappreciated victories in public health in the last half third of the 20th century was the steady expansion of universal newborn genetic screening. Here, too, we owe a great deal to Dr. Robert Guthrie. Starting about 1962, he became a tireless advocate before state legislatures and departments of public health for the enactment of mandatory newborn screening laws. Several clinical geneticists who knew him have regaled me with anecdotes concerning how Guthrie traveled from state to state with a bottle of bourbon in his suitcase, a beverage that he found effective in getting his message across in evening discussions with some elected officials. He was helpful. Between 1962 and 1968 *every* state in the United States enacted laws requiring that newborns be tested for PKU.

Most of these laws were constructed to allow scientific experts in the states to decide whether or not to add other disorders to the testing menu. For a few years in the 1990s, I was a member of the advisory group that evaluated proposed genetic tests for the state of Massachusetts. Today, pursuant to these laws many states screen for about 40 rare genetic disorders, and the number should grow dramatically in coming years. In many cases the first line of response to discovering a child with an inborn error of metabolism is immediately to place him or her on a special diet that averts mental retardation and, not infrequently, death. These disorders are individually rare, but collectively common. About one in 1000 children is born with one of them. Over the last 50 years mental retardation has been averted in tens of thousands of children with orphan genetic diseases thanks to a combination of scientific advance and wise public policy.

Neural Tube Defects

In Ireland in the 1960s about one in every 150 newborn children was afflicted with a neural tube defect (NTD)—at the time the highest incidence of this severe birth defect anywhere in the world. NTDs include a variety of disorders that arise because of the failure of the tissue that is destined to form the spinal cord to form a completely closed *tube*. As gestation continues, a fetus with a NTD may develop one of several different structural problems ranging from a very small, protected lesion at the base of the spine to a profoundly devastating lesion that causes major brain malformations. A critically important fact to understand if one is struggling to prevent these birth defects is that the embryonic neural tube normally closes at about 21 days after conception—*before many women realize that they are pregnant.* Depending on where the failure of the tube to close occurs and the amount of nerve tissue exposed to the outside world, affected children may be diagnosed either with *anencephaly* (in which the brain is severely damaged) or *spina bifida* (in which the lesion is limited to a portion of the spinal cord, the lower the better). About 40% of affected infants have anencephaly (and usually die soon after being born) and about 60% have spina bifida. In the 1960s NTDs were so common and so severe that in some countries they constituted a national public health problem, one that demanded vast resources for treatment and care.

Although there are a few chromosomal and rare single-gene disorders that cause NTDs, in the vast majority of cases, we do not yet understand why a particular fetus develops one of these malformations. But epidemiological data show that powerful genetic risks exist. Although the background risk that any healthy woman will give birth to a child with a NTD is about one in 1000, if that woman has given birth to one child her risk of bearing an affected child in the next pregnancy jumps to about 1:25, a 40-fold increase. If she has given birth to two affected children, the risk for the next pregnancy is about 1:10. If a man or woman who was born with spina bifida procreates, the chance that the fetus will also have a NTD is also ~4%, a huge step-up in risk. Over the last 40 years, scientists have documented a number of other risk factors. Hispanic women are at higher risk for bearing children with NTDs than are white women; African-American women are at lower risk than are white women. Obese women, women with diabetes, and women who must take antiseizure medicines are at increased risk. Each of these factors might reflect underlying genetic or environmental risks, probably both.

As is often the case with birth defects, the incidence of NTDs varied widely among countries, from as much as one in 1000 across the United Kingdom to only one in 10,000 births in parts of Asia. Generally speaking, it was most common among northern Europeans, especially those of Celtic origin. Nobody knew what caused it, but everyone recognized that there must be genetically disposing factors because the single most powerful predictor of risk was a positive family history for the disorder. Part of the mystery of NTD is that throughout the world it affects more girls than boys with a ratio of about 1.2:1. If we assume that about one in 1000 children is destined to be born with a NTD that means that each year in the United States and Europe about 10,000 babies are born with anencephaly or spina bifida, about 100,000 babies in a decade. Of course the worldwide numbers are far higher.

Over the last four decades two remarkable developments have significantly reduced the number of children born with NTDs in the United States. The first was the discovery by the British scientist Nicholas Wald in 1974 that a biomarker called α-fetoprotein (AFP) that could be easily and inexpensively measured in a woman's blood was higher than normal in pregnancies in which the fetus was afflicted with a NTD. Especially in the United Kingdom, several groups quickly undertook large studies of AFP in maternal serum (in affected fetuses this protein can leak through the lesion into the mother's circulatory system) and were able to show that elevated levels of AFP (often double the normal range) were strongly associated with a markedly increased risk (but not a certainty) that the fetus was afflicted. The actual diagnosis is usually confirmed by repeat testing and by ultrasound studies that can visualize the lesion. During the mid-1970s proponents of population-wide maternal serum AFP testing published arguments in British medical journal *The Lancet,* favoring what is essentially a eugenic approach—that affected fetuses should be identified early enough that the mothers might be given the option to terminate the pregnancy.

In 1977 a major U.K. collaborative study unequivocally established the value of a national screening program. Less than a decade after this discovery, largely because the National Health Service could efficiently introduce the screening test into routine prenatal care, there was a sharp drop in the number of births of children with NTDs. In the United Kingdom, most women who were offered the AFP test took it, and most women who learned that they were carrying an affected fetus did terminate the pregnancy. In 1999 Wald reported that in just 20 years the annual number of live births of infants with NTDs in the United Kingdom had decreased by 95%!

In the United States screening for fetuses with NTDs lagged behind the British program until May of 1987 when the American College of Obstetricians and Gynecologists (ACOG) issued a medical liability alert, warning that a physician who failed to offer the test could be sued if a child was born with a NTD. In effect, ACOG was declaring a new standard of practice that its members would ignore at their peril. The alert had a profound impact on obstetric practice. By 1990—just 3 years later—the vast majority of pregnant women in the United States were being routinely offered maternal serum AFP (MSAFP) screening. Although the data in the United States are not as comprehensive as those in the United Kingdom, it appears that over the last 20 years the number of children born with NTDs in the United States has declined by >30%. The mandatory fortification of the food supply with folate is almost certainly the major cause of the decline. Many young pregnant women do not opt for NTD screening and among those who do that learn the fetus is affected, many do not terminate the pregnancy.

The implementation of MSAFP screening was not without its problems, the most important of which is that the screening algorithm was set so that the test would miss few affected pregnancies. Because the variability in the test results, for every true positive result that the test reported, it also called out about 10 that would turn out to be *false* positive results. Almost every woman who tested positive had to be promptly retested. Fortunately, on subsequent testing about 90% of those women turned out *not* to be carrying a fetus with a NTD. Even though the odds favored them, many of the women who were initially informed that the result was positive lived in a state of profound anxiety during the 10–14 days that it took to repeat the test and obtain the results. I well recall some of them telling me that the fear generated by a positive result on the initial screen cast a pall over the rest of the pregnancy. Over the years the quality of the test has improved, but it still generates many false positive results.

The second major development that contributed to the decline in births of children with NTDs was the discovery that women who regularly ingested the B vitamin folic acid (folate) could reduce their risk of bearing an infant with an NTD by about 50%. In the early 1960s Bryan Hibbard, an epidemiologist working at the University of Leeds, began to study pregnancy outcomes in poor women with the hope of identifying environmental risk factors that might be corrected with simple public health interventions. In 1964 he reported the results of his detailed analysis of 1484 low-income women in Liverpool with an unusually high level of problems in their pregnancies, noting that among other dietary deficiencies, the women had low levels of folate in their urine. This led him in 1965 to hypothesize that folate (which is an essential cofactor for many biochemical reactions in human cells) deficiency could cause birth defects, especially those involving the central nervous system. During the late 1960s and 1970s an increasing number of retrospective studies and uncontrolled trials were conducted that seemed to confirm Hibbard's hypothesis. In 1968 Hibbard and his colleagues showed that by providing folate supplementation throughout pregnancy to women who had in the past given birth to a child with an NTD they could reduce the expected recurrence risk by ~70%. But, it was not possible to generalize from this happy finding to make a claim about the benefits of folate to all pregnant women.

It was not until the Medical Research Council in England and the Public Health Service in the United States conducted prospective trials that physicians fully grasped the magnitude of the preventive benefits of folate. In 1991 Wald and others showed with prospective studies that women who were consuming folate before and when they got pregnant had a sharp reduction in risk for bearing a child with a NTD. In 1992 the U.S. Public Health Service (USPHS) recommended that all women who were planning a pregnancy or who could become pregnant consume 400 μg of folate a day.

In the United States about one-half of pregnancies are not planned. Study after study showed that despite major educational efforts, many fewer than half of the women of child-bearing age were consuming folate *supplements* (although they were easily available in multivitamins) and that among those that were, many were taking only about half the folate needed to reduce the risk of NTDs should they become pregnant. This disappointing fact generated prolonged, often passionate debate that reached to the highest levels inside the FDA.

After years of study, in 1998 the FDA took the extraordinary step of mandating that cereals be *fortified* with at least 140 μg of folate per 100 g of grain (essentially a serving of breakfast cereal), in effect altering the food consumed by everyone to ensure that they got the vitamin to fertile women. Of course, it was not the first time that a vitamin has been routinely added to the food supply. Since the 1920s, many nations have added iodine to the food supply to prevent thyroid disorders. During World War II, disturbed by the nutritional deficiencies of many army recruits, the United States pushed for supplementing flour with thiamine, niacin, and riboflavin. Most commercially sold milk is fortified with vitamin D (to prevent the now rare bone disease called rickets).

The decision by the FDA to mandate the fortification of the nation's flour supply with folate—a decision that Commissioner David Kessler called the most difficult of his tenure—satisfied neither its proponents nor its critics. The proponents rightfully pointed out that a steadily growing body of evidence argued for fortification at much higher levels than the FDA recommended. Critics argued that the decision to fortify was in effect an uncontrolled experiment on the American people. Because fortification affects foods that almost everybody eats almost every day, it was possible that the program was creating a new health risk for the entire population.

That is why the new *fortification* rule only mandated that cereal makers provide about 25% of the folate intake recommended to avert NTDs. This reflected an uneasy compromise. No one knew whether regular exposure to high levels of folate might pose other kinds of risks. When one fortifies the nation's food supply with a chemical, even a tiny increase in risk for some serious disorder such as cancer could more than offset the public health gains of averting the births of about 1000 children with NTDs. Further, it was already known that taking high levels of folate could mask signs of an uncommon blood disease caused by vitamin B_{12} deficiency. Part of the logic for limiting the mandatory fortification to relatively modest levels was that women also consume folate when they eat green, leafy vegetables and some fruits. But experts estimated that the mandate resulted in most young women consuming about one-half the daily intake needed to avert NTDs should they become pregnant.

Soon after the FDA made its decision, Chile and Canada adopted a folate fortification policy, but, remarkably, most nations, including those in Western Europe, did not. This is especially troubling in the face of powerful evidence of efficacy, as evidenced by public health efforts in Chile. From 1967 to 1999, the incidence of births of children with NTDs in Chile was steady at 17 cases per 10,000 births. One year after Chile adopted a folate fortification program, the blood levels of folate in young women tripled. In the years since then the incidence of live births of infants with both anencephaly and spina bifida decreased by about 50%! Put another way, many children who would have been born with a NTD have been born without the congenital defect. Prenatal screening and abortion played only a small role in this sharp reduction in incidence. Over the last few years many cereal makers in the United States have increased the levels of folate to those (400 μg per serving) that scientific studies suggest could deliver significant prevention from NTDs, but some still do not. As of 2012 about 50 nations have decided to require that flour or cereals be supplemented with folate. Inexplicably, no European nations have yet mandated folate supplementation of food.

Although the government regulations were issued about 15 years ago, the fortification debate has continued. Among the leading proponents for increasing the levels of fortification is Dr. Godfrey Oakley, former director of the Birth Defects Division of the United States Centers for Disease Control and Prevention in Atlanta. In 2006, Oakley coauthored an editorial in *Pediatrics,* in which he asserted that in the 15 years since the emergence of a scientific consensus in 1991 that folate supplementation averted NTDs, failure to aggressively push fortification and supplementation programs had resulted in the births of about 3 million babies throughout the world with congenital malformations that could have been easily prevented. Although the overstated claim reflects the passion of the advocate, the surprisingly slow adoption of a public health policy that offers immense benefits at reasonable cost surely has led to the birth of many thousands of affected babies in nations that have the resources to educate and supply folate to young women.

To this day, scientists remain uncertain as to why increased levels of folate in pregnant women's blood reduces the incidence of NTDs, but the evidence that it is does is irrefutable. Although the magnitude of the benefit varies considerably across populations (possibly because of differing genetic backgrounds or unknown environmental factors), dutiful and consistent consumption of 400 μg of folate a day on average reduces the background risk of bearing a child with spina bifida or anencephaly by at least 30%–50%.

Let us make sure we grasp how profound a statement this is. Consider the United States. If there was no screening and the general risk in the population for giving birth to a child with a NTD before the use of folate was one in 1000, some 4000 affected babies would be born each year. If proper folate consumption reduced that number by 50%, then only 2000 affected babies would be born each year and 2000 babies who would have developed a NTD would be born without that disability. Over a 20-year period that means that the simple act by women who could become pregnant of taking a daily supplement would result in the birth of 40,000 normal babies who might otherwise have had burdensome birth defects. Because the United States is one of the countries that requires the *fortification* of cereals with folate, it has achieved significant reductions in NTDs, but nowhere near 50%. Only the regular ingestion of supplements guarantees that a woman will maintain a protective level of folate. If all women in the United States who were able to become pregnant took 400 μg of folate a day, the number of children born with these defects could be cut substantially below current levels.

Our approach to averting the births of children with NTD is maddeningly contradictory. Few would oppose having women who may become pregnant take a daily vitamin supplement to reduce by half the risk of a serious birth defect in the fetus should they become pregnant. But many will be uncomfortable with a screening program that is intended to identify affected fetuses early enough to permit an abortion. This is especially so because many children with an NTD other than anencephaly are of normal intelligence (although often burdened with serious medical problems such as the inability to walk and poor bladder control). It is not possible during pregnancy to predict the future intellectual status of a fetus with a low spinal lesion, so the mother and father confront a troubling existential uncertainty. In deciding to terminate, they must acknowledge that a significant fraction of fetuses with similar lesions would be born destined to have normal or near normal intellect. Women in the United States terminate pregnancies in which the fetus has been found to have spina bifida at a sharply lower rate (about 60%) than do women in Europe (about 80%). Of course this difference reflects the influence of many different factors. But one can surely infer that in the United States we should redouble our efforts to educate women about the protection that consumption of folate confers on the fetus and strive to convince as many women who could become pregnant as possible to consume 400 μg of folate each day.

Despite the incomplete success of the folate fortification programs, they still constitute a great victory in the war to conquer orphan disorders. Just as a low-phenylalanine diet averts mental retardation after birth, a high-folate diet averts congenital malformations before birth. Because neural tube defects occur about 10 to 20 times more often than does PKU, a highly effective folate supplementation program prevents about fivefold to 10-fold as many cases of NTDs than does a newborn screening program designed to avert mental retardation due to PKU.

Although they represent two of the great early successes in the field of clinical genetics, neither the increase of population-based newborn screening to detect inborn errors of metabolism nor the use of folate to reduce the risk of spina bifida is closely tied to the rise of that new medical discipline. To trace the trajectory of what today has become an immensely powerful set of technologies orchestrated to develop therapies for hundreds of orphan disorders, it will help to offer some historical context. No medical discipline arises because of the efforts of a single person. Many factors—the growth of a research mindset at a nation's medical schools, the availability of governmental and private funds to support that research, discoveries in basic science that open the door to new therapeutic possibilities, and lobbying efforts by families affected by genetic disorders are among the most obvious—contributed to the rise of clinical genetics. Still, in this field, the contributions of a few people certainly drove it forward. In providing some historical background, I have chosen mainly to highlight the work of several scientists who virtually everyone working in clinical genetics would agree played critical roles.

2

The Rise of Medical Genetics

False Start

AFTER THE PRINCIPLES OF MENDELIAN INHERITANCE were rediscovered in 1900, their immense relevance to agriculture and animal husbandry rapidly drove the study of genetic science forward. Much of the great early work in genetics was concerned with improving crop yields and enhancing desirable characteristics in cattle and poultry. Plant and animal breeders were essentially studying what in the modern era we call quantitative trait loci (QTLs), and they were using statistical methods to infer gene action.

As I discussed in Chapter 1, the most important early contributor to *medical* (also called clinical) genetics was Archibald Garrod, the English physician who described several single-gene disorders and who coined the term, "inborn errors of metabolism" in 1908. He is without question the founding father of human biochemical genetics. But his contributions to understanding biochemical aspects of disease did not stimulate explosive growth in clinical research. Indeed, during much of his career, clinical genetics was misdirected or suppressed by flawed enthusiasm for eugenics, a movement that began in England and the United States in the 1880s. Medical genetics did not really emerge as a legitimate applied science until after World War II.

Eugenics, a neologism coined in 1883 by Francis Galton, a talented Victorian scientist who was enamored of statistics and who had published a book called *Hereditary Genius* in 1869, derives from the Greek for "well born." Throughout the western world, the elegant (and deceptively) simple principles of Mendelian inheritance were rapidly incorporated into intellectual discourse. In particular, some leaders of the "Progressive" movement that held sway in England and the United States in the late 19th and early 20th centuries saw genetic knowledge as a tool to help manage social problems. "Mendelism" arrived at a time when there was already substantial societal concern about hereditary diseases, especially those that affected human cognition and behavior. Thus, it was hardly surprising that after the basic sciences (such as plant breeding) led to the rediscovery of Mendelian laws, progressives would seek to apply them to problems in the human condition. During the years from 1905 to 1925, a group of highly motivated scientific eugenicists (an oxymoron to be sure), many of whom were closely linked to the Eugenics Record Office at Cold Spring Harbor on Long Island in New York, published a series of monographs that through the extensive study of extended families purported to show that many people who had complex phenotypes such as mental retardation, mental illness, alcoholism, and epilepsy were afflicted primarily because of a hereditary defect in the germ plasm—what we would today call single-gene disorders.

Charles Davenport, a Harvard-trained biologist who was both an excellent scientist and a deft fund-raiser, was among the first to apply Mendelian analysis to human disorders and conditions. From his perch as Director of the Genetics Laboratory at Cold Spring Harbor (an independent research institute largely funded by wealthy families), he began as early as 1905 to foster efforts to collect massive amounts of data from families in which particular phenotyptes seemed to be heritable. Thanks to a major financial gift from the family of Mrs. E.H. Harriman (the widow of a railroad magnate), in 1910 he created the Eugenics Record Office (ERO), a research institution that existed until World War II. In 1911 Davenport published *Heredity in Relation to Eugenics*, the first major compendium of human diseases and conditions linked to genetic factors. He described more than 100 disorders that he concluded were monogenic. In many cases, such as achondroplasia, Huntington's chorea, and alkaptonuria, he was correct. In many others, such as narcotism, pauperism, and criminality, he was gravely mistaken.

Soon after creating the ERO, Davenport recruited Harry Hamilton Laughlin, a young midwestern schoolteacher with a passion for eugenics, to lead the operation, which he did with a passion. Besides his interest in human clinical genetics, Laughlin was instrumental in driving eugenical thinking in broad areas of national social policy. For example, he lobbied many state legislatures to adopt laws permitting the sterilization of the mentally ill and the mentally retarded, and he testified at length before Congress on hearings concerning a proposed law to limit immigration. That law, which was enacted in 1924 and remained the major statute on the subject until 1968, implemented a quota system that restricted immigration by people thought to be of less desirable stock (such as those from eastern Europe) and imposed more generous limits on immigrants from England and Germany. At their height, sterilization laws, which were upheld by the Supreme Court in 1927, were on the books of more than half the states. Under their imprimatur, officials sterilized about 60,000 institutionalized mentally retarded persons, both men and women.

During the 1930s, the spread of eugenic programs slowed. One important, but all too quiet, braking force was the deepening understanding among basic scientists that the expression of complex traits was influenced by many factors, and that most did not manifest simply because of a single mutant gene. More obvious was the ghastly racist invective emanating from Nazi Germany, which, in 1935, would lead to a vast enterprise that sterilized at least 400,000 persons with various diseases as a prelude to the murder of millions because of their heritage.

Reviewing the earliest work in medical genetics leads one to conclude that for most of the first half of the 20th century, the quest to understand rare genetic disorders was carried on by a small cohort of deeply committed academic scientists and physicians, the latter often specialists, such as neurologists and ophthalmologists. During the late 1920s and 1930s, a few academic physicians began to advocate that lectures in medical genetics be added to the always overcrowded medical school curriculum. Among the first three were Madge Macklin, who gave a few lectures as early as 1926 at University of Western Ontario; Laurence Snyder, who began giving lectures at Ohio State University in 1933 and who published the first textbook in the field in 1941; and William Allan, who developed some lectures in 1936 at Bowman Gray School of Medicine. In 1940 he planned the first Department of Medical Genetics at Bowman Gray, but the project stalled because of World War II. In 1952, another leading human geneticist, Lee Dice, a biologist at the University of Michigan, asserted that Snyder had done more than any other person to coalesce this new medical discipline.

In the late 1920s, a fruit fly geneticist and future Nobel laureate, Hermann J. Muller, showed that radiation causes mutations, a finding of immense intellectual as well as practical importance. Muller strongly suspected that mutations caused by exposure to background radiation and chemicals were the source of new genetic diseases. But he was not a clinician. Unfortunately, he also (for a short time) became enamored with a dystopian form of eugenics that contemplated controlled breeding to improve the overall intelligence of the human species.

What was the state of medical genetics before World War II? There was not yet a department of medical genetics at any medical school in the world. There were no journals devoted exclusively to human genetic disease, no academic societies focused on advancing the field, and only a single, small textbook to support a course. Ironically, given the horrors that unfolded in Nazi Germany, many of the leading clinicians who showed special interest in heritable human disorders were Germans—in particular, ophthalmologists. So, despite the efforts of a few visionary academics, it is fair to say that a discipline of medical genetics did not exist before World War II.

Founding Fathers

During the period from 1941 to 1945, the demands placed on academic medicine by World War II effectively halted most non-war-related research activity. But, beginning in 1946, in part influenced by the birth of the Nuclear Age and an intense concern for the impact on humankind of deleterious mutations caused by radiation, medical genetics began to grow rapidly. In 1948, interested academic physicians, having reached a critical mass, incorporated the American Society of Human Genetics (ASHG), ever since then the most important organizational entity in this ever-growing field.

Although arbitrarily drawing lines is fraught with error, I believe that the birth of the field of *medical* genetics—the moment when physicians began to offer help to patients dealing with orphan genetic disorders in specialized clinics—can be pinpointed fairly well. It was the day in the spring of 1946 that Dr. James V. Neel (1915–2000), a *Drosophila* geneticist turned physician, became director of the "heredity clinic" at the University of Michigan School of Medicine in Ann Arbor. Thanks to a grant from the Horace Rackham School of Graduate Studies, the clinic had opened its doors in 1941, but there had been little activity during the war.

Neel was, unquestionably, a founding father, but he was not the only founding father of this new clinical field. Within the next few years, physicians at the Bowman Gray School of Medicine, the University of Utah, the University of Texas, the University of Toronto, and the Hospital for Sick Children in Montreal also founded medical genetic clinics. But, among Neel's competitors for the title of "Founding Father," only Victor McKusick (1921–2008), who originally planned to be a general practitioner in his beloved home state of Maine, has an equal or better claim. The exigencies of the war permitted McKusick to leave Tufts before completing an undergraduate degree to enroll in The Johns Hopkins School of Medicine, from which he graduated in 1946, and he never left. There were, of course, other physicians and scientists who made important early contributions to the field of medical genetics—James Crow, H. Bentley Glass, Kurt Hirschhorn, William Schull, Barton Childs, Alex Bearn, Clarke Fraser (a Canadian), Arno Motulsky, and Alfred Knudson being among the more notable. But the two men I have singled out for special attention in this chapter are certainly exemplars of the founding cohort. Looking back over my years in clinical genetics, I take pleasure in realizing that I met all of these fine people and got to know most of them.

A child of the Great Depression, when he graduated from high school, James Neel, whose father died when he was 12, had no other option but to attend the College of Wooster (Wooster, Ohio) located a few blocks from his home. There he fell under the influence of a young geneticist named Warren Spencer, who, in 1935, pushed him toward graduate school at the University of Rochester, largely because it was the academic home of one of the country's greatest geneticists, Curt Stern, who had recently fled from Nazi Germany. Although his PhD thesis on genetic control of certain aspects of development in fruit flies went well, by 1939 Neel's thoughts were turning to the possible applications of genetic knowledge to human disease. At the time—partly because of the stigmatization that human genetics suffered because of the misguided eugenics movement that had dominated early thinking in the field—this was a bold move. In 1939 few academic scientists were working on problems in human inheritance. When, in his final year as a graduate student at the University of Rochester, Neel asked Stern if he could organize a seminar in human genetics, the older man readily agreed. When Stern scoured the scientific literature for what he considered solid papers, he found only 20 that he felt had conducted research on the inheritance of a human trait with anywhere near the rigor typically found among scientists working with fruit flies and corn.

Although Neel became ever more deeply interested in medical genetics and was sure a medical education was in his future, he spent the next 3 years as a postdoctoral fellow, working in some top laboratories and making important discoveries about the environmental causes of mutations. In mid-1942 he returned to the University of Rochester to study medicine, where (because of his doctoral work and the accelerated training necessitated by World War II) he was able to complete his undergraduate medical work in just 2 years. Ironically, given his abhorrence of the eugenics movement, in 1943 Neel (while still a medical student) published his first paper in clinical genetics, positing that being a redhead was the consequence of having inherited two recessive alleles (later proven to be true), based on research he did in the Eugenics Record Office archives at Cold Spring Harbor.

After spending his intern year at the University of Rochester Hospital, Neel had a stroke of luck that shaped the rest of his life. Just a few years earlier, Lee Dice, an embryologist at the University of Michigan, had obtained a grant to fund a research program in human genetics. The war years made it difficult to staff new medical programs. In 1945 Dice offered the job to Neel (still a medical resident at Rochester), who promised to come as soon as he completed his training and military service. In 1948 James Neel, age 33, became the director of one of the first two heredity clinics (as they were then called) in the United States. The choice of such a young physician was significant, in part, because Neel made a major contribution to understanding the inheritance of an orphan blood disease then known by several different names—familial microcytic anemia, target cell anemia, and thalassemia—while he was still in his senior year of medical school.

The word "thalassemia" is fashioned from the roots of two Greek words: "*thalassa*," the word for sea, and *anemia*, the word for blood. George Whipple (a future Nobel laureate who had studied Greek in high school) coined the term when he was the Dean of the University of Rochester. It had been long known that this strange blood disease was found most commonly in people who had a Mediterranean heritage. The condition was first well described clinically about 1925 by Dr. Thomas Benton Cooley, one of America's leading pediatricians (France awarded him the Legion of Honor for his work in Paris during World War I overseeing the care of thousands of children), who was at that time Chief of the Children's Hospital of Michigan. Even though he was deeply occupied by his administrative duties both in Michigan and as President of the American Pediatric Society, Cooley decided to train himself in childhood blood diseases, then an almost unexplored field. During the period from 1921 to 1925, he encountered four children in his Detroit clinic with severe anemia and who also had dramatic changes in the shapes of their facial bones. These children soon died. In 1925 he published this small series of cases, giving the disorder the name "erythroblastic anemia," correctly recognizing that the misshapen bones were caused by the dramatic expansion of the bone marrow space—as if the body was trying desperately to make needed red blood cells.

When Neel saw his first patient with what was then called Cooley's anemia, as any medical student would do, he consulted a hematology textbook. He was struck by the author's observation that each of the parents of children with Cooley's anemia often seemed to have signs of a much milder anemia (characterized by smaller than normal red blood cells). In an epiphany that he remembered his whole life, Neel immediately surmised that Cooley's anemia was a rare recessively inherited disorder. He and a fellow medical student named Bill Valentine set out to see if they could confirm the pattern of inheritance.

On their precious off-duty hours, they combed hospital records to identify at-risk families and rode the city bus lines to visit them, take a full family history, and draw blood for study. They eventually examined one or more affected children who had been diagnosed within the last 15 years in 11 families—all of Italian origin and all hailing from southern Sicily! The resulting paper, "The hematologic and genetic study of the transmission of thalassemia (Cooley's anemia, Mediterranean anemia)," was published in a leading medical journal in 1944. Neel named the severe form of the disease (in which the child had two recessive alleles) "thalassemia major" and the mild carrier form (just one recessive allele), thalassemia minor, terms still in wide use today. Their work also unearthed enough information to calculate the incidence of babies born with the disease and the frequency of the abnormal allele in the Italian population. Their calculations that about one in 25 persons in southern Europe carried a single copy of the recessive gene and that about one in 2400 babies born to couples of Italian origin were born with thalassemia, turned out to be quite accurate.

World War II and its aftermath almost completely severed communications between scientists living within the United States and those in Italy. I believe it likely that Neel did not then know of studies similar to his that were being conducted in Italy in the mid to late 1940s by Ezio Silvestroni and Ida Bianco, two hematologists at the University of Rome, who also deduced that persons with benign microcytic anemia carried a single mutant allele, and that if a person was born with two copies, he or she would have severe thalassemia. More than 60 years later one can still occasionally find complaints in the scientific literature that these two fine Italian scientists were never awarded adequate credit for helping to prove that thalassemia is caused by a rare single-gene disorder.

Clinicians who were struggling to help children with thalassemia major—which if untreated, is uniformly fatal in childhood—benefitted greatly from advances in blood transfusion medicine that were in part driven by World War II. Beginning about 1950, physicians began to treat young children with thalassemia by giving them regular blood transfusions, a strategy that quickly extended life expectancy, often by many years. In addition, if the diagnosis was made early in life and if hematologists carefully monitored the children to prevent them from ever becoming too anemic, many fewer developed misshapen bones or experienced unusually short stature. This is because regular transfusions suppress the body's effort to sharply increase red cell production, which is what expands the bone marrow space and causes bony abnormalities. Unfortunately, frequent transfusions have the terrible side effect of overloading the body with too much iron, a condition that damages many organs, including the heart, and which is also eventually fatal. Beginning about 1970, hematologists—especially in Italy where the disease is relatively common—began to treat affected children with bone marrow transplants. We will return to these advances in a later chapter.

The first research task that Neel set for himself when he joined the faculty of the University of Michigan was to study families with sickle cell anemia (SCA), another orphan disease that was relatively commonly seen in African-Americans, and that he suspected might also be caused by two recessively inherited genes. For three decades (1945–1975), studies of SCA were crucial to framing an approach to understanding molecular aspects of all genetic disease.

It is no surprise that the oldest medical literature concerning SCA derives from Africa. During the 19th century, some Africans called the widely recognized disorder "ogbanjes," a term loosely translated as "children who come and go," because of the high childhood mortality rate. The earliest use of the term "sickle cell" was by Dr. J.B. Herrick, a Chicago cardiologist (1861–1954), and his intern Ernest Edward Irons (1877–1959), who saw the abnormally shaped, long curved red cells under the microscope when he examined a blood smear taken from Walter Clement Noel, a 20-year-old black dental student from Grenada (and who they erroneously diagnosed with "muscular rheumatism," probably because of the painful joints that are a hallmark of the disease). The earliest use of the term to describe the disease itself may have been in 1922 when Verne Mason, a physician at The Johns Hopkins Hospital, used it for that purpose.

Neel believed that understanding of SCA was overly influenced by the fact that physicians usually cared for sickle cell patients only when they were hospitalized with a severe pain crisis. Seeking to better understand the natural course of the disease, he collaborated with Dr. Wolf Zuelzer, a pediatric hematologist at Children's Hospital, to study every family in Detroit's large black population in which at least one child had SCA. Neel developed a laboratory method to induce red blood cells in persons who carry one copy of the gene for SCA to make the abnormal shapes for which the disease is named. Among the first 42 parents (who each carried just one abnormal gene) of children with sickle cell disease, he found that he could induce sickling in the blood of each, but that it was *extremely difficult to induce sickling in the blood of their unaffected, noncarrier* children. By taking a careful medical history and performing laboratory tests on blood samples, Neel showed that SCA was an autosomal recessive disorder, and that being a carrier (having just one abnormal allele) was a benign condition. This work, which was published in *Science* in 1949, has long been considered a key paper in the emergence of medical genetics. Later, Neel learned of a little known paper that reached similar conclusions about the inheritance pattern of the disease, which was published contemporaneously by a military doctor named E.A. Beet in East Africa. In reaction, Neel, who was a fair, but highly competitive, person, was quick to point to an even earlier paper of his in which he had strongly surmised that SCA was an autosomal recessive disorder.

Just a few months later another paper appeared in *Science* that Neel (with whom I occasionally talked at scientific meetings) recalled as having read with greater intensity than any other among the thousands he read in his lifetime. Studying their migration in an electrical field, a team led by Linus Pauling (later *twice* a Nobel laureate) showed that the hemoglobin molecules of persons with SCA moved differently than did those from persons without the disease. The only logical conclusion was that some aspect of the protein was altered in persons with the disease. In this paper, Pauling became the first to use the term "molecular medicine," so it may fairly be considered to mark the birth of that field, which is now in an explosive growth phase. Seven years later, Vernon Ingram (1924–2006), a German who fled from the Nazi regime to England as a teenager and went on to become one of the world's most famous protein chemists, used a technique called paper chromatography to prove that the change in the movement of hemoglobin molecules taken from persons with SCA in an electrical field was due to the substitution of a single amino acid (valine for glutamic acid) in the sixth position of β-hemoglobin protein. That is, SCA is caused by a tiny change in a giant protein!

Throughout the 1950s, Neel expanded his studies of sickle cell disease. As he looked further and heard from more physicians around the world who were caring for patients who had abnormal hemoglobin but for whom the illness did not fit the standard descriptions, Neel realized that the clinical severity of hereditary anemia varied widely. Collaborating with the group that had first studied hemoglobin in an electric field, Neel identified a new variant dubbed hemoglobin C (so called because it was the third form to be found) and showed that children with SC anemia (one copy of each mutation) were not as severely ill as those born with classic sickle cell anemia. Sensing that a whole new world of medical research was opening, many scientists reoriented their careers to study the genetics and chemistry of hemoglobin molecules. Today, scientists have identified scores of different mutations in the hemoglobin molecules of humans.

In addition to his many contributions to human genetic research, Neel realized early that, even if there were few or no efficacious treatments for the steadily growing number of disorders that were clearly attributable to mutations in a single gene, prompt diagnosis would be of immense importance to the parents of an affected child and to their close relatives. For parents who gave birth to children with severe autosomal recessive disorders like thalassemia and SCA faced a one in four risk that any subsequent pregnancy would also result in the birth of an affected child. Neel, a great teacher, became convinced that academic medical centers needed to train specialists in clinical genetics. He was among a few dozen physicians and scientists who in 1948 founded the ASHG, which today is the preeminent organization of its kind in the world.

In the early 1950s he substantially expanded the activities of Michigan's heredity counseling clinic, which had been contemplated in 1941. About the same time (1943), Sheldon Reed, a PhD human geneticist, began working at the newly formed Dight Institute for Human Genetics at the University of Minnesota in Minneapolis. Reed (who I knew well and who once discussed this with me) was uncomfortable with the fact that his patron, Dight, a wealthy Minneapolis businessman, had been an ardent eugenicist. Partly to distance his work in the "heredity clinic" from the horribly tainted notion of eugenics, Reed embraced the term "genetic counseling," which had been proposed by Lee Dice, the third president of the newly formed ASHG. At the group's fifth annual meeting in New York City in 1952, Dr. Dice chaired a discussion on the challenges of "genetic counseling" that was subsequently published in the society's journal.

During his long, distinguished career Neel made many important contributions to genetics, especially in the areas of radiation biology and population genetics. His work in radiation biology began in November of 1946 when as a young army officer he was appointed to a small team to assess the long-term health effects on the population exposed to radiation after the atomic bombs were dropped on Hiroshima and Nagasaki. During the 6 months that the team spent there, Neel (working closely with a fine population geneticist named William Schull) focused on developing a system to quantify the frequency of birth defects and on studying the many cases of childhood anemia that were probably due to mutations caused by the radiation. He spent most of the next 2 years of his military service living in Nagasaki. Although primarily concerned with the impact of the radiation on pregnancies, he also helped initiate a long-term effort to monitor the population for cancer. His postwar studies in Japan led to a lifelong relationship with the nation. In the early 1960s he conducted studies to evaluate the impact of consanguineous marriages (usually between first cousins) on the risk for genetic disease, an indirect method of assessing the prevalence of disease-causing mutations in the human gene pool.

The work that Neel pioneered in radiation biology stimulated a major change in his career. It drove him from a primary interest in clinical genetics to a much deeper interest in understanding the gene pool—the sum of all genes among all individuals in a population. For more than 3 decades (1960–1990) he would primarily explore the gene pool through cultural, clinical, and biochemical studies of two isolated human populations: the Xavante in Brazil and the Yanomami in interior Venezuela. In effect, he was taking on one of the great questions first articulated by Darwin: How does natural selection work? His first major paper on this effort was an 88-page summary of his research published in 1964, a work he regarded as a pilot study to guide more sophisticated efforts in the future. In the 1960s he spent extended time in the interior of the Amazon basin, understandably with some trepidation. Neel once told me that he had undergone a "prophylactic" appendectomy in Ann Arbor to eliminate one possible threat to his life—acute appendicitis—in a region without any access to medical care! It is appropriate that when Neel published his autobiography that he called it *Physician to the Gene Pool*!

Neel conducted one of his earliest studies in human population genetics in Michigan. One of the big questions he wanted to answer was to understand the spontaneous mutation rate among genes in the human genome. One early example of this work involved a dominantly inherited disorder called aniridia, a congenital form of severe vision impairment due to the failure of the formation of the iris. In 1958 Neel assigned the problem of determining the exact number of patients with aniridia among the population of Michigan to a young physician named Margery Shaw, who had studied genetics at Columbia and Cornell before going to medical school at the University of Michigan. Neel reasoned that because the severe eye disease was easy to ascertain, if the total number of cases could be counted, Shaw could infer the mutation frequency of the gene and how likely it was that affected persons reproduced (a measure of "selection pressure" against the mutation).

Shaw, who had just finished her internship at the University of Michigan, spent the better part of 2 years on this project, the results of which were published in the *American Journal of Human Genetics* in December of 1960.

I met Margery Shaw (1923–2012) 13 years later in the spring of 1973 at the University of Texas in Houston, where I spent 2 years working in her laboratory (which focused on human cytogenetics, the study of chromosomes). Dr. Shaw, who is one of only two women who has served as president of both the ASHG and of the Genetics Society of America, was my first and most important mentor in human genetics. She often regaled me with humorous anecdotes about her life. My favorite concerned the aniridia study. In one small town that shall remain unidentified, she found several teenagers with aniridia—an unexpectedly high number given the rarity of the condition. Because none of the parents were affected, these children all seemed to have the defect because of a spontaneously arising mutation in the egg or sperm cell. On mathematical analysis, the odds of this having occurred were astronomical. A few weeks after her first visit, Dr. Shaw returned to reexamine the issue. One day she pulled into a gas station and to her immense surprise recognized that the handsome man who pumped the gas had aniridia. On questioning him, Shaw found the solution to the problem. The charming fellow acknowledged that he could have fathered one or more of these children.

This is one of the earliest examples of how nonpaternity can confound efforts to study human genetics. With the advent of DNA technology, it is now easy to address the problem of possible nonpaternity in such studies. But DNA tests do not help us resolve the ethical issues that arise when a clinician discovers what had been unknown or well-hidden evidence of nonpaternity. This is one clinical dilemma that physicians are likely to face with greater frequency in the future. Studies repeatedly suggest that ~2%–3% of individuals are not the offspring of the acknowledged father.

While Neel was building his career at the University of Michigan, Dr. Victor McKusick was following a different path at The Johns Hopkins School of Medicine in Baltimore. One-half of a set of identical twins, McKusick, who was born in rural Maine and schooled in a one-room schoolhouse by the same teacher for 8 years, was drawn into medicine because of an illness he suffered as a teenager. In 1937 he developed an abscess of his left axilla and an ulcer on his right elbow, both of which would not heal. After local care came up short, he was transferred to the Massachusetts General Hospital where physicians eventually cultured a fastidious form of streptococcus, and successfully treated him with a 10-week course of sulfanilamide.

McKusick, who attended a high school that had not offered even introductory science courses, matriculated at Tufts University with his eyes on medical school. Because of the disruptions of World War II, some medical schools began accepting undergraduates after just 3 years of study. In 1943 he applied and was admitted to Johns Hopkins Medical School after just $2\frac{1}{2}$ years of undergraduate work at Tufts, thinking that 4 years later he would be doing his internship at a hospital in his beloved Maine. But, the core tenet of the medical school—that good patient care depends on rigorous clinical research—seduced him. After graduating from Johns Hopkins Medical School in March of 1946 at age 24, he stayed on to do his residency in medicine. When he completed it in 1948, he joined the faculty and never left. A quarter century later (1973) he was named the William Osler Professor-in-Chief of Medicine at Hopkins, arguably the most honorific position in medical education in the United States (Osler was one of the founding physicians of the medical school).

McKusick's fascination with genetics began early. In June of 1947, while still an intern, he cared for a teenager named Harold Parker who had a bizarre constellation of findings—a large number of intestinal polyps (that required repeated surgery to avert the threat of cancer) and melanin freckles on the inside of his lips. Over the course of the next 2 years, McKusick evaluated four similar patients. Learning that a Boston physician named Jeghers had also been "collecting" similar patients, McKusick proposed that they combine their 10 cases and write up a detailed analysis of the condition. In 1949 they coauthored a paper (published in *The New England Journal of Medicine*), comprehensively describing this *monogenic* cancer syndrome. Some years later it acquired the eponym of Peutz–Jeghers syndrome in honor of the Dutch physician who first pointed out some of the clinical signs in 1921. In trying to understand what caused the odd constellation of findings in this disorder, McKusick was influenced by Bentley Glass, a prominent basic scientist who also worked at Hopkins. Glass dissuaded him from attributing the findings to the effects of two closely linked genes, instead asserting that *pleiotropism* (varying signs caused by the pathological effects of a single mutant gene) was far more likely to be the unifying explanation.

After his internship, McKusick focused on cardiology, itself a new specialty. From 1948 to 1950 he directed the nascent cardiovascular unit at the Baltimore Marine Hospital, where he helped to develop tools for analyzing heart sounds. He was among the first physicians to perform cardiac catheterizations, and he may have been the first to reverse an abnormal rhythm with a drug (lidocaine). On July 1, 1950, McKusick returned to Hopkins to complete his residency, and in 1952 he stayed on as an assistant professor of medicine on the Osler service. In addition to his regular clinical duties, he pursued a new interest that had emerged from his study of heart disease. Although the mechanisms were at best poorly understood, cardiologists knew that a number of uncommon causes of severe heart disease arose because of disorders involving the connective tissue (which is crucial to maintaining the strength and distensibility of the large blood vessels). About 1952, McKusick embarked on a comprehensive study of five such disorders, of which a disease called Marfan syndrome was the centerpiece. In 1956 he published *Heritable Disorders of Connective Tissue*, an achievement that certainly makes him the father of genetic cardiology.

Marfan syndrome is a dominant disorder in which the underlying gene mutation harms many different organ systems—another example of pleiotropy. Affected persons have—to mention a few of many findings—unusually long arms, remarkably flexible joints, and eye lens abnormalities, and they were, in the 1950s, at great risk for sudden death from dissecting aortic aneurysms (because of a weakened vessel wall). The disease is named for a distinguished French pediatrician, Jean Antoine Marfan, who was (among other achievements) a pioneer in treating children with tuberculosis and in the role of nutrition in preventing and ameliorating disease. In 1896 he published a report of a 5-year-old girl who manifested most of the features mentioned above. It is unclear as to whether the child actually had this disorder or one of several disorders that are similar to it. But, in 1931 a Belgium physician named H.J.M. Weve, in describing a patient with similar stigmata, coined the eponym, and it stuck.

According to McKusick, medical genetics became "institutionalized" at Johns Hopkins on July 1, 1957. That was the day that the 36-year-old McKusick on the invitation of his boss, Dr. A. McGehee Harvey, became the Director of the Moore Clinic, a once famous institution that had originally focused on treating syphilis, but that had been languishing. McKusick convinced Harvey that the Moore Clinic should refocus to study hereditary disorders. In the ensuing 15 years, a position at the Moore Clinic became one of the most sought after training spots in the world for those interested in medical genetics. Among them, the talented young doctors who came to work with him described and delineated several dozen new diseases. For McKusick it was a period of immense productivity in a career that would include more than 800 publications!

Of his many contributions to medical genetics, McKusick is probably best known for creating the first comprehensive catalog to correlate single-gene alterations with particular human phenotypes. This monumental project, which was conceived about 1960, grew out of a monthly journal club held at McKusick's home in which the research fellows would have to bring index cards on which they had written summaries of the newest findings concerning patients with rare single-gene disorders that they had evaluated at the clinic. For the first decade or so this ever-growing, multivolume work was called "Catalogs of Autosomal Dominant, Autosomal Recessive, and X-Linked Phenotypes." It was first published as a book with the title *Mendelian Inheritance in Man* (MIM) in 1962. Online access became available in 1987. The 12 editions that McKusick published from 1966 through 1998 constitute a detailed history of progress in human genetics, especially gene mapping and clinical delineations of thousands of phenotypes. The latest online version of MIM describes nearly 10,000 phenotypes that have been firmly associated with, or are suspected of being caused by, a variation in a single gene! Even today, it is rare to walk into the office of a clinical geneticist and not glimpse at least one hardcover edition of MIM on a shelf. In addition to online MIM (OMIM), there is now a cognate version for companion animals called OMIA, which lists all known single-gene disorders in dogs, cats, horses, cows, and other husbandry animals.

One exceedingly important, if serendipitous, event in McKusick's career was his long-term relationship with the communities of Old Order Amish in Lancaster County, Pennsylvania and other areas, a connection that he began with them in 1963 and that lasted until his death. Several forms of dwarfism and other monogenic disorders are present among them at a much higher frequency than in the general population. As an honor for their intense efforts to study the genetics of dwarfism, Victor and several of his fellows were made honorary members of The Little People of America! His research led to the delineation of several rare genetic skeletal disorders.

McKusick became involved in gene mapping (locating the gene for a particular disorder to a particular segment of a particular chromosome) early. Once a technique called somatic cell hybridization became available in the 1970s, efforts to map human genes accelerated. Along with Frank Ruddle at Yale, McKusick was a leader in organizing this effort and in running periodic meetings to update the scientific community. In the current era of low-cost DNA sequencing, such efforts seem antiquated, but they constituted a tremendous advance at the time.

McKusick was eager to stay abreast of new technologies, so it is not surprising that he became an early, ardent supporter of the Human Genome Project in the mid-1980s. As DNA analytical techniques improved and could be applied at reasonable cost to clinical research, in 1986 he became coeditor of a new journal called *Genomics* to publicize those advances. In 1988 he agreed to serve as the first president of HUGO, an international body created to help coordinate what would be, given the state of technology, a massively challenging effort—to sequence the human genome. As an international statesman of human genetics, he was a superb choice. In 2008 Dr. McKusick won the Japan Prize, that nation's highest scientific award, for his contributions to "Medical Genomics and Genetics." During the 1950s and 1960s the medical centers at the University of Michigan and The Johns Hopkins University were the two top sites in the United States to train if one wished to become a medical geneticist.

Although she did not achieve the stature or influence of Neel or McKusick, Madge Thurlow Macklin has a fair claim to being a "founding mother" of medical genetics. Born in Philadelphia in 1893, Madge Thurlow earned her BA at Coucher College in Baltimore in 1914. After a year or two of political activism in the suffragette movement, she enrolled in The Johns Hopkins School of Medicine, which awarded her an MD in 1919. It was at Johns Hopkins that she met her future husband, a Canadian named John Macklin. In 1921, when he took a professorship at the University of Western Ontario in London, Canada, they married and she followed him there, spending the next 5 years acting as his research assistant. By 1926 Madge Macklin had become interested in genetics and embarked on a path to study the influence of genes on risk for cancer. About then she also began advocating for adding genetics to the medical school curriculum, an advocacy that would continue unabated for the rest of her life.

Macklin's star would shine more brightly in the firmament but for her early and persistent advocacy of eugenics. From about 1925 to about 1945 she was active in the Eugenics Society of Canada, serving once as its president. She openly advocated the forced sterilization of mentally retarded persons, until the madness of the Nazi regime ended that strain of thinking in North America. In 1946 Macklin separated from her husband and moved to Ohio State University where she pursued a distinguished career in breast cancer genetics until she died in 1962. One measure of her importance is that in 1959 she became the first woman to serve as president of the ASHG, 5 years after Neel and 15 years before McKusick.

Another physician who nearly matches the level of contribution made by Neel and McKusick is Arno Motulsky. Born in Germany in 1924, Motulsky enjoyed a happy childhood until 1933 when the Nazis came to power. As antisemitism grew progressively more violent, the family sought to leave

Germany. In May of 1939, Motulsky and some of his relatives (along with 1000 other Jews who had obtained coveted visas) departed for Cuba on the SS *St. Louis*. But while they were on route, the Cuban government revoked the visas. The United States refused to admit the travelers, and the ship returned to Europe. Just 2 days from arrival, the passengers learned some countries would give them asylum, and Motulsky wound up in Brussels where he eventually obtained an American visa. But the Germans soon overran Belgium, and Motulsky began a life of internment in various camps. Eventually, he was able to leave Belgium, and via several stops, he arrived in Chicago in 1941 where he was reunited with his father.

He started working as a laboratory technician and taking pre-med courses. At age 20, he was drafted, an event he later recalled as one of the luckiest of his life because the U.S. Army sent him to medical school at the University of Illinois. He did his residency at Michael Reese Hospital in Chicago where he became fascinated with hematology, which in turn drew him to medical genetics. In 1953 Motulsky took his first academic job—as a hematology instructor at the University of Washington, where he remained throughout his career. He began offering lectures in medical genetics that year. Motulsky, who served as President of the ASHG in 1977, is credited with founding the field of pharmacogenetics—the study of how one's genes influence drug metabolism—because of studies that he initiated about 1957. Today, the study of how variations in genes shape the body's responses to small molecules is a core aspect of drug development. Among the many people that Dr. Motulsky trained is Joseph Goldstein, with whom he initiated studies of lipid metabolism. Goldstein, later working closely with Dr. Michael Brown, made the challenge of understanding lipid metabolism his life's work. In 1985 Goldstein and Brown shared a Nobel Prize for their success in elucidating the nature of the lipid receptor.

In 1948, inspired by rapid advances in the field of genetics, a small group of physicians and scientists incorporated the ASHG, which they intended to foster research into their field. They also promptly created the *American Journal of Human Genetics* as a forum for publishing promising research results. The group elected Hermann J. Muller its first president. Muller, the polymath who had studied with the great Thomas Hunt Morgan (the first gene mapper) at the "fly lab" at Columbia University, had recently won (in 1946) the Nobel Prize in Physiology or Medicine for his discovery in 1927 that radiation caused mutations in genes. Initiating a tradition of having the president deliver and publish an annual address, Muller wrote a classic paper entitled "Our Load of Mutations," one of the first to argue that mutations were a much more common cause of disease than most scientists thought. To appreciate Muller's profound insights, one should remember that at the time, scientists still debated whether DNA *or* protein was the hereditary material, and that it would be another 5 years until scientists deduced the structure of the double helix. In 1954 Neel served as the sixth president of the ASHG. Sheldon Reed served as president in 1955, and Victor McKusick served as president in 1974. From a few dozen members in 1948, the ASHG has grown into a flourishing organization with more than 6000 people regularly attending its annual meeting.

During the 1950s and 1960s the study of human genetics grew steadily. The following advances are a few examples of that growth. In North America and Europe more and more academic physicians decided to focus their work on genetic disease, and the number of heredity clinics grew rapidly. In 1952, a young Chinese researcher named T.C. Chu, who was one of the first scientists to work at the great M.D. Anderson Cancer Center in Houston, perfected a method to better study human chromosomes. In 1953 James Watson and Francis Crick deciphered the helical structure of DNA (a discovery that did not lead to advances in clinical medicine until many years later). In 1956, the year that scientists finally established the correct number of human chromosomes (46) in a cell, the First World Congress of Human Genetics was held in Copenhagen. Also in 1956, at the University of Michigan, Neel created the first *Department* of Human Genetics in a medical school, an event that was replicated at several other universities the following year. In 1958 the French clinical geneticist Jérôme Lejeune proved that Down syndrome was caused by the presence in cells of an extra chromosome 21, setting off a frenetic (and highly successful) search for other correlations between chromosome variations and disease. In 1962 the English geneticist Mary Lyon hypothesized that in women one of two X chromosomes is inactive in each cell, a finding that helped explain X-linked disease and that opened up new thinking about gene regulation. In 1968 a Danish scientist named Torbjörn Caspersson discovered a way to use dyes to reliably (if crudely) map human chromosomes, in effect increasing our powers of resolution more than 10-fold.

Yet, despite these impressive scientific discoveries, the field of medical genetics for some time remained a relatively small discipline that did not even have its own "institute" at National Institutes of Health (NIH), the source of most funds for academic clinical research. It was not until 1989, largely in response to the birth of the ambitious Human Genome Project, that the NIH finally created a National Center for Human Genome

Research, calling on James Watson to be its first director. The federal government did not elevate the center to an independent institute (thus finally giving it parity with 26 other institutes) until 1997. Also slow to come was the recognition by the American Medical Association (AMA) that clinical genetics deserved to be a bona fide specialty. The American College of Medical Genetics was incorporated in 1991, and the AMA finally recognized the field as a specialty in 1996.

Currently, there are about 40 training programs available in the United States to physicians who seek to specialize in medical genetics. Despite the stunning technological advances that promise to drive genetic information into every medical specialty, the profession remains comparatively small. Only 216 individuals passed a board examination in medical genetics in 2013. This may be in large part because medical genetics has always been practiced almost completely within academe. In addition, it may reflect the growing familiarity in all areas of medicine with how to collect and analyze genetic information.

Genetic Counseling

The field of medical genetics expanded rapidly during the 1970s, and by the late 1970s there were more than 100 clinics in the United States. Essentially all of them operated in academic medical centers under the direction of a professor of pediatrics or medicine. After the initial challenge of attempting to make an accurate diagnosis of an orphan disorder in a child was completed, the next task was often to explain future reproductive risk to the parents. Of course, in the frequent cases in which a definitive diagnosis could not be made, estimation of recurrence risk for future pregnancies was fraught with uncertainty. Depending on the diagnosis and the parents' understanding of statistics, discussions about reproductive risk and other topics often required at least an hour for each couple. To meet the demand for the counseling services required much more time from attending clinicians than they (who were often more interested in laboratory research) could allocate. Sometimes nurses helped handle the workload.

Although others may have sensed in the late 1960s that the time was ripe to create a new health care profession focused on the delivery and interpretation of genetic information, many experts give credit to Melissa Richter, who at the time directed the Program for Continuing Education at Sarah Lawrence College in New York, for initiating an effort that led to the first master's degree program in genetic counseling. In 1968, Richter, in collaboration with Joan Marks who became the director of the program, worked out the structure of a program that would focus on traditional classes in biology, social work, and psychology in the first year, and in the second year place students as interns in genetics clinics in the New York City area. The first students in genetic counseling started in the Sarah Lawrence Program in the fall of 1969. The idea caught on; over the next decade about a dozen similar programs opened their doors.

The number of master's degree programs has enjoyed slow, but steady, growth since then. In 2014 in the United States there were 25 fully accredited programs with several others under development. For more than a decade only colleges and universities in the United States operated master's degree programs, but today they can be found (albeit in small numbers) around the globe. Canada has three, England has two, Japan has six, and there is at least one university-based program in many other nations.

By the mid-1970s the academic leaders of genetic counseling, of whom the most influential was Joan Marks, began to discuss the need to make the infant profession more robust. In 1977, after a meeting attended by nearly 100 genetic counselors, which generated a spirited debate, the group created the National Society of Genetic Counselors, which in 1979—despite some resistance from physician-geneticists who at first questioned the competence of a graduate of these programs to offer adequate counseling—became the official organ of the profession. Over the ensuing years the National Society of Genetic Counselors (NSGC) did much to develop the profession of genetic counseling. In 1993 it created the American Board of Genetic Counseling, and developed a certification examination (an examination for which one could only sit if he or she had completed a master's degree program), launched its own journal, worked to achieve recognition and respect from physician-geneticists, and lobbied for adoption of state licensing laws. State licensure is the key to professional independence. It provides an irrefutable legal basis for engaging in private practice, and is crucial to enable third party billing of insurers. Unfortunately, thus far only a handful of states have enacted such laws, and very few genetic counselors have a robust private practice.

Nevertheless, in 2014 there were only about 30 accredited genetic counseling programs in the United States and three in Canada. In the United States there are now about 3000 board-certified, actively practicing genetic counselors. Those who have remained in the profession (which is more than 90% female and is characterized by substantial interruption of professional careers to have children) earn a median salary of about $65,000.

Although most counselors are employed by medical schools or in large obstetric group practices, a growing number (~10%) now find employment at companies that are offering a rapidly expanding number of genetic tests. For example, Genzyme, until recently the nation's largest genetic testing company, once employed nearly 100 counselors. Industry-based counselors often interface with physicians more than they do with patients. One of their main jobs is to explain and interpret very esoteric test results to physicians who often know little about them.

Extraordinarily rapid advances in understanding genetic predisposition to cancer, developmental disabilities such as autism, and common disorders such as heart failure and diabetes suggest that a demand for genetic counselors will grow substantially. By one count the number of diseases for which there are genetic tests has tripled since 2001 to more than 2500 today. The big insurance company, United Health, has predicted that revenue from all genetic tests in the United States will soar from $10 billion in 2010 to $25 billion in a few years. Some concern has been voiced that genetic counselors working for laboratory testing companies face an inherent conflict of interest. I disagree. The core tenant of the profession is to remain absolutely neutral in discussion with patients and doctors about the need for genetic testing, the implications of the test results, and the right course of action. This is in large part because in its early days nearly the entire work of the profession involved the sensitive topic of prenatal diagnosis.

What impact have genetic counselors had on the burden of genetic disease? For most of the profession's 40-year history most counselors have spent much of their professional time interpreting the results of prenatal testing and providing support to women and their partners as they confront deeply troubling information. The two most common scenarios involve testing women (for most of the past four decades defined as those who would be 35 or older at delivery, but now enlarged to include all pregnant women) for the risk of carrying a fetus with Down syndrome (trisomy 21) or some other chromosomal error or advising them about their option to be screened to identify those who are carrying fetuses with neural tube defects.

There is no method I know of to measure the impact of genetic counseling on decisions to continue or terminate a pregnancy depending on the results of a prenatal test. In the United States, over time there has been a steady annual decline in the number of live births of children with Down syndrome compared with the number that would be expected purely on the basis of the number and age distribution of pregnant women. However, the actual number of children born with Down syndrome each year in the

United States has not changed much. The ever-larger number of women (nearly 20% of those who give birth) who delay pregnancy until their late 30s (when the risk of having a child with Down syndrome is much higher) offsets the decline in numbers that one would expect from those women who choose prenatal testing and terminate on learning that the fetus is affected (currently estimated in the United States to be ~75%). The fact that the absolute number of births of children with Down syndrome has not risen over the last decade (as expected because of population increase) suggests that most women who learn that they are carrying a fetus with trisomy 21 terminate the pregnancy.

Given their commitment to value neutrality, one would expect that genetic counselors have had no discernible impact on this trend. It is possible that many genetic counselors hold a private bias in favor of termination, but that is not likely to influence the choices made by their patients. Rather, it is the emergence of the testing as a standard part of prenatal care that has made the difference. More than 30 years have elapsed since it became a standard of care in the United States for a doctor to inform a pregnant woman about risks of certain birth defects and the availability of tests to discern them. Today, a physician who does not discuss the availability of prenatal screening runs the risk of a malpractice lawsuit should the woman give birth to a child with one of the disorders in question. Advances in testing technology and a fear of legal liability are among the forces that are driving the growing use of testing and the decline in live births of children with Down syndrome and other abnormalities. Data collected from 21 registries in Europe (where there are relatively few genetic counselors) including 6 million pregnancies show a picture that is quite similar to that in the United States.

Given the ubiquity of the test, over the years the most common task for genetic counselors has been to educate women about the availability of a screening test to detect fetuses with neural tube defects (NTDs) and to inform and counsel them about the results. These disorders arise around the 21st day of gestation due to a failure of the spinal cord to seal properly. Depending on the location of the lesion along the spine, the condition is associated with mild or severe disabilities. In general, higher lesions (those closer to the head) cause much greater disability than do lower lesions. There are several forms of NTDs, ranging from fatal anencephaly to a mild lesion of the lower spine that is compatible with a quite normal existence. Unfortunately, the current biochemical tests do not assess severity of the NTD.

Beginning in 1987, it became standard practice in the United States and Europe to offer all pregnant women a screening (nondiagnostic) test for

NTDs. If the test is positive it is followed up with other tests, especially biochemical analysis of amniotic fluid, to make a firm diagnosis. Genetic counselors have been and still are called on to manage this screening and counseling task. Here the observations that I made in regard to testing for Down syndrome apply. Counselors do not attempt to influence choice, and there is no evidence that they have shaped pregnancy outcomes.

Although millions of screening procedures have been performed to identify women who are at risk for giving birth to an infant with a NTD, the reduction in live births of affected children in the United States does not seem to have been achieved as much by testing and pregnancy termination as by the widespread use of folate acid supplements. Surprisingly, in certain areas of the world the live births of affected children who were diagnosed early enough in pregnancy to permit a legal termination are on the *rise.* For example, in New South Wales over the last 20 years the number of women who have terminated an affected fetus has declined from 82% to 62%. This may reflect parental reaction to the physician's inability to predict the severity of the condition, a greater tolerance for and acceptance of persons with disabilities, or a greater confidence that needed social services will be available.

In the United States today, every medical school and some large hospitals operate medical genetic clinics. The menu of cytogenetic, biochemical, and DNA tests available to diagnose children born with rare disorders has soared. For example, as we probe the human genome, we are finding many germline mutations that increase the risk for developing cancer later in life. Many academic oncology centers now include clinics where genetic counselors counsel men and women about their genetic risks for developing cancer. Genetic testing is rapidly altering the field of cancer medicine. All cancers are clonal in origin; that is, they arise from mutations accumulating in a single cell. Thus, all cancers are in the most fundamental sense genetic disorders. We are learning that each tumor has its own genetic profile. It will soon be routine for an oncologist to order a DNA analysis of tumor tissue to assess if it contains mutations that might make the patient fare better on one drug or combination of drugs than another. In 2013 Foundation Medicine, a tumor DNA profiling company based in Cambridge and the largest among the first generation in this new field, went public and earned significant valuation from investors. The first realization of the dream of "personalized medicine" may well be in cancer medicine. Ten years hence it is quite possible that genetic counselors will be working as often in cancer clinics as they do in the traditional setting of reproductive risk.

3

Blood

Transfusions

THE POLYMATH CHRISTOPHER WREN (who among other accomplishments was the architect who designed St Paul's Cathedral in London) is often cited as being among the first to have experimented with blood transfusions—between dogs—in 1659, just 30 years after William Harvey became the first person to accurately describe the circulation of the blood. Others give credit to an Oxford physician named Richard Lower, whose work in transfusing dogs in the mid-1660s is more thoroughly documented. The earliest well-documented record transfusion of blood into a *human* that I know is a report in 1667 from physicians at the Royal Society in England who paid Arthur Coga, a "mad" Cambridge student, 20 shillings to agree to permit them to transfuse sheep blood into his body (an experiment that would never be approved by an ethics committee today!). Some physicians apparently thought that the transfusion might cool the fire in his brain. They must have transfused very little ovine blood, as Coga survived. He later wrote an essay in Latin saying that the experience had improved his madness (perhaps the earliest example of the placebo effect), but some who knew him said that was not the case.

Over the next 150 years, the idea of blood transfusion engendered much speculation and not a little humor. For example, one wag wondered what would happen if a Quaker donated blood to an Anglican Archbishop. Would the ardor with which the church father adhered to the Creed diminish? Would blood from high-minded people improve the behavior of reprobates? Could judicious use of blood transfusions reduce criminal behavior? Many physicians, for whom the letting of blood was an essential element of the therapeutic arsenal, were predisposed to believe that just as reducing blood volume might relieve congestive heart failure and kidney disease, the of blood might act as a general tonic to improve everything from muscle strength to dementia. Fortunately, it only took a few misadventures to establish that one could *not* use animal blood to treat human illness (a death rate approaching 90% is fairly convincing).

During the early 19th century, physicians occasionally tried human-to-human transfusions, but with only limited success. Credit for the first *successful* transfusion to humans is often given to James Blundell, a British obstetrician who in 1818 allegedly saved the life of a woman having postpartum hemorrhage by taking blood from her husband and immediately injecting it into the woman. Twenty-two years later Blundell also participated in the first documented transfusion to treat a boy with hemophilia. Although physicians were astute enough to realize that a donation from one family member to another was preferable to using an unrelated donor, success (defined here as *merely surviving* the transfusion) was still a matter of luck. It depended on the chance that the recipient had a blood type that was compatible. Only ~15% of the English population at the time had "O negative" blood, which we now recognize as safe to give to anyone.

The first great advance in developing safe methods for donating blood to patients was the discovery by Karl Landsteiner (1868–1943)—then a young, assistant professor of pathology at the University of Vienna—in 1900, of the ABO blood groups. Deeply interested in biochemistry, after he graduated from medical school, Landsteiner spent 5 years in chemistry laboratories in Zurich and Munich. He then returned to Vienna and began to work in the new field of immunology, which was emerging from the triumphant advances in understanding bacterial infections. In 1900 Landsteiner showed that when a blood donation from one human to another failed, a microscopic examination of the clumped red cells looked the same as when blood was given from one species of animal to another. This observation laid the foundation for the notion of "compatibility" and laid the groundwork for transfusion medicine. Landsteiner posited that some sort of reaction to "nonself" caused the sudden release of vast quantities of hemoglobin from billions of exploding red blood cells, and that this explained the jaundice and shock that one saw in patients in whom transfusions failed. His observations did not immediately garner much attention. But, over the next 8 years he meticulously worked out and named what we know today as the ABO blood group system. In 1930 Landsteiner, who had moved to New York to work at the Rockefeller Institute for Medical Research, was awarded the Nobel Prize for this work.

In one of the first great fusions of genetics with physiology, in 1907 Reuben Ottenberg (1882–1959), a graduate of Columbia University's College of Physicians and Surgeons who was to spend his entire career at Mt. Sinai Hospital in New York City, observed that Landsteiner's ABO blood groups obeyed the rediscovered Mendelian laws of inheritance. This led him to perform the first human-to-human transfusion based on tests designed to *exclude* genetically incompatible donors. The next transformative advances—in 1915 and 1916—involved the first use of anticoagulant agents to preserve donated blood. This ended the short-lived era in which blood was transferred directly from donor to recipient as vein-to-vein transfusions and led directly to the first blood depots—banks of stored blood kept close to the front lines in World War I.

Each year in the United States today, physicians use about 10 million units of blood to treat victims of trauma, manage blood loss in major surgery, treat the anemia associated with cancer, and help patients with rare hemoglobin disorders. From World War I to the conflict in Afghanistan, donated blood has saved tens of thousands of soldiers. Much less well known, but equally important, was the development between the late 1930s and early 1950s of a new therapeutic strategy to save the lives of thousands of children with β-thalassemia: a rigorous regimen of regular blood transfusions.

Beginning about 1950, when physicians came to understand that an immensely confusing set of overlapping clinical observations about apparently different forms of anemia were actually observations about a single monogenic disorder that Jim Neel had described as thalassemia major, the stage was set to develop a much deeper understanding of the disease. During the middle of the 20th century, regardless of the name it carried, it became clear that one intervention was the key to helping affected children—regular transfusions. Fortunately, the availability of safe transfusions grew exponentially in the decade after World War II.

The term "blood bank," so familiar today, was coined in 1937 when Dr. Bernard Fantus, at Cook County Hospital in Chicago, launched the first such entity in the United States. World War II radically altered the priorities of American medicine, but in late 1945 many of the advances first used in the battlefield were brought home. The American Association of Blood Banks was created in 1947, and by 1950 there were 1500 hospital-based blood banks in the United States and 31 American Red Cross regional blood centers. Fifteen years later there were 4400 blood banks and 55 Red Cross blood centers. Although at a less dramatic pace, blood banking grew rapidly in Europe as well.

Perhaps the most astounding aspect of β-thalassemia is the frequency of the mutated alleles. In Italy, during World War II, Silvestroni and Bianco showed that overall, about one in 20 Italians carried a disease allele, but there were regions such as Ferrara where as many as one in 10 persons was a carrier. Studies in other parts of the world subsequently showed that mutations were extremely common from Italy to India to Southeast Asia. Rough calculations suggested that literally millions of humans carry one mutation for this rare disease and that each year at least 100,000 children—mostly in third world countries—are born with thalassemia major. During the 1950s the vast majority died before the age of 5.

But, why are these mutations so common? In 1949 J.B.S. Haldane, who by then was a leader in the field of population genetics, proposed that those who carried one copy of a mutation in the hemoglobin gene were protected from becoming ill from malaria. In his words, "The corpuscles of the anaemic heterozygotes are smaller than normal, and more resistant to hypotonic solutions. It is at least conceivable that they are also more resistant to attacks by the sporozoa which cause malaria." His now famous hypothesis was informed by studies conducted first in Italy as early as the 1920s and throughout the world during the 1940s that showed that the frequency of thalassemia alleles in human groups was highest among those living in the world malarial belt (lands near the equator).

Haldane led such an interesting life that he merits a brief detour. The son of an academic biologist, in mid-childhood he was already working in his father's laboratory, often as a guinea pig in experiments of human physiology. As a young adult, he continued experimenting on himself, once drinking dilute hydrochloric acid to study its effects on his muscles! A large, fearless man, during World War I he single-handedly carried out daring attacks on the enemy, leading his general to call him the "the bravest and dirtiest officer in my army." After the war, Haldane returned to academic life. A gifted student who was already a master of four languages, he went on to make important contributions in chemistry, genetics, and mathematics. He is most remembered for his applications of mathematics to the field of population genetics. At one point in life he claimed to be one of only three men who understood the mathematics of Darwin's theory. Haldane loved to speculate about the impact of science on the future of society. He wrote a science fiction novel, *Daedalus; or, Science and the Future,* that predicted test-tube babies, and which was the basis for Aldous Huxley's much more famous novel, *Brave New World.* An outspoken communist, Huxley eventually retired to Sri Lanka where he worked to bring science education to the third world.

β-Thalassemia

As with the struggle to conquer PKU, the quest to save children from the ravages of β-thalassemia has more than a few heroes, but also as with PKU the story of the quest to conquer β-thalassemia can be told around the work of a few key people. No man is more closely associated with the story than is the British physician, David Weatherall, whose lifelong work on the disease began when he met a young Nepalese girl named Jaspir Thapa (much of the next two pages is drawn from Weatherall's highly readable "biography" of thalassemia). In 1958, Weatherall, a newly qualified doctor just 25 years old, arrived in Singapore to begin his 2 years of national service. He was assigned the weighty task of being the medical officer in charge of the children's ward at the Alexandra Hospital (a colonial era institution that is today a first-rate teaching hospital). Jaspir, the daughter of a Gurka soldier, was a patient. She had been diagnosed with severe, chronic anemia since infancy and was being kept alive by frequent blood transfusions.

Weatherall had no idea of the cause until a biochemist at the hospital named Frank Vella showed him a paper published a few years earlier under the then odd title of "Mediterranean Anemia in Thailand." In the late 1950s, western physicians still largely believed that the disease described by Cooley in 1925 and studied by Neel in the late 1940s was found almost exclusively in people born in the Mediterranean basin or in Africa. After studying Jaspir's medical history and examining her blood, the two men wrote up a case report that appeared in the *British Medical Journal.* They had no idea at the time that they were describing a disorder that afflicted many tens of thousands of children in Southeast Asia and India. Weatherall spent his second year of national service in Taiping near the border with Thailand. In his free time, he used a car battery and filter paper to rig up an electrophoresis machine, to enable him to identify individuals with thalassemia (hemoglobin of people with thalassemia moves differently in an electric field than does that of unaffected persons).

After being discharged from the military in 1960, Weatherall, now hooked on studying the genetics and physiology of blood disorders, went to Johns Hopkins University for further study with McKusick's group. There, he joined with John Clegg and Michael Naughton, two talented protein biochemists, to better understand how the recently discovered pairs of α- and β-globin peptide chains that comprise a single molecule of hemoglobin were synthesized. Several years of hard work resulted in proof that thalassemia major was caused by a failure of cells to produce adequate amounts of the β chain, leading to an often profound excess of the α chains, which wound up as harmful deposits in the cells. Their paper, published in *Nature* in 1965, is one of the landmark papers in the birth of molecular medicine.

Weatherall had found his life's work. He returned to the medical school at Liverpool, England, where he wrote *The Thalassemia Syndromes*, the first comprehensive text on these conditions, and started a hematology research group that later moved with him to Oxford. Over the years, he and his team made a string of elegant discoveries showing that tiny genetic deletions caused many different forms of thalassemia. They were also among the first to show that microdeletions of chromosomes could cause mental retardation. In 1970 he and his team showed that babies who were stillborn because of a severe blood disorder had a genetic disease due to which they could not make any α chains. In 1971 they discovered a rare form of α-thalassemia brought about by mutation that causes the DNA to make a protein that is much too long! During the 1970s, they also proved Haldane's malaria hypothesis when they compiled definitive proof that *carriers* of another form of thalassemia arising from mutations in the genes that code for the production of the α chains are strongly protected from malaria caused by *Plasmodium falciparum*.

Perhaps Weatherall's most remarkable achievement unfolded during the late 1960s and 1970s as he and others realized that, although rare in Europe, the genetic thalassemia syndromes constituted a huge public health problem in Asia, the unexpected and astounding dimensions of which came into focus after the World Health Organization engaged Weatherall to lead an extensive fact-finding study. Over the ensuing decades, Weatherall was foremost among many hematologists who advocated for creating clinical partnerships with hospitals in the world's poorest countries to bring optimal care to afflicted children. For many years Weatherall focused his energy on supporting a clinic in poverty-stricken Sri Lanka, an immense challenge when one contemplates the logistics of collecting and storing the needed blood, closely following the children's clinical course, and managing the iron overload that is an inevitable consequence of the transfusions and, itself, a frequent cause of death. In 2010 Weatherall won the highly prestigious Lasker–Koshland Award, "[f] or 50 years of international statesmanship in biomedical science—exemplified by discoveries concerning genetic diseases of the blood and for leadership in improving clinical care for thousands of children with thalassemia throughout the developing world."

Beginning about 1925, the year that Cooley compiled a comprehensive description of the disorder, the history of efforts to conquer thalassemia major can be (not too arbitrarily) divided into several periods. Until the early 1960s, the mainstay of therapy was periodic blood transfusions. But, during the 1940s and 1950s few physicians were giving careful attention to the optimum use of this therapy. Key questions remained unanswered. The most important was, what minimal level of hemoglobin should physicians strive to maintain in affected children? No one knew. Approaches varied, but most hematologists were letting children become severely anemic (say to 5–6 g/L) before they transfused them. In poor areas of the world this policy was driven by a lack of regular access to blood donations; in rich countries, it was driven by a growing awareness that the accumulation in the tissues of excess iron caused by frequent transfusions posed grave risks to the patients.

In 1963 at a meeting sponsored by the Cooley's Anemia Foundation in New York City, Dr. Irving Wolman, who cared for patients at the Children's Hospital of Philadelphia, presented a small study that showed that as compared with children who were transfused only when symptomatic, children in whom doctors consistently kept the hemoglobin level above 10 g/dL grew taller, had less enlarged livers and spleens, had more normally shaped bones, and had fewer fractures. His work confirmed that children with poorly managed thalassemia major often developed profoundly misshapen bones (because of expansion of the bone marrow space, as the stem cells whose job it is to make red cells receive biochemical alerts about the body's low level of hemoglobin and desperately attempt to overcome it). Wolman acknowledged the risk of iron overload, but argued that a higher level of hemoglobin would suppress the body's tendency to absorb iron from food (a response to chronic anemia) so that the net increase in iron might be acceptable given the clinical benefits he anticipated.

Unavoidably, studies of how children respond to therapy for a chronic disease take many years. Wolman persisted; 6 years later (1969) he reported to the same group on the progress of the 17 children he had maintained at a comparatively high level of hemoglobin. By now there was no doubt that they had less bony deformities, fewer fractures, and less enlarged livers and spleens. However, six of the 17 children had died of causes directly attributable to excess iron, which damages many organs, but is particularly harmful to the heart. For the children to benefit from maintaining a high level of hemoglobin, physicians needed a way to limit the accumulation of iron that was an unavoidable consequence of so many transfusions. Thus, another key problem to solve was to find a way to chelate the body's stores of iron (molecules that bind to metals and remove them are called chelators).

By this time, researchers had acquired a fairly sophisticated understanding of how humans acquire, use, recycle, and store iron. There are only ~5 grams (about one-third of an ounce) of iron in the body of a healthy adult. About 70% of this is in the circulating red blood cells (iron atoms are essential for their work), ~20% is bound to special storage proteins, and ~10% is in all the other cells. A child with thalassemia who is receiving 10 transfusions each year gains about an extra 2 g of iron; over 10 years he or she accumulates about four to five times the total body stores of iron that he should have. This volume overwhelms the body's mechanisms for eliminating iron and much of it winds up in tissues where it should not be.

In the early 1960s scientists at the Swiss Federal Institute of Technology and at the CIBA pharmaceutical company in Zurich began studying molecules called sideramines, which some bacteria use to capture the iron they need to live. This led to the development of the first iron-chelating agent approved for human use, a compound called desferrioxamine (also called deferoxamine). Unfortunately, the drug could not be given by mouth and intramuscular injections were painful. To reduce the body's iron stores, patients would have to take desferrioximine every day, a very difficult regimen. Early trials of the chelator were limited by the patients' ability to tolerate the drug, and the reductions in body iron were not impressive. However, some clinical researchers persisted.

Good news came in 1974 when a team at the Great Ormond Street Hospital in London reported on a 10-year study in which they had combined intramuscular injections with intravenous administration when the children received their blood transfusions. Although the study was small, there were fewer deaths from iron overload in the treated group than the untreated group. A few years later, a group in Boston showed that by giving desferrioximine by continuous intravenous therapy (over 24 hours), they could sharply reduce body iron stores. But how could this finding be adapted for regular clinical use?

At Oxford Dr. Martin Pippard had the simple, but clever, idea of doubling the dose while halving the infusion time. It worked well. Although it was hardly optimal for the patient, Pippard's finding ushered in a new clinical era. Patients could be maintained at high levels of hemoglobin without being poisoned by iron if they were given intravenous chelation therapy at home while they slept. It takes a long time to show that a chelating therapy is safe to administer each day for years, that it can reduce iron levels that have been accumulating for many years, and that organs will remain healthy. But, by about 1995 several clinical teams had shown that despite the obviously difficult problems in complying with nightly intravenous therapy, long-term chelation clearly led to a significant reduction in deaths and disability from iron overload. The success stimulated several pharmaceutical companies to begin research programs to develop an iron-chelating drug that was longer acting and could be taken by mouth.

Improved transfusion regimens and the advent of chelation therapy eventually provided the answer to another difficult question. Many children with thalassemia developed very large spleens (photos of children with thalassemia major taken in the 1920s show children with spindly bowed legs, oversized jaws, and swollen abdomens that one associates with profound malnutrition). Large spleens worsened the anemia in two ways—more red cells were destroyed there, and large spleens increased fluid volume. When surgeons removed these big spleens, the anemia often improved somewhat, but the children were now at greater risk for serious infection. Debate swirled for several decades as to the value and proper timing of such surgery, but with the advent of aggressive transfusion management and chelation the need for this operation fell sharply.

The twin advances of careful titration of hemoglobin levels throughout childhood and aggressive chelation therapy have had a positive impact on life expectancy. Because thalassemia major is rare in the United States (about 1500 patients), one can find better data in Italy, which in the immediate post–World War II era was the only developed nation that had both a reasonably large population of affected patients and reasonably good transfusion services. In 1998 a team at the University of Ferrara reported on its study of 1146 patients with thalassemia major born between January of 1960 and 1988. For the cohort born in 1970–1974, the doctors found that if transfusions were readily available, 82% of patients survived to age 25. This was a big advance, but the patients had numerous serious side effects from the iron overload that was a consequence of regular transfusions. About 80% of both sexes had diminished sex hormones and ~6% had serious cardiac problems.

Seven years later, another Italian group reported that 65% of patients in a long-term study were alive in mid-adulthood. This improvement in survival was largely due to much careful management of body iron stores. Another major study compared the survival of 977 patients born before 1960 with that of 720 patients born after 1970. It found that the more recently born patients—those treated from an early age with both regular transfusions and an iron-chelating agent—had a dramatically better ($p < 0.0005$) rate of complication-free survival. Data from a large patient registry in the United Kingdom indicate that between 1970 and 1990, life expectancy of children born with thalassemia major jumped from 17 to 37 years. By 1999, thanks largely to the aggressive use of newer chelating agents to reduce iron poisoning of the heart, life expectancy had soared; between 2000 and 2008 the annual death rate had decreased by ~70%. Together, the two interventions were changing the nature of thalassemia major, converting it from a fatal disease of childhood to a serious chronic disease of adulthood.

What is it like to be a child with thalassemia major who has been cared for since infancy at a *first-rate* thalassemia clinic such as the one led by Dr. Elliott Vichinsky at the Children's Hospital and Research Center in Oakland, California? If today you met such a child on a playground, you would likely have no idea that he or she had a serious disease. Because children are diagnosed at birth (due to newborn screening), they can be cared for from infancy by experts. It is clear that when that happens, the children enjoy much healthier lives. The 56-page Standard of Care Guidelines at Oakland begins by saying "Treatment for thalassemia has dramatically improved. Patients should live full lives with careers and children of their own." But it then cautions that "Unfortunately, many patients die prematurely of morbid preventable complications." In the United States and Europe and in some other parts of the world, thalassemia major, once a fatal childhood disorder, is now a manageable chronic disease akin to type 1 diabetes. But we are not yet able to declare victory. The challenges of lifelong management are demanding indeed for the patient, the family, and the clinical team. For many adult patients compliance with the rigorous management regimen is a daily challenge, frequently not met.

And, sadly, these great advances are unavailable to many children in the world's poorest countries. It is painful to report that although the average life expectancy of an affected child in the United States and Europe now may be more than 50, worldwide the median life expectancy may still be as low as 7 and certainly not much more than 10. In too much of the world the fate of affected children is not much better than that day in 1958 when Dr. David Weatherall first met Jaspir. If children born with β-thalassemia do not have access to regular transfusions—the case in much of Africa and Southeast Asia—they die. One of the most difficult challenges that the field of clinical genetics faces is how to provide new—and expensive—lifesaving therapies to children everywhere.

Sickle Cell Anemia

Because the study of sickle cell anemia (SCA) has played a critical role in the maturation of our understanding of the molecular basis of genetic disorders, it is fitting in a chapter on blood to go deeper into this disorder that affects some 30,000 people in the United States and many hundreds of thousands around the world.

Despite the great advances in understanding the pathophysiology of SCA and in diagnosing the condition in infancy, our ability to care for people with the disorder has remained woefully inadequate for decades. When I was a house officer at Boston City Hospital in the early 1980s, we regularly saw young adults in the emergency room who were in terrible pain because of the small infarcts in their bones or lungs caused by the stickiness of the oddly shaped cells in the capillaries and other small vessels. We provided intravenous hydration and pain medications, but little more. Many young adults with SCA looked chronically ill. They usually walked with difficulty (because of injuries to their joints), showed signs (such as a drooping cheek) of having had a stroke, were unable to work because of chronic pain, and suffered—understandably—from depression. Until the 1980s, ~10% of children born in the United States with SCA died in childhood, often because of pneumococcal pneumonia to which they are predisposed. Overall, during the last quarter of the 20th century in the United States, about one-half of all patients died before reaching the age of 40. Sadly, severe, recurring pain crises drove many more men with inadequate health care to drug addiction, further compounding their hard lives. The first drug (hydroxyurea) shown to improve the health of patients with SCA was approved for that purpose in 1995.

What is it like to be a patient with SCA in the United States today? Who better to answer that question than a young person with the disease? I first met Helen Sarpong in 2011 when I was seeking a patient to act as a teacher to the staff at bluebird bio, a company that was working to develop a gene therapy for treating patients with SCA. I met Helen through Dr. Martin Steinberg, the director of the outstanding Sickle Cell Disease Clinic at Boston Medical Center, after I asked him to connect me with a patient who would be comfortable talking about her disease to a room full of strangers.

Helen, a willowy black woman then in her 30s, has velvety skin, a captivating smile, and a cheerleader's personality. Recently married, she bragged about her husband and shared with me their dream to some day open a bakery (both graduated from a culinary school) in Somerville, a town that borders Cambridge, just north of Boston. When I met with her to plan the company visit, I quickly realized that she was quite an expert on her disorder. For example, she reminded me that most babies born with SCA are healthy in the first months of life. This is because humans have evolved a complex system for carrying oxygen to cells. In the first few months a slightly different protein called fetal hemoglobin does much of the job. Helen also recalled that when she was an infant in 1976 most babies with SCA announced that they had a problem through abdominal pain, swollen, painful hands or feet, or pneumonia. This is no longer the case. Since about 2000, all babies born in the United States are tested for SCA; parents learn that their babies have SCA *before* they undergo a medical crisis.

Most of the problems in sickle cell disease arise because of one of two phenomena. During their trip through the body's thinnest blood vessels, the misshapen red cells tend to clump together and block the passage, depriving the nearby tissues of vital oxygen, causing pain and swelling, and killing cells. Sickle cells also tend to self-destruct (hemolyze), especially in their travels through the spleen, so patients sometimes become severely anemic, occasionally rather quickly.

As with so many patients with SCA, the full impact of the disorder on Helen can only be understood in the context of her family life and of the resources available to her. During the first 2 years of her life, Helen lived in New York City. After a doctor told her mother that she would likely soon die of her disease, Helen's mother sent her to Ghana to live with her grandparents and extended family. But, in her words, "Despite the odds against me, I continued living, albeit suffering greatly without any hospital care or medication."

Helen returned to the United States in 1985 at the age of 10. This was the first year that she benefitted from treatment of pneumonia with penicillin. Helen recalls this as the era in which "the nightmares of midnight emergency room trips and month-long stays in the hospital began." Her worst encounter with the medical system happened when she was a college student in Florida at an affluent private hospital in Naples. She recalls, "I had to beg the ER doctors and nurse to please look up information on SCD online just to get help, but instead they locked me in a room and told me to be quiet because my loud crying was scaring the other patients!"

After college Helen moved with her boyfriend, Johnathan, to Boston "solely based on the fact that there was a Comprehensive Sickle Cell Center at Boston Medical Center. The one thing I *always* tell other folks I know in Boston who have sickle cell is to avoid community hospitals," she said. "You can be almost certain that the doctors in the ER know little about the disease. Most of the time, despite the fact that you are writhing in pain, they think you are faking it to get drugs."

Helen turned 40 in May of 2015, a milestone she trumpets, as when she was young she was told she would likely not live that long. Sickle cell disease continues to injure her. One reason why she and her husband have yet to open the bakery they dream of owning is that she now has avascular necrosis in her spine and both hips. In 2014 she underwent a total hip replacement and spinal surgery. Helen's upbeat spirit and her supportive husband are critical to the success with which she has battled her disease. Women with SCA face special risks in pregnancy. Nevertheless, Helen and her husband, working closely with expert doctors, now have two healthy children!

In the larger cities today, clinics that are dedicated to helping patients in crisis have reduced hospitalizations. Specialists are available who can help control pain without increasing risks of addiction. Hydroxyurea, which induces production of fetal hemoglobin to well above normal levels, has reduced the need for transfusions. But SCA is still a deadly disease. As patients enter adulthood, available services fall away, unemployment is a huge problem, and medical care moves from being proactive to reactive. Nevertheless, consistent clinical management from early childhood with a strong focus on maintaining a healthy lifestyle has extended the median life expectancy from about 15 years in 1973 to about 50 years in 2015 (with women tending to outlive men). This is great progress, but much remains to be done.

Given the large patient population and huge unmet medical need, it will be no surprise to learn that many scientists at academic medical centers and at biotech companies are working assiduously to develop new therapies for patients with SCA. The website, Clinicaltrials.gov, lists nearly 300 open research trials. Unfortunately, few of them appear to be focused on breakthrough technologies. Many are studying the impact of combining a second drug with hydroxyurea. But because hydroxyurea is a potentially toxic compound originally developed to kill cancer cells, the long-term effects of its use in children (despite reassuring studies thus far) remains a cause for concern. Currently, the more innovative efforts range from harvesting and treating the patient's *stem cells* with lentiviral vectors that can then be returned to the body to populate the bone marrow and produce normal hemoglobin to efforts to develop small molecules that will be safer and more efficacious than hydroxyurea in up-regulating the long-term production of fetal hemoglobin. A particularly exciting effort is being driven by a California-based company called Global Blood Therapeutics that is developing an orally available small molecule to modify the affinity of hemoglobin S for oxygen. It is among several that intend to develop novel small molecules to reduce the polymerization that leads to the stickiness of the cells. Early studies in rodent models have generated signs of efficacy, but human trials will not occur for a couple of years. Given the immense effort under way to harness novel technologies to treat SCA, it is likely that several breakthrough therapies will emerge over the next decade.

Factor VIII

Hemophilia A, one of the best known of the several thousand rare single-gene disorders, may have introduced the idea of inherited disease to humanity. The earliest written references to a familial bleeding disorder are found in rabbinical texts from the second century of the Common Era. For example, Rabbi Judah the Patriarch exempted a woman's third son from being circumcised if her first two sons died of bleeding. Even more insightful was Rabbi Shimon ben Gamaliel, who forbade a boy to be circumcised if the sons of his mother's three elder sisters had died after circumcision. In the 12th century, Maimonides forbade the circumcision of a boy born to a woman married for the second time who had lost sons to a bleeding disorder during her first marriage. Thus, at least by then, learned men recognized that the bleeding problem was hereditary, that it affected only boys, and that the risk was transmitted through the mother—the hallmarks of a highly penetrant, X-linked single-gene disorder!

Detailed case reports of families with hemophilia began to appear in the 18th century. According to Laureen Kelley, the president of the Save One Life Foundation, the first person in North America to be diagnosed with the disorder was a boy named Appleton who lived in Ipswich (Massachusetts) about 1660. In 1803 one American physician traced his affected patient's family back nearly a century to a woman who had settled in Plymouth, New Hampshire. In the early days this disorder went by several different names, but by 1828 a leading British treatise on blood chose—oddly enough—to call it haemophilia (literally, love of blood). William Osler, Physician-in-Chief at the Johns Hopkins School of Medicine, was among the first (1894) to publish (in his textbook of internal medicine) a robust argument that hemophilia was a sex-linked disorder in which girls were unaffected. He worked this out by studying extended families more than a decade before Mendel's laws of inheritance were rediscovered.

It is likely that at the start of the 20th century more was known about hemophilia than any other genetic disorder. For example, in 1911 two British physicians published a major text on the disease and cited nearly 1000 references and 200 documented pedigrees! In England immense medical and popular interest in hemophilia was driven by the birth of Queen Victoria's eighth child, Leopold, Duke of Albany, who was affected. Because the disease was unknown among either side of Victoria's ancestry, Leopold's disorder arose because of a new mutation in a sperm cell from Victoria's father, Edward, Duke of Kent (some have raised the issue of possible nonpaternity, but it has never been proven). Leopold suffered many severe bleeding episodes in his life, but grew to adulthood and married, before dying at age 31 from a bleed in his head caused by a fall.

In an era of large, interconnected families, the hemophilia gene spread quickly through the royal families of late 19th century Europe. Quite possibly, the most famous orphan disease patient in history was Alexis, the only son of Leopold's niece, Alexandra, and Tsar Nicholas II of Russia, who was born in 1904. Anguish over Alexis' many bleeding episodes and frustration over the inability of physicians to help the boy led the royal family into the clutches of a visionary monk named Rasputin who came to exert great influence, probably distracting the Tsar from the many national problems that would culminate in the Communist Revolution.

Even though little was yet known about the devilishly complicated process of blood coagulation, by 1910 it was understood that hemophilia was caused by the failure of blood to clot. About then scientists who were studying different fractions of plasma found evidence to suggest that persons with hemophilia had reduced levels of what they called prothrombin. That erroneous conclusion dominated thinking until 1935 when a brilliant scientist named A.J. Quick discovered that neither reduced levels nor malfunction of prothrombin caused the overly long blood clotting time in patients. Just 2 years later a research group at Harvard discovered that normal human plasma carried a large molecule other than prothrombin, just a tiny amount of which will clot the blood of hemophilic patients. They named it "antihemophilic globulin." Over the ensuing 20 years, further research led to the discovery of many other proteins that kept the body's clotting system in balance, and it was established that most people with classical hemophilia lacked adequate amounts of a protein called by convention Factor VIII.

Still, there was no treatment. Patients with the most severe form of the disorder suffered severe recurrent bleeding episodes, especially into their joints, that crippled them in childhood. They typically died in adolescence or young adulthood, often after a fall. The long list of substances used in an effort to treat hemophilia between 1900 and 1935 reflects the limited understanding of the disease. One of the first substances shown to have any therapeutic value was a snake venom that was such a powerful clotting agent that it could only be used topically.

In the first half of the 20th century, research physicians were perplexed by the variation in the severity of hemophilia among patients. In those days before DNA testing, *Drosophila* and corn geneticists had already come to realize that the manifestation of a genetic condition could vary significantly by the location and type of mutation within a gene, but human geneticists (who could not conduct breeding experiments) remained uncertain of the relevance of that finding to human genetic disorders.

Over the last 40 years by painstakingly studying the relationship between particular mutations, the expression and function of the damaged proteins, and the patients' clinical condition, hematologists have learned a great deal about Factor VIII. This protein is so efficacious that boys who have as little as 3%–4% of the normal amount have only a relatively mild bleeding disorder. However, among patients with <1% of protein activity the disease is life threatening. This finding is generally true for many other rare genetic disorders that arise because of a loss of function of a protein. In general, evolution seems to have provided us with a buffer; we can usually function well even if a key protein is functioning at only 10% of the normal range, and sometimes we can do well with even less.

The second half of the 20th century witnessed extraordinary advances in developing therapies for hemophilia. To appreciate those advances, we have to turn back the clock to about 1950. At that time the key treatment for hemophilia was whole blood transfusions in response to a crisis. But this posed great difficulties. The blood-banking system was in its infancy, and many community hospitals did not yet store blood. In those days patients with classical hemophilia could suddenly become severely ill (e.g., from a spontaneous hemorrhage into a joint space). One could not assume that a bleed could be managed with admission to a large urban hospital. Severe bleeds might well require more blood of the desired type than was available.

Far more challenging was the problem that a unit of whole blood actually contains very little Factor VIII (the lifesaving protein that had not yet been discovered). It was both difficult and dangerous to give multiple transfusions to small children who were bleeding into their joint spaces and muscles. There was a thin line, indeed, between getting enough Factor VIII into them and throwing them into pulmonary edema because they received more fluids than their hearts could handle. Among severely affected patients, especially active children from 5 to 12, bleeds (which often started with the slightest trauma such as a fall on the playground) were very frequent. Some parents spent 50–100 nights (this is not a misprint) a year in the hospital as doctors and nurses tried to stop the bleeding that, once started, might last for many days. Some years later, the discovery that units of fresh frozen plasma were safer and more effective to use than units of whole blood because they greatly reduced the risk of volume overload, constituted an important advance in treatment, but it only partially reduced the number of lonely nights in the hospital that parents spent watching their children in pain (bleeding into joints is intensely painful).

Among the many articles and books written by dedicated parents striving to draw public attention and support to an orphan disorder that has incapacitated their children, none is more moving than *Journey* by Robert and Suzanne Massie. As has been repeatedly the case in the quest to treat orphan disorders, a single family often galvanizes interest in driving improvements in research, patient support, and public education; in the case of hemophilia, it was Robert and Suzanne Massie. Published in 1975, *Journey* recounts their unending battle to preserve their son's life. Bobby was born in 1956, and as the first 18 years of his life—the period covered in the memoir—was a period of stunning advances in the care of hemophiliacs, the book is an immensely valuable and moving public record of scientific progress (as well as of disappointing clinical setbacks). I met Suzanne Massie in 2011 at a dinner to raise funds to help care for patients with hemophilia in the world's poorest countries, at which she graciously signed my copy of her book. I know she will not mind if I briefly recapitulate the extraordinary battle that she and her former husband waged on behalf of her son.

In 1956, Robert, a Yale graduate who had studied at Oxford on a Rhodes Scholarship, and Suzanne, the daughter of a Swiss diplomat who had earned her degree at Vassar, both in their mid-20s, had recently married and were just embarking on promising careers in journalism. On August 17, Suzanne gave birth to a son. The doctors and nurses noticed right away that the infant had a large hematoma on the back of his skull and some bruises scattered about his body, but it had been a difficult delivery requiring forceps, and bruising was not so unusual. As was the routine back then, mother and child stayed in the hospital for a few days. On the second day after his birth, Bobby was circumcised. When he was handed back to his mom, he was still bleeding, but the nurses merely assumed a little vein had been cut and dressed the wound. In due course mother and child went home, and the bruises resolved. Five months passed. One day Suzanne noticed another rather large bruise and showed it to the pediatrician. Now concerned, the doctor drew some blood for laboratory tests and became more worried when Bobby bled for a long time after the needle stick. The results showed highly abnormal clotting times, and the pediatrician immediately referred Bobby to New York Hospital. A day or two later a hematologist informed Bob and Suzanne that Bobby had hemophilia. Suzanne recalls the moment this way: "[A]s surely as if we had been abandoned on the bleak surface of the moon, our lives had changed. We had no idea what lay ahead."

Naturally, Suzanne and Bob had lots of questions: Who is the best specialist? What are the treatments? What are the risks? How serious is it? Is bruising dangerous? What if he bleeds into his head? Will he have normal intelligence? Will he grow normally? The list grew steadily, but the list of answers did not keep pace. Realizing how crucial it was to protect her little boy from even minor trauma, Suzanne gave up her job and was rarely away from her son. Not surprisingly, she recorded her new life in a diary. For the first few months, other than bruising easily, Bobby did pretty well. But just before his first birthday he started to bleed badly into his pelvis, and his scrotum greatly expanded as it filled with blood. They rushed him to the hospital where he was kept overnight for his first blood transfusions. Back then rigid protocols forbade parents to be with children while they were undergoing medical procedures, so Suzanne and Robert spent long nights in grim waiting rooms just waiting.

During the next two years of Bobby's young life, he bled frequently and needed many transfusions. The worst episode began with a seizure—a warning sign that Bobby was bleeding in or just outside of his brain, a potentially fatal complication. Fortunately, it stopped. As Bobby grew older his need for transfusions exceeded the highest level that any doctor had dared to estimate. In 1965 it was 73; in 1966 it was 92; in 1967 it was 107. The costs, most not covered by insurance, were exorbitant. Bobby spent more time out of school than in school. As often happened to patients with hemophilia, big bleeds into his knee joints made him unable to walk. Bobby spent 7 years of his childhood in a wheelchair. Once when he was asked what the worst part of his disease was, he skipped right over the endless visits to hospitals, the ferocious pain that comes with bleeds into the hips and knees, and the fear that engulfed his family. What he wished the most, he said, was to be able to walk. For a child growing up in an era before society resolved to reduce the physical barriers that so compromised people with disabilities, the profound social isolation that Bobby experienced—the inability to go to school because there were too many stairs, the inability to play sports or go to birthday parties, the difficulty in nurturing friendships—caused him much greater unhappiness than did hemophilia itself.

Desperate for better therapies, the parents scoured the medical literature, called doctor after doctor, and grasped at every newspaper report. For a brief period one family of a child with hemophilia touted the benefits of peanut flour, and Bob Massie searched the nation until he found a supplier in Georgia that was willing to send him a 50-pound bag of the stuff. But it tasted awful and it did no good. Years later a controlled study confirmed that it was useless. Understandably, anguished parents continued to pursue many false leads (another common experience across many orphan disorders).

Especially frustrating to parents was that, even among experts, opinions about the best way to treat hemophilia varied greatly. Convinced that bleeding into the joints was the worst complication a boy could face, at the first sign of such trouble one renowned hematologist in New York required complete immobilization of the limb for months. Another equally renowned expert believed that joint *mobility* was the key to protecting long-term use of the limb. Even more troubling was that no two experts seemed to agree on how many transfusions should be given, what the blood volume in each transfusion should be, when they should be started, or how long they should be continued. Unfortunately, there was good reason for uncertainty. No one yet knew that the amount of the key protein—soon to be called antihemophilia factor, or AHF—in a unit of blood or plasma varied significantly. Given this and that the severity of hemophilia naturally varied among boys, it became difficult to choose a transfusion regimen. Doctors were just guessing.

Strangely enough, although the economic burden of caring for little Bobby had kept Bob and Suzanne on the edge of poverty, it was their intimate familiarity and fascination with hemophilia that ultimately made them wealthy. During Bobby's early childhood, his dad worked as a journalist for several of the nation's weekly magazines. His job was to write long feature articles that would entice the readers, so it is hardly surprising that he often proposed a story about hemophilia to his editors. They usually nixed the idea as not being likely to appeal to a broad enough audience. Nevertheless, Bob kept doing research on the disorder, work that led to a deep fascination with the impact of hemophilia on the royal family and on the chaotic politics in Imperial Russia in the early 20th century. One day, after a long discussion with Suzanne, Bob agreed to take a big chance. He quit his job and began work on a book about that era. Three years later (1967) his historical work, *Nicholas and Alexandra,* won him instant acclaim and propelled them into a level of wealth they had never envisioned.

Throughout their lives, both Robert and Suzanne remained professionally focused on studying Russian history and the modern Soviet State. In 1981 Robert won a Pulitzer Prize for his biography of Peter the Great, and in 2011, he published a biography of Catherine the Great. Suzanne became an important advisor on Russian culture to Ronald Reagan. It was she who taught Reagan to use the famous phrase, "trust but verify" in Russian. Thirty-five years later, *Journey* is still the best single memoir of its kind, a work that stimulated other parents to form advocacy groups for hemophilia, including the Save One Life Foundation that is dedicated to helping children with hemophilia in third world countries and to helping local nongovernmental organizations (NGOs) develop systems to improve the care of these boys.

In 2011, at the fund-raiser sponsored by the Save One Life to honor Suzanne Massie (who herself founded the Firebird Foundation, the first not for profit in the Soviet Union dedicated to caring for children with hemophilia), I had the privilege of hearing Robert recount his remarkable journey. In his unassuming way, he provided a coda to *Journey*. The struggle with hemophilia never ends. As a young adult, Robert developed hepatitis from a blood transfusion. It became so incapacitating that he was "out of commission" for 7 years. But his journey, which was marked by several brushes with death, ended in triumph. Faced with imminent death from a liver that was failing because of a viral infection, he underwent a liver transplant that cured both the infection and the hemophilia (the new liver made lots of Factor VIII). A graduate of Princeton, Yale Divinity School, and Harvard Business School, Robert has been a lifelong critic of disparity in health care and took a (failed) run at the Massachusetts Senate seat vacated by the death of Ted Kennedy.

The period covered by *Journey* (1956–1974) was a time when the dream of everyone touched by the disease was that scientists would figure out how to isolate and concentrate large amounts of AHF and deliver it at low cost to patients so that they would not suffer the horrible consequences of repeated internal bleeding. There was really good reason to hope. In the late 1940s a team in the United States managed to concentrate it to about fivefold the level in human plasma, albeit at a prohibitive cost. At Oxford University, Rosemary Biggs and R.G. Macfarlane were able to concentrate it more than 100-fold from animal blood, but that product could not be used in humans because it would cause severe allergic reactions. In Sweden, Dr. Birger Blombäck and Dr. Margareta Blombäck had managed to purify human AHF nearly 20-fold, but here again the cost was too great to be commercially viable in a big nation like the United States.

By 1963, in the United States the standard treatment for hemophilia was to administer fresh frozen plasma (FFP). It required about 250 milliliters of plasma to deliver about 250 units of AHF. Bobby and the boys his size often needed 6 units at a time. It typically required an overnight stay in the hospital with doctors closely monitoring fluid volumes to deliver enough FFP to stop a bleed. For large, intractable bleeds the patient might need scores of units of FFP and stay in the hospital for a couple of weeks. Then, in 1964 a dream came true.

There are many heroes in the history of hemophilia, but I believe the greatest is Dr. Judith Pool, the woman who that year showed that by simply freezing and then slowly thawing and decanting plasma taken from normal donors, Factor VIII activity could be highly concentrated in the "sludge"—the last, most viscous portion of the fluid at the bottom of the bag. Pool was born Judith Graham in New York City in 1919. Even in high school she showed much promise in science. She majored in physiology at the University of Chicago, where while an undergraduate, she married Ithiel Pool. Judith Pool graduated with honors in 1939 and stayed on to earn her PhD in the same field in 1946. During that time she worked in the laboratory of Ralph Waldo Gerard, and in 1942 coauthored a paper on the first successful use of a microelectrode to study the electrical activity of a single cell. In 1950 Gerard was nominated for a Nobel Prize for this line of research. Many have claimed that Pool did not get the recognition she deserved for her contribution to this work. In 1942 Judith followed her husband to Hobart and William Smith College in upstate New York (where she continued her thesis work). In 1952 she again followed her husband—to Stanford—where she soon obtained an appointment as a research associate and decided to study the physiology of coagulation. The couple divorced in 1953. She published her first paper on coagulation in 1954.

When Dr. Pool embarked on her research as a young PhD at Stanford, it was like being dropped into a mountain range knowing there was gold in the hills, but not knowing where. When she started working with FFP, she was shocked by how little AHF it actually contained. She spent months trying to figure out where the AHF was hidden and in developing a more precise measurement of its concentration. One day after studying hundreds of units of FFP, she compared the AHF levels in the first drops of a sample going into a patient to that in the last drops. The concentration near the end was much higher, but the fluid was cloudy and full of threadlike material. It did not even look safe to put in people. To her vast surprise when she studied this sludge, it was chock full of AHF. Working intensely over the next few weeks, by varying the way she handled the FFP, Dr. Pool determined that if you slowly thawed FFP from 0°C to just 5°C and decanted it, that a soft white residue at the bottom was highly concentrated AHF. Although she privately called the residue sludge, she knew that it would not fly, so she gave it the much more scientific name of "cryoprecipitate."

Almost overnight this material revolutionized the care of patients with hemophilia, saving many lives and greatly extending life expectancy.

In two seminal papers that appeared in *Nature* in 1964 and *The New England Journal of Medicine* in 1965, Pool showed that by using her "cryoprecipitate," one could deliver about 10-fold the amount of Factor VIII than could be obtained from transfusions. The standard method for treating hemophilia immediately changed. By 1968 virtually every large blood bank in the United States and Europe was preparing cryoprecipitate for patients with hemophilia. An added benefit of the system she developed was that the units of blood from which she extracted the cryoprecipitate could be reconstituted and reused in anyone (other than hemophiliacs) for transfusions. In the ensuing decades, many other clinical uses for Factor VIII would be discovered. The process for purifying Factor VIII was quickly industrialized, and several companies (notably Baxter) began to produce vast quantities of it in a highly purified form that could be reconstituted in a tiny vial of fluid and easily administered. This permitted another great advance. Doctors could now train parents to use a simple intravenous line to deliver cryoprecipitate at home. The future for patients with hemophilia was looking much brighter.

During the 1970s there was every reason to believe that a major therapeutic victory had been won in the battle against hemophilia. Cryoprecipitate was saving lives and reducing crippling joint disease. Many scientists were working on further purifying the concentrate. If they could produce enough of it, the door to preventive treatment would open. If treatment were started early, most patients might be able to live normal lives. One problem that remained was that some patients developed powerful antibodies ("inhibitors") against the proteins in the donated concentrates, which necessitated giving massive amounts of cryoprecipitate, at astronomical cost, to overcome the neutralizing antibodies they were making. To this day, it may cost more each year (more than $1,000,000) to treat patients with hemophilia who have developed these inhibitors than it does to treat patients with any other disease.

But just a few years after it appeared that we had conquered hemophilia, a devastating new complication loomed on the horizon. About 1970, pharmaceutical companies began to purify Factor VIII from pools of plasma collected from large numbers of donors. This permitted the development of products that had sufficient concentrations of the protein to permit patients to undergo major surgery and—for the first time—regularly take the product to *prevent* major bleeds. But within a few years it became obvious that pooled plasma carried a higher risk of infection to the patients. A new form of incurable viral hepatitis called "non-A, non-B" (now called hepatitis C) infected patients, carrying with it a vastly increased risk of developing liver cancer. But that was only the beginning.

In 1982, about 6 months after the first description of a patient with AIDS was published, came the first report of a hemophiliac patient with this mysterious new disease. At that time and for some years thereafter, there was no simple method to screen blood donors for being HIV carriers. The horrible fact is that for years persons with hemophilia who were regularly taking the new glycine precipitated concentrates were regularly exposing themselves to the risk of HIV infection. I was an intern at Boston City Hospital in 1981–1982, the year that medicine began to grasp that an AIDS epidemic was starting to unfold. A crude mnemonic that my fellow medical residents and I used that year to remember who was at risk for AIDS was "4H." The Hs stood for heroin addicts, homosexuals, Haitians, and hemophiliacs. Dirty needles, anal intercourse, and repeated exposure to unscreened blood products explained three of the Hs. The unfortunate inclusion of Haitians (many of whom had immigrated to Boston) was because the island was burdened by extreme poverty, lack of access to clean needles, and widespread prostitution.

The 1980s was a horrible period for people with hemophilia. Just when it seemed that medicine could bring the disease under control, persons with hemophilia became the cohort that was most completely devastated by the HIV pandemic. It turned out that the companies that manufactured the concentrates (and therefore required huge amounts of "donated" blood) were purchasing much of it from the highest risk populations—drug abusers and prisoners. For example, some companies set up plasmapheresis facilities in Haiti and other third world nations. Because the final concentrate products were made from pools of thousands of donors, the chance that a dose of concentrate carried HIV was high.

In 1983 the National Hemophilia Foundation estimated that the life expectancy of a hemophilia patient had, since the advent of cryoprecipitate therapy, increased from 42 to 63 years. Just 6 years later the median age at death had decreased again—to 40. In less than a decade, more than one-half of the roughly 15,000 hemophilia patients in the United States became HIV positive and about one-third of them died of AIDS, a percentage that rivals the destruction wrought by the Black Plague in 13th century Europe. The story was the same throughout much of the rest of the world. The death rate among hemophiliacs from AIDS in France was even higher. The companies that had harvested AHF from tainted blood products were vilified (as well as inundated with lawsuits from persons with hemophilia that led to huge settlements). Not until 1985, when it was shown that HIV could be destroyed by heat, did changes in manufacturing processes eliminate the risk of HIV infection in donated blood or blood products.

During this terrible time, patients with hemophilia who did not die of AIDS suffered in other ways. Dr. Holbrook Kohrt, a hematologist at Stanford who does research on cancer, recently recalled his childhood experiences with hemophilia for *The New York Times*. These include neighbors who opposed blood transfusions on religious grounds and who told his parents they would go to hell for using them to keep him alive, and discrimination against him during his high school years by students who conflated hemophilia with AIDS. Perhaps most painful for him was remembering a summer camp for kids with hemophilia that closed after several years because so few of the campers remained alive.

But the 1980s was also a period of extraordinary advance in molecular biology. The new ability to clone (isolate) genes of interest for intense study became the basis for the biotech industry. In 1984, finishing in a dead heat, groups at the Genetics Institute in Cambridge and at Genentech in South San Francisco isolated the Factor VIII gene, setting the stage for mass producing it in cultures of Chinese hamster ovary cells. Just 3 years later, recombinant human Factor VIII was ready for clinical trials. The new products were free of virus and highly efficacious. Since 1986 virtually no person in the United States with hemophilia has developed a severe infection from a blood product! Today >90% of patients in the United States with hemophilia are treated with a recombinant (cloned) product that poses no risk of HIV infection. Life expectancy has soared—approaching 70—and few affected children experience anywhere nearly as much pain and disability as did Bobby Massie. Because of the availability of safe and easy to administer Factor VIII, many boys now take the protein on a regular preventive basis.

As is the case for any rare disorder, it takes a highly experienced team of health care providers to render the best possible care of people with hemophilia. In 1973, the National Hemophilia Foundation (founded in 1948) started lobbying the government to fund a national network of treatment centers. Today there are 141 federally funded treatment centers and programs throughout the nation. In 2011, the federal government (through Maternal and Child Health Bureau [MCHB] and the Centers for Disease Control and Prevention [CDC]) allocated about $12 million to support these centers. A study conducted by the CDC of 3000 persons with hemophilia has shown that people who follow a rigorous program of preventive care at one of these centers are 40% less likely to die of a complication related to the disease and 40% less likely to be hospitalized because of bleeding.

What lies ahead for patients with hemophilia? Probably the major weakness of recombinant Factor VIII therapy is its short half-life—about 12 hours—which means many patients must inject themselves at least every other day. Because scientists now have access to large quantities of the purified protein, they can study it in new ways to see if by slightly tweaking its structure they might be able to increase its half-life. In the spring of 2014 the Food and Drug Administration (FDA) approved Eloctate, a drug developed by Biogen Idec for this purpose. Scientists developed Eloctate by *fusing* the Fc portion of a certain kind of immunoglobin G (IgG1) with a modified (B domain deleted) Factor VIII protein to increase its half-life in the circulation. Most patients with Hemophilia A who use Eloctate as a strategy to prevent bleeding will need to undergo an IV infusion only once every 4 days. Proteins are large and have complicated structures, but we know a lot about some of them, and it is often possible to figure out which parts (called domains) do the most critical work. The development of Eloctate is an example of how protein engineering can sometimes improve on nature!

The next great advance in treating people with hemophilia is likely to be gene therapy. Scientists began trying to cure hemophilia in animals about 1990. The discovery by Jim Wilson and his colleagues at the University of Pennsylvania that a type of adeno-associated virus called AAV8 preferentially enters liver cells (the cells that make Factor VIII), which made it the vector of choice for carrying a normal copy of the gene, has spurred much research. During the 1990s Katherine High, a hematologist who had trained at the University of North Carolina, one of the nation's top centers for research in clotting disorders, emerged as a key leader in gene therapy for hemophilia. In 1998 her research team showed that it could use gene therapy to cure the disease in Irish setters, the animal model that was most similar to the human condition. But despite heroic efforts by several research teams, they could not replicate the success in humans. Only in 2011 with the report from London that gene therapy had been used with partial success to treat Hemophilia B (Factor IX deficiency), a less common and somewhat less severe clotting disorder, was hope restored. I will recount this part of the hemophilia story in the gene therapy chapter.

In the United States today there are about 20,000 persons with hemophilia, and about 400 affected babies are born with it each year. Thanks to early diagnosis, preventive treatment, and careful attention to the risk of infection, our newest patients will lead far less difficult lives than has Bobby Massie. There is an excellent chance that they will be among the first generation to be cured of the disease. But, together, children with hemophilia in the United States and Europe constitute only 15% of the world population. For the rest, life remains grim. Experts estimate that as you read these words there are about 400,000 boys in the world with hemophilia, a number that does not make the disease seem so rare. Only ~25% of those boys receive any regular treatment, and many die young. In 2012 in the Philippines, nine boys died because they had no access to clotting factors. In Africa less than half the boys with classical hemophilia live to the age of 10.

The many technological advances that have driven the rapid growth of the biotech industry and enabled so many new approaches to curing orphan diseases, astounding efforts the telling of which occupies most of the rest of this book, have also driven great advances in the diagnosis of genetic diseases. Because our testing and screening abilities have thus far outpaced our drug development efforts and because genetic testing opens doors that often permit the avoidance of having children with certain genetic disorders, these tools occupy a unique place in genetic medicine. No discussion about efforts to develop new therapies for rare disorders is complete without an evaluation of the future of genetic testing, including the degree to which it will be used to avoid the births of affected infants. In the next chapter I review genetic testing.

4

Genetic Testing

Avoiding Disease

IN A BOOK WRITTEN ABOUT THE STRUGGLE to develop breakthrough treatments for children with rare genetic conditions, a chapter devoted to genetic testing may seem out of place. But for the last 40 years, our ability to diagnose fetuses with some chromosomal and some single-gene disorders and our ability to warn couples about the risk they face of bearing children affected with severe monogenic disorders have outpaced advances in therapy. Of course no one should equate avoiding the conception or birth of an affected child with a therapy. Yet, since about 1970, advances in testing have averted the births of tens of thousands of children with serious genetic and chromosomal disorders. Genetic testing is widely used and will be much more widely used in the future to identify individuals and couples at risk for bearing children with severe incurable disorders and for screening fetuses early enough to permit pregnancy termination.

For those who practice clinical genetics, the results of genetic tests often raise ethical dilemmas: Given the variations in which any one syndrome can present, how can one predict quality of life? What constitutes a severe disorder? What are the chances that new treatments might emerge in the near future, and how should that thinking be factored into genetic counseling? As I shall discuss more fully in a later chapter, the rapid growth of clinical DNA sequencing ensures that these questions will entangle ever more families. For the foreseeable future, giving women the opportunity to screen for and avoid the birth of a child with a serious genetic disease will remain an essential element of the practice of clinical genetics.

During the 1970s, the field was dramatically changed by two new technologies: (1) the introduction of *amniocentesis* (the use of a long needle to aspirate amniotic fluid and capture fetal cells suspended therein), a technique used primarily to inform women whether or not they were carrying a fetus with a chromosomal disorder; and (2) *carrier* testing (usually based on biochemical analysis of levels or properties of particular proteins) to identify couples who, although themselves healthy, were at risk for bearing a child with a particular single-gene disorder. Amniocentesis followed by the analysis of fetal chromosomes has been used primarily to test fetuses to determine if they have an extra chromosome number 21 (Down syndrome). Carrier testing programs initially focused on three relatively common single-gene recessive disorders—sickle cell anemia (SCA), β-thalassemia, and a fatal childhood neurological disorder called Tay–Sachs disease—each of which only occurs when a baby inherits one copy of a mutated gene from each parent. If two persons—each of whom carries a mutation for the same genetic disorder—marry, in each pregnancy there is a one in four chance that the infant will be affected with the disorder. As I shall recount, the first programs of population screening to alert persons to a reproductive risk involved profoundly different orphan diseases predominantly affecting persons in different cultures and led to markedly different outcomes.

Amniocentesis and Prenatal Diagnosis

Occasional efforts to obtain amniotic fluid to assess the fetus can be traced back at least a century, but sustained, sophisticated studies began in 1961 when a New Zealand physician showed that sequentially measuring the level of a chemical called bilirubin (which is elevated if many red cells are breaking down) in amniotic fluid accurately predicted the severity of Rh disease, a potentially fatal immunological disorder caused by antibodies generated by the pregnant woman against the red blood cells of her fetus. This disease manifests because the father has contributed an Rh-positive gene to the fetus that is expressed on its red cells and that an Rh-negative mother's immune system recognizes as foreign. Today, in the developed world, fetal Rh (standing for rhesus because the basic scientific work was performed in rhesus monkeys) disease is rare because doctors can give pregnant women who are at risk a drug called RhoGAM that prevents the mother's antibodies from reaching the fetus' red cells.

Amniocentesis was perhaps first used to diagnose a genetic condition in 1960 when a couple at risk for bearing a child with hemophilia A sought it to determine if their fetus was male or female. As hemophilia is X-linked, only males are affected. The first effort to use gender identification to assess a fetus for risk of X-linked Duchenne muscular dystrophy occurred in 1964. For the last four decades, fetal sex determination has been used as a method to screen pregnancies in couples (who are ethically comfortable with doing so) at risk for bearing children with severe X-linked disorders. Each time, the couple faces not one, but two, coin tosses. The first is as to gender. Female fetuses will either be unaffected or be healthy carriers, but in either case will not have the X-linked disorder that runs in the family. If the fetus is male, the couple must face another coin toss. Depending on which X-chromosome they inherited from their mother, one-half the males will be normal; the other half will have the disease. Before the advent of molecular techniques that permitted a definitive diagnosis, many at-risk couples faced the wrenching choice of whether or not to abort male fetuses rather than take the 50% chance that a live-born infant son would have the disease. Fortunately, that era is over.

A major breakthrough occurred in 1966, when Mark Steele and W. Roy Breg, Jr. (among others) showed that one could culture fetal cells found in amniotic fluid and study their chromosomes (a process called karyotyping). This grew out of the discovery in the mid-1950s of a new method to count human chromosomes by swelling white blood cells with hypotonic solution. When the cells popped open, the chromosomes would scatter. One could then apply a fixative to the glass slide on which this was done, find so-called metaphase spreads under the microscope, and count and photograph the chromosomes. I well remember the moment (in the fall of 1973) that I stared through a microscope on the fifth floor of the MD Anderson Cancer Center in Houston (where I was starting graduate work in the laboratory of Dr. Margery Shaw) and counted and photographed my chromosomes. In these days of automated DNA sequencing, I find it hard to believe that the next step in my analysis was to use household scissors to cut the image of each chromosome from the microphotograph (as a child might cut little paper dolls) and align them in pairs by size (the lower numbers indicating larger chromosomes) on a piece of white cardboard.

In the late 1960s and early 1970s human cytogeneticists, especially Torbjörn Caspersson (1910–1997), discovered how to use dyes to provide deeper resolution of chromosome structure (a process called *banding*), permitting the identification of more subtle abnormalities such as duplications of genetic material. For about 25 years, high-resolution banding of chromosomes, which in the best hands could identify up to 850 bands in the human karyotype, was our best tool for identifying abnormalities. Today, this technique is fading away as DNA array testing and DNA sequencing (discussed later) offer much deeper resolution.

In 1968, Dr. Henry Nadler reported the first successful diagnosis of a fetus with Down syndrome. In 1970, he and his colleagues published a large series of cases in *The New England Journal of Medicine,* showing that amniocentesis posed low risks to the fetus and that one could construct a karyotype that accurately diagnosed Down syndrome and other disorders caused by the presence of an extra chromosome. It had been known since the studies by an English physician named Lionel Penrose in the 1930s, that the single biggest risk factor for bearing a child with Down syndrome is advanced maternal age. Only about one in 1500 pregnant women in their mid-20s gives birth to a child with Down syndrome; but among pregnant women who are 35, the number increases to about one in 300. And among pregnant women in their early 40s, about one in 50 is carrying an affected fetus. In 1973, when a major study in Canada showed that the risk of losing a pregnancy after undergoing amniocentesis (via spontaneous abortion) was only about one in 300, it quickly became the standard practice in obstetrics to inform women who would be 35 or older at their delivery of the availability of a test to determine whether the fetus had an extra chromosome 21. Coincidently, the mid-1970s marked the start of a steady increase of the number of women who were opting to delay having children until they were in their mid-30s. Using age 35 as the agreed on trigger to offer amniocentesis and fetal chromosome analysis meant that each year in the United States doctors alerted about 250,000 pregnant women to their increased risk of bearing a child with Down syndrome. Soon, ~70% of such women were opting for the test, and ~80% of those who learned they were carrying a fetus with Down syndrome aborted the pregnancy. The rapid increase in use of amniocentesis was in part driven by *numerous* lawsuits in which women who had not been offered a test to screen a fetus for Down syndrome and gave birth to an affected child successfully won substantial damages for the tort of "wrongful birth." These lawsuits, which were consistently upheld on appeal, established a new standard of care in obstetrics.

As there were always far more pregnant women below the age of 35 (each of whom carries a small risk in each pregnancy that her fetus will have Down syndrome or some less common chromosomal disorder) who were *not* being screened, widespread use of age-related amniocentesis and fetal chromosome analysis did not substantially reduce the total number of babies born with Down syndrome each year in the United States. But, as more and more women delayed childbearing, and as more and more women in their 30s chose amniocentesis for fetal chromosome studies, targeted abortion did avert what might have been a substantial increase in the number of live-born affected infants.

During the last 30 years, the approach to screening for fetuses with Down syndrome was changed dramatically by the advent and steady improvement of a screening test that was performed on *blood* drawn from the pregnant woman. Because such noninvasive testing was relatively low-cost, posed no risk to the pregnancy, and actually reduced the number of women to whom amniocentesis would be recommended, obstetricians quickly became comfortable with it. In 1984, scientists reported that in pregnancies in which the fetus had Down syndrome the pregnant woman had reduced levels of α-fetoprotein in her blood. In 1987 other researchers showed that women carrying affected fetuses also had elevated levels of human chorionic gonadotropin (HCG), a second readily measured protein, in their blood. The timely measurement of these two chemicals in combination with the measurement of estriol levels became known as the "triple test." By 1995 the test was being ordered in ~60% of all pregnancies. By 2000 it had become the standard of care to offer this screening test to *all* pregnant women, regardless of age. The test was sensitive enough to identify ~60% of the affected fetuses among those tested. About this time, fetal ultrasound was added to the screening protocol. This was because of its ability to diagnose edema on the back of the fetal neck, a sign strongly associated with Down syndrome. This raised the sensitivity of the screening protocol to 74%. In time, the addition of a fourth biochemical test (called PAPA) brought it to 80%. But, the test also generated many false-positive results. Among the women who had a positive biochemical test and therefore were offered amniocentesis and fetal karyotyping, only a minority actually turned out to have a fetus with Down syndrome. This high false-positive rate created much anxiety among women who would later learn that they were not carrying a fetus with an extra chromosome 21.

In January of 2007 the American College of Obstetricians and Gynecologists (ACOG) issued *Practice Bulletin Number 77*, announcing a new standard of care that embraced a two-tiered screening system. It recommended that during the first trimester, *all* pregnant women (regardless of age) should consider using the noninvasive maternal blood tests (those that measured chemicals that, depending on how far their levels in maternal blood deviated from normal, indicated an increased risk) and ultrasound screening for increased risk for Down syndrome. Those who scored positive should then consider (depending on gestational age) either chorionic villus sampling or amniocentesis and fetal karyotyping to obtain a definitive answer.

The decision to recommend offering the screening test to all pregnant women (and the acceptance of that standard by the insurers that must reimburse for it) elicited a bitter critique from the National Down Syndrome Congress and other support groups that argued that screening devalued their children. But the practice of offering the test to all women quickly took hold.

In the years since ACOG issued its new standards, what has happened to the incidence (births each year) of Down syndrome? No one has precise statistics. It appears that the biochemical screening test is being offered to almost all pregnant women, and that far more than one-half of them are taking it. However, among those who receive a positive result, many do *not* choose amniocentesis to confirm that positive screen, and among those who do opt for fetal karyotyping and learn that the fetus does have an extra chromosome 21, only ~70% end the pregnancy. Perhaps because persons with Down syndrome are much more visible in society today compared with in past decades, they are no longer perceived to have significant disabilities. At any rate, available data indicate there has actually been a tiny *increase* in the number of infants born with Down syndrome each year in the United States and Europe over the last two decades (largely driven by the increased median age of pregnant women).

Today, a powerful new technology is again rapidly changing the current situation. In 2012, a company in California named Sequenom began offering a blood test of the *mother* to determine if her fetus has an extra chromosome 21 (the most common such chromosomal condition, excluding those involving the X or Y). This test relies on the astounding ability to find and capture fragments of fetal DNA circulating in the mother's blood. Because there is extremely little fetal DNA to retrieve, the fact that this new test (marketed by Sequenom as MAT21) is highly accurate is extraordinary. In 2011 the company published a paper showing that it correctly identified 209 out of 212 fetuses with Down syndrome, a success rate of ~99%. In the same study, it incorrectly identified a fetus as having Down syndrome in just three of 1471 cases. These numbers are superior to the existing biochemical tests and ultrasound (which can detect nuchal translucency, which is more common in fetuses with Down syndrome). The MAT21 test and tests like it (being marketed by companies including Verinata and Ariosa) are growing quickly. Their routine use is likely to sharply reduce the number of amniocenteses (about 200,000) performed each year in the United States. Given that the test is not perfect, physicians will still recommend amniocentesis and karyotyping for the women who test positive on the Sequenom test and others like it. About 100,000 pregnant women used Sequenom's test during its first full year (2013) in the marketplace. During 2014 about 300,000 pregnant women used its test or similar ones offered by competitors. That is still only a small fraction of the 4 million women who give birth in the United States each year. Given that ACOG has recommended that prenatal testing be offered to women regardless of maternal age, I expect that (despite their high price tag) tests like those developed by Sequenom will soon be used by many more pregnant women.

The MAT21 test is a harbinger of much more comprehensive tests to analyze the genetic status of the fetus. Several companies are trying to develop technologies to detect and capture *whole* fetal cells circulating in the maternal bloodstream. This approach is potentially far superior to that embraced by Sequenom. If one can capture a fetal cell (rather than merely fetal DNA fragments), one can avoid the knotty problem of contamination caused by maternal DNA, which is many-fold more common. Access to DNA harvested and amplified from fetal cells permits one to run a vastly greater number of genetic tests on the sample. I suspect that it will be some time before this becomes the basis for a new standard in prenatal screening. Even when the technology is ready, genetic experts will have to decide what findings in widespread DNA analysis are clinically meaningful. For example, there are scores of small deletions and duplications scattered with various frequencies among human chromosomes, but that are not yet clearly associated with serious abnormalities. Fetal cell analysis will raise immensely complicated questions for those who counsel the woman about her results. It will take some years to reach a consensus on which findings are sufficiently associated with abnormalities to be viewed as appropriately triggering a discussion that includes the option of pregnancy termination. Still, biochemical screening of pregnant women to identify those at increased risk for having a fetus with Down syndrome will soon be history.

The advent of prenatal diagnosis challenged deeply held values and generated much ethical debate. In an era in which there were exceedingly few good therapies for single-gene disorders and none for chromosomal disorders (which involve abnormalities in number of hundreds of genes), the development of techniques that let couples avoid the births of children with serious untreatable diseases held great appeal for some persons. For others, it seemed dangerously similar to the eugenic mindset that had reached its apotheosis in the monstrous crimes in Nazi Germany. Indeed, after the United States Supreme Court held in 1973 in *Roe v. Wade* that the Constitution protects a woman's right (until the point at which a fetus might survive outside the womb) to terminate a pregnancy, Jérôme Lejeune, the French scientist who discovered that Down syndrome was caused by an extra chromosome 21, became an ardent, lifelong *opponent* of prenatal diagnosis for the purpose of aborting affected fetuses. He was horrified by what he considered a devaluation of human life, and he was convinced that targeted abortion would dissuade researchers from studying how to improve the lives of children with Down syndrome.

Unfortunately, today—more than 40 years since prenatal diagnosis became a part of obstetric care—the terrible dilemma faced by women who want to have a child, but learn that the fetus has a serious genetic disorder, continues to cloud a growing number of pregnancies. This is because the diagnostic reach of prenatal diagnosis is broader than in the past, and because many women demand and most physicians feel the obligation to provide them access to various screening tests, even though the chance that they will result in a positive diagnosis is quite low. In addition to testing for chromosomal disorders, doctors can order other tests (such as for Fragile X syndrome, the most common form of X-linked inherited mental retardation).

And this is just the beginning. The number of disorders for which it will be possible to screen will explode in number, probably in less than a decade. DNA sequencing technologies, which are highly accurate and soon will be astoundingly inexpensive, will provide women with the option of screening for literally hundreds of orphan conditions. The rapidly increasing ability to assess the small amount of free fetal DNA (or the even more rare intact fetal cells) that leaks across the placenta into the maternal circulation and analyze it for virtually any known genetic condition will create a vexing dilemma. For although almost everyone would agree that certain genetic disorders cause severe, heartbreaking disability, there are many other disorders about the burden of which reasonable people might sharply disagree.

In the near future women will have to wrestle with the question as to whether a newly discovered variation in fetal DNA may lead to medical problems severe enough to justify ending a wanted pregnancy. Given the vast difference between the tiny number of serious orphan disorders for which we have effective therapies and the large number of those disorders for which we could theoretically identify affected fetuses sufficiently early in pregnancy to legally terminate the pregnancy, for the foreseeable future many pregnant women will confront that difficult choice.

Carrier Testing

During the 1970s, three distinctly different population-based screening programs were developed to identify persons who were born with ("carried") a single copy of a disease allele that although harmless in such individuals, represented a serious reproductive risk if two carriers with mutations in the same genes married. Each program targeted a serious orphan disorder for which there was either no treatment or inadequate treatments. Each anticipated a future in which it will someday be possible to greatly reduce the number of children harmed by orphan genetic disorders merely by avoiding their conception or birth. Each had a profoundly different impact on the population to which it was provided.

In 1972, a team led by Michael Kaback, a clinician and a biochemical geneticist then at The Johns Hopkins University School of Medicine, developed a serum assay that could reliably identify persons who carried a disease allele that if present in two copies in a fetus caused a fatal neurological disorder of childhood called Tay–Sachs disease. The disorder, which manifests by about age 2, is caused by a defect in an enzyme called hexosaminidase A, which renders the cells incapable of breaking down and recycling molecules called gangliosides. As they accumulate in the cells, they render them dysfunctional. Tragically, children with Tay–Sachs disease lose their ability to talk and walk and go blind before they die, usually by age 5.

Although the disease can affect children in any ethnic group, the causative mutations are most common in Ashkenazi Jews among whom about one person in 30 is a carrier (normal, but carrying one mutated gene). This means that about one in 900 Ashkenazi Jewish couples had a one in four risk that each pregnancy would result in the birth of a child with Tay–Sachs disease. In the United States and Europe >90% of all children born with this disorder were within Ashkenazi Jewish families. Because an affected child may not show evidence of the disease for the first couple of years of life, before the advent of carrier testing many at-risk families had more than one affected child.

The Ashkenazi Jewish population in the United States quickly embraced the test that Kaback and his team developed. Without benefit of state funding or support, community-based programs that were often held at temples on Saturday morning collected blood samples and sent them off for testing. In the Orthodox community, which favors arranged marriages and forbids pregnancy termination, Chevra Dor Yeshorim (The Committee to Prevent Jewish Genetic Diseases) amended traditional matchmaking activities to include "compatibility testing," essentially requiring that all young people be tested to see if they carried a Tay–Sachs mutation. Simply by doing this, over the years the community has *averted* many hundreds of marriages between carriers.

In less than a decade, Tay–Sachs screening became an accepted part of health care for Ashkenazi Jews. In the two decades from 1974 to 1993, carrier screening for Tay–Sachs disease became a global practice. More than 1,000,000 persons of Ashkenazi ethnicity were tested, and more than 36,000 carriers were identified. The programs found 1056 couples at increased (one in four with each pregnancy) risk. Prenatal testing was provided in 2416 high-risk pregnancies, and nearly all affected fetuses were aborted. By the end of 2006, worldwide nearly 2,000,000 people had been tested for Tay–Sachs carrier status, and more than 65,000 carriers had been identified. About 1500 at-risk couples were monitored (many during several pregnancies), resulting in the abortion of nearly 800 fetuses and the *live* birth of about 2400 unaffected children. In 2007 less than 10 children with Tay–Sachs disease were born to Jews in North America; about 15 affected children were born in other ethnic groups. In a single generation the birth of children with Tay–Sachs disease to Ashkenazi couples in the United States fell by $>90\%$.

In any human group in which the dominant practice is for its members to marry within the group (endogamy), rare recessive mutations that arise spontaneously are more likely to be maintained and become relatively more common over time. Several rare, severe, untreatable single-gene disorders are more common among Ashkenzi Jews. Since the pioneering work of Michael Kaback and his colleagues, other scientific teams have developed tests for many of these other disorders. Today, a number of leading medical centers operate laboratories that offer testing for the "Jewish panel," which from a single blood sample determines whether a person carries any one (or more than one) of about 10 serious autosomal recessive disorders that are more common among Ashkenazi Jews than among other groups. Over time, widespread testing coupled with avoidance of pregnancy or abortion of affected fetuses will follow the historical trajectory of testing for Tay–Sachs carrier status.

All screening programs struggle to define what constitutes success. Is the main goal to educate individuals about reproductive risks, or is it to reduce the number of individuals who are born with the severe disorder? It is hard to ignore the eugenic thesis implicit in carrier testing. In effect, such programs are premised on the belief that certain diseases are so burdensome and so lacking in efficacious treatment that it is preferable not to conceive individuals who would be affected and to abort affected fetuses when possible. In the case of Tay–Sachs disease—a truly horrible disorder that causes much suffering and early death—it is possible, except in regard to those persons who oppose abortion for any reason, to accept this blunt premise. But, the other two large, population-based, carrier screening efforts—for SCA and for β-thalassemia—targeted disorders that are treatable, albeit with significant limitations, and do not typically affect brain function.

SCA is—as orphan diseases go—quite common. About one in 10 African–Americans carries a copy of the causative mutation—which substitutes one amino acid for another at the sixth position in the hemoglobin β gene. Persons with classical sickle cell disease have a chronic anemia, but their more severe clinical problems arise from the stickiness of the deformed cells that causes frequent, extremely painful occlusive events in the small vessels and capillaries, especially the lungs. People with sickle cell disease are at risk for many problems, including strokes, kidney disease, and pneumonias. In the early 1970s, the life expectancy of a person with sickle cell disease (even one receiving state-of-the-art care) was about 28 years. In that era, about 30,000 Americans had sickle cell disease. Even though the affected population was small compared with, say, persons with cancer, when considered from the perspective of lost years of life, sickle cell disease, which manifests in infancy, has a significantly greater impact on public health than did many of the less common cancers.

Although there were some earlier calls to do so, one crucial event in the increase of sickle cell screening was an article, "Health Care Priority and Sickle Cell Anemia," written by a black physician named Robert Scott that was published in *The Journal of the American Medical Association* in 1970. Almost certainly enabled by the successes of the Civil Rights Movement, Scott's article emphasized both the magnitude of the clinical problems faced by affected patients and paucity of federal funding devoted to research on the disease. He then asserted, "the most important factor which would make a preventive program effective would be the early knowledge of trait carrying (trait means having one copy of the mutation). If screening were offered before marriageable age those who were found to carry an abnormal hemoglobin gene could be counseled to be sure that their mates were tested at the time of marriage. . . ."

The article, accompanied by a supportive editorial advocating for the creation of sickle cell testing and education programs in neighborhood health centers across the nation, garnered much attention. Within months an African–American man working in the Nixon administration had drafted language favoring sickle cell screening programs that became part of a major address on health care policy that the President gave in February of 1971. Nixon called on Congress to appropriate $5 million to initiate the effort. Within weeks, Senators Ted Kennedy and John Tunney were drafting a bill that would be enacted in May of 1972 as the National Sickle Cell Anemia Control Act.

During the period 1970–1972, 13 states enacted SCA testing laws. Well-meaning legislators in states with large black populations passed a variety of laws, each intending to corral as many young people as feasible for testing. Most of these laws targeted either (1) entry into the public schools, in effect saying a black child needed proof of having been tested before he or she could enter a certain grade level, or (2) young couples about to marry (or both). Unfortunately, the legislators who drafted the bills moved too quickly; in many cases they failed to get sufficient input from the black community, especially concerning the mandatory nature of the proposed laws.

Even before the new federal bill became law in mid-1972, both black physicians (like Jim Bowman at the University of Chicago whose daughter many years later would be in President Obama's inner circle) and community leaders decried mandatory testing, criticized the failure to educate parents of black children, and dismissed as impossible the idea of counseling a 6 year old about the consequences of being a carrier of SCA. The state laws were rarely, if ever, enforced. Instead, state programs had no choice but to embrace public education, voluntary testing, and genetic counseling if they wanted a slice of the federal financial pie that was available to compliant states under the national law.

Not surprisingly, as one set of well-intentioned black leaders challenged another set of well-intentioned black leaders, confusion (sometimes bordering on chaos) spread. Some egregious legislative missteps, such as enacting laws that required proof from self-identified African–Americans of having been tested for sickle cell trait as a condition of obtaining a marriage license, stimulated allegations of racial discrimination, and evoked comparisons with the lurid eugenic theories embraced by the Nazi government from 1934 to 1945. Although many sickle cell clinics were created with funds made available by the federal law, unlike the Jewish community did with Tay–Sachs testing, the black community did not embrace carrier testing. By 1975, it was clear that comparatively few adults were going to seek voluntary screening for sickle cell trait.

In the late 1980s, however, population-based testing for sickle cell carrier status underwent a sudden change. It had been known for many years that infants and very young children with sickle cell disease were at high risk for bacterial pneumonia; ~10% of affected infants died of pneumonia in the first year of life. Too often, they were diagnosed with sickle cell disease only after they were admitted to the hospital with pneumonia. The development of low-cost tests to detect sickle cell disease by examining how a small sample of blood moved when placed in an electrical field (hemoglobin electrophoresis) paved the way to add sickle cell disease to the list of disorders in *newborn* screening programs. Although some states began screening in the early 1980s, it was the publication in 1987 of a National Institutes of Health (NIH) consensus statement favoring the practice that drove universal screening. By the mid-1990s, 42 states had added sickle cell disease screening; today all states provide it.

One consequence of expanded newborn screening for sickle cell disease was that the test also identified infants who were carriers. Because one in 10 African–American babies is a carrier, the new test posed a challenging problem for state health departments usually operating with very limited budgets. In most states the finding that an infant had sickle cell disease triggered an effort to ensure that he or she received rapid and thorough follow-up from expert physicians. The discovery that a child was a carrier merely resulted in a letter being sent to the mother and (if there was one) the pediatrician of record. Subsequent research has indicated that until recently, children who were found to be carriers through newborn screening often did not know their status later in life.

In England, newborn screening to detect babies with SCA and β-thalassemia is firmly embraced by the National Health Service. During the period 2009–2010 more than 650,000 people were tested for SCA and more than 16,000 were identified as carriers. About half of their partners were screened, which led to the ascertainment of 1130 high-risk pregnancies. About 35% of those pregnant women chose to undergo prenatal testing of the fetus. Scotland adopted a national screening program in 2010. Similar efforts are proliferating throughout the world.

There have been few efforts to study whether 25 years of identifying almost all babies with sickle cell trait at birth has had any impact on reproductive behavior in the United States. For some time, however, the standard of care in obstetrics has been to alert African–American women to the possibility that they are carriers of sickle cell trait. Unfortunately, this discussion usually occurs after they are already pregnant. Some women agree to be tested, and a positive result triggers advice to test the partner as soon as possible. However, in the United States prenatal diagnosis and abortion of affected infants is uncommon.

The number of children detected each year in the United States with sickle cell disease is available for the period from 2001 through 2009. The data suggest that carrier testing is modestly reducing the number of live-born affected children. In 2001, among 4.05 million births there were 1213 children with sickle cell disease. In 2005, among 4.19 million births there were 1047 such children. In 2008, among 4.30 million births there were 1033 affected kids, and in 2009, among 4.18 million births there were 887 affected children. Thus in the 5-year period from 2005 to 2009, there was about a 15% reduction in the number of live births among virtually the same number of overall births. Of course this surmise ignores other important possible influences such as the decline in births among black women. However, these data do suggest that compared with voluntary community-supported Tay–Sachs screening (which has resulted in a 90%–95% reduction in annual live births of affected children), sickle cell carrier screening has had, thus far, a minor impact on reducing the number of live-born children with this serious, life-threatening disorder.

The third major screening program that emerged in the 1970s targeted β-thalassemia. As I discussed earlier, until about 1950 classical β-thalassemia (caused by the inability of blood stem cells to make normal adult hemoglobin) was almost uniformly fatal in childhood. In the 1950s in those countries where it was possible to do so, regular (about monthly) blood transfusions extended patient life expectancy to about 20 years, but at great cost. Chronic blood transfusions greatly overload the body's ability to handle iron. The resulting iron overload damages many organs, especially the heart, usually leading to death in the 40s. In effect, patients receiving chronic transfusion therapy died of the treatment, not the disease. Even though the advent in the past 10–15 years of new drugs to remove excess iron from the patients has been tremendously beneficial, the disease is currently curable only with a bone marrow transplant (which itself carries about a 10% risk of death).

Antonio Cao, the physician who pioneered population screening for β-thalassemia, traces its origins to the Fourth International Congress on Birth Defects, which was held in Vienna in the fall of 1973. It was there that he heard Michael Kaback report on his early efforts in Tay–Sachs screening and met Dr. Y.W. Kan, who was developing the molecular techniques that would make prenatal diagnosis of β-thalassemia possible. A few months after the meeting, Dr. Cao moved to a new position as Director of Pediatrics at the University of Cagliari in Sardinia, where he had been born. He immediately realized that in Sardinia, where one in every 250 babies was born with β-thalassemia, it was a major public health problem with more than 100 new affected babies arriving each year, each of whom required vast health care resources. In that era, even with the availability of the first chelating drugs, life expectancy was not much more than 20 years. Over the next few years, Cao established a close collaboration with Kan (who worked in San Francisco) and characterized the type of mutations in the Sardinian population and the methods of prenatal diagnosis.

In the mid-1970s, Cao led a massive public education effort concerning the inheritance of β-thalassemia and the possibility of avoiding the birth of affected children. Rather than screening the entire population, each time he identified a carrier of the disorder, he focused testing on that person's extended family. By doing so, he managed to identify the vast majority of carriers on the island by testing only ~15% of the population. Within a decade or so he and his team had identified ~80% of the predicted 2700 at-risk couples in Sardinia. Once couples knew they were at one in four risk for having an affected child, the vast majority chose testing. The percentage climbed even higher when use of a new sampling technique called transabdominal chorionic villus biopsy (a needle is used to aspirate from a uterine cell layer that contains tissue of fetal origin), which posed less threat to the fetus, became available. By the end of 1992, Cao's team had tested 3968 pregnancies and diagnosed disease in 1023 fetuses. Nearly all (99%) of the couples with affected fetuses terminated the pregnancy. In 1992, only four babies in Sardinia were born with β-thalassemia; the birth incidence had *decreased* from 1:250 to about 1:4000. This 95% reduction in births of babies with β-thalassemia in Sardinia has remained steady for the last 20 years, making it, arguably, the most effective targeted genetic testing program in the world.

Programs modeled on Sardinia have been initiated in many other countries over the last decade. In Greece, data from a national registry shows a marked reduction in births of children with β-thalassemia over the last 3 years. In Saudi Arabia, where consanguineous marriages are common, between 2004 and 2009 a mandatory premarital screening program increased the voluntary cancellation of planned marriages due to risk of bearing children with β-thalassemia by more than fivefold (about one-half of at-risk marriages were canceled). In Pakistan, the offer of free prenatal diagnosis for at-risk couples in Multan, a region of four million people, led to 105 tests. Each of the 23 couples that found they were carrying an affected fetus terminated the pregnancy. Since 1998, a pilot project in Zhuhai in Guangdong Province in China has identified 67 couples of which both members were carriers for β-thalassemia. The vast majority of pregnancies were terminated. Although it will probably take many more years to become widely established, it appears that targeted testing leading to prenatal diagnosis and selective abortion will drive a major reduction in the number of affected births in those countries that can afford to deploy such programs.

However, it is quite unlikely that programs modeled on the β-thalassemia program will be deployed for many other autosomal recessive disorders. For example, as long ago as 1993, in a speech given on his acceptance of the prestigious William Allan Award from The American Society of Human Genetics for his work on β-thalassemia screening, Cao opined that it would soon be possible to use a similar strategy to sharply reduce the number of children born with cystic fibrosis (CF). But, even though a test to screen for carriers of CF became routinely available about 1992, there has not yet been a significant reduction in incidence. Why? As rare genetic diseases go, β-thalassemia and CF pose roughly similar burdens. Both are early onset, require frequent hospitalizations, and (until recently) may be offered only supportive therapies. Neither severely affects cognition. Persons with either condition face the risk of substantially reduced life expectancies. Wisely, since its creation more than 50 years ago, the Cystic Fibrosis Foundation (CFF) has consistently devoted its considerable funds and influence to developing therapies for the disease. When carrier screening and prenatal diagnosis became widely available in the 1990s, the Foundation did not alter its position. In its view, avoiding the births of children with CF could indirectly harm efforts to undertake game-changing research. The CFF has never deployed its significant resources to support carrier testing or programs premised on prenatal diagnosis.

In the not too distant future—perhaps 10 years from now—physicians will routinely offer all pregnant women the option of a noninvasive screening test that assesses *hundreds* of orphan genetic disorders, including CF. For both carrier testing and prenatal diagnosis are soon to be revolutionized by the extraordinary advances in DNA sequencing. The first "consensus" sequence of a human being was completed in 2001 by several teams of scientists, most working with federal funds that eventually totaled ~$2.3 billion. Today, the cost of generating the complete DNA sequence of a human being is less than $5000. This constitutes a 500,000-fold decline in cost over a decade. Advances in sequencing technology continue unabated, and it is likely that several years hence the cost of sequencing will be less than $500. The low cost will remove what has been the main barrier to widespread uptake of the technology to date. Once sequencing has become highly accurate, highly automated, and astoundingly inexpensive, pressure to adopt carrier and prenatal screening across a wide array of tests will be intense. Indeed, an early version of a universal sequencing test could be possible today. At least five companies are currently offering or preparing to offer noninvasive prenatal testing. They will do so by capturing the very tiny amount of free fetal DNA that escapes through the placenta and travels in the mother's bloodstream. In China, a company called BGI (previously Beijing Genomics Institute) is already performing about 1000 such tests a day! In just a few years the use of amniocentesis to obtain fetal DNA for diagnosis will give way to safer, noninvasive sampling.

Based on current data, ~80% of women will accept the offer of broad screening for hundreds of disorders, and ~80% of those women who learn they are carrying a fetus affected with a severe disorder will terminate the pregnancy. Why 80%? At least 20% of pregnant women will refuse prenatal testing for religious or ethical reasons. Why 10 years? Why not sooner? It is highly likely that a low-cost, highly accurate test coupled with strong interpretive algorithms will be available long before then, but there are many logistical, economic, and ethical challenges that will slow large-scale deployment of such a test. As usual, well-educated affluent women will use and benefit from the test first.

The population-based carrier screening programs that I described are harbingers of far more comprehensive screening tools. In many respects DNA-based testing is easier to perform than a host of biochemical assays. Several young companies are preparing to launch direct to consumer DNA-based carrier testing for scores of autosomal recessive, single-gene disorders. Given that such tests are most valuable when they assess carrier status before one has chosen a partner (i.e., among teenagers and young adults), it would not surprise me if test results were commonly listed on one's Facebook page or other social media sites. At the very birth of molecular medicine—when Linus Pauling and colleagues deciphered the biochemical difference that identified carriers of SCA, he proposed that carriers determine some means to alert potential partners to their status as a way to dissuade carrier–carrier marriages! Mark Zuckerberg has provided them with an easy way to achieve the same goal.

Low-cost, accurate carrier screening becomes progressively more valuable as the diseases in question become progressively rarer. Many orphan disorders are so uncommon that there is little or no economic incentive for industry to attempt to develop new therapies for them. As Antonio Cao showed in Sardinia, widespread use of carrier testing for rare disorders can quickly and sharply reduce the births of newly affected infants. For the many really rare orphan disorders, carrier testing and avoidance of affected pregnancies will often be the most efficient choice.

The advent of genetic testing technologies triggered important social policy questions. One of the most important was how to make sure that testing and counseling were made widely available to women and families regardless of socioeconomic status. In 1976, Senators Jacob Javits and Ted Kennedy drafted and guided a bill called the National Genetic Diseases Act (NGDA) into law. Essentially, the bill offered significant funds to states that developed testing and counseling programs based on principles such as voluntary participation and access to genetic counseling. Ultimately, 34 states developed or expanded programs that met the guidelines of the NGDA, and more than 130,000 women received genetic counseling under the state programs.

Beginning around 1990, there was growing concern among medical geneticists and genetic counselors, patient groups, and others that the ability to screen people for risk of developing a genetic disease or giving birth to a child with a genetic disorder would compromise access to or the premiums charged for health insurance coverage. Over the last 20 years a majority of the states enacted laws to specifically avert that possibility. In 2008, the United States Congress passed and President George W. Bush signed the Genetic Information Nondiscrimination Act (GINA), which provided a fairly comprehensive protection against this risk, as well as risk in employment discrimination on similar grounds. As there was little documented evidence of insurance discrimination based on genetic testing before GINA, it is unlikely that the law will have a discernible impact. It may help to encourage otherwise reluctant persons to have needed testing, but that will be difficult to determine.

Preimplantation Genetic Diagnosis

Perhaps the most exciting advance in genetic testing is the rapidly expanding use of preimplantation genetic diagnosis (PGD), a technology that combines in vitro fertilization with DNA testing to help couples who know they are at high risk for bearing a child with a single-gene disorder to avoid that possibility without having to consider the termination of a wanted pregnancy. PGD is also widely used to screen for chromosomal disorders such as trisomy 21 (Down syndrome) in infertile women who undergo in vitro fertilization (a quality control check on a comparatively common problem) and, rarely, in situations in which the parents of a child who has been diagnosed with a severe disorder for which the only therapy is bone marrow transplantation choose to conceive a human leukocyte antigen (HLA)-matched younger sibling (often called a "savior sibling").

PGD for orphan disorders became possible because of the nearly contemporaneous emergence of two technologies: in vitro fertilization (IVF) and the polymerase chain reaction (PCR). The successful use of IVF in humans (beginning in 1978) capped three decades of studying conception and early development in frogs and many other species. Today it is a major clinical activity through which many thousands of children are conceived and born each year. The PCR, which is surely one of the greatest advances in the history of biotechnology, enables researchers to greatly amplify particular stretches of DNA for study. Kary Mullis, the unconventional California molecular biologist who conceived the basis for the technology, has famously recounted that he did so on a single evening in the spring of 1983 while on a moonlit ride with his girlfriend in his silver Honda Civic on a California highway. Scientists at the Cetus Corporation where Mullis was working—in particular, Henry Ehrlich—did much to refine the technology, which is now fundamental to the performance of an untold number of DNA-based tests. Mullis was awarded a Nobel Prize in 1993.

PGD requires the study of multiple human embryos that have been fertilized outside the uterus. Typically, the woman undergoes ovarian stimulation, the physician harvests multiple eggs, and they are fertilized in vitro with her husband's sperm. The next challenge is to capture a cell from the developing embryo. The most widely used approach is called cleavage stage biopsy. Embryos are cultured until they reach their third division (day 3 after conception). A technician uses one of several techniques to pierce a thin structure called the zona pellucida and then uses an ultrafine pipette to aspirate one or two blastomere cells for analysis. This procedure is typically repeated on several embryos. Technicians then isolate the DNA from the cell and use PCR testing to diagnose the fetus—that is, to determine if the blastomere cell under study is derived from an affected or an unaffected embryo. A British scientist, Alan Handyside, was the first to combine IVF with PCR to permit a couple at risk for a monogenic disease to initiate a pregnancy that they knew could not be affected. The report, published in 1990 in *The New England Journal of Medicine,* tested eight-cell embryos at risk for an X-linked disorder for the presence or absence of DNA from the Y chromosome. The physicians only implanted female embryos in the mother's womb.

Over the last 25 years, PGD has steadily become widely used both to screen for chromosome disorders in infertile women using IVF and for couples facing a high risk of bearing a child with a severe monogenic disorder. Countries have diverged widely in their approach to regulating both IVF and PGD. In 1990, Germany enacted a law that effectively banned the test, but in 2011 it relaxed that position. England has exercised the most far-reaching and thoughtful oversight. In the 1990s, Parliament created a new agency, the Human Fertilisation and Embryology Authority (HFEA), and set it up as an independent authority to oversee the practices at infertility clinics, especially in regard to the disposition of human embryos. Thus far, HFEA has required review of PCR-based testing of blastocysts for new single-gene disorders on a disease-by-disease basis. In the United States, there is no federal and little state regulation of PGD. Although IVF and PGD appear to be safe, it is impossible to rule out problems that might not emerge in individuals until years or decades later.

Much experience has confirmed that the removal of a single cell from an eight-cell human embryo for PCR-based testing does not pose ascertainable risk to the development of that embryo if it is subsequently transferred to a woman's uterus and a pregnancy ensues. However, although there is no doubt that the vast majority of PGD analyses leads to the transfer of an unaffected embryo into the woman's uterus, testing errors can occur. The most current data on this question probably comes from a retrospective study of and reanalysis of more than 1000 untransferred embryos stored at six major European fertility clinics. Retesting showed that 94% of embryos were correctly classified at the time of PGD (a finding that must be improved upon). More important, the sensitivity of the testing was >99%, meaning that the testing team failed to identify an affected embryo in <1% of cases. However, it may be that, given a history of greater regulatory oversight in Europe than in the United States, a similar survey in the states would yield less reassuring numbers.

PGD can now be used to assess virtually *any* single-gene disorder. One of the largest IVF clinics in the United States, the Virginia Center for Reproductive Medicine, discloses it has used the technique to help couples at risk for more than 200 different disorders. In recent years, couples in which one partner has or is at one in two risk for developing a severe *dominantly* inherited disorder, especially Huntington's disease (HD), have become frequent users of PGD.

The history of HD is one of the most intriguing in medicine. During the 1860s, George Huntington, the teenage grandson and son of physicians who practiced on Long Island, often accompanied his father on his rounds. Among the patients were several families in which many members suffered for several decades and eventually died from a neurodegenerative disorder, the hallmark of which was chorea, a term used to describe uncontrollable movements of the extremities. Although the disease was sufficiently common to have been recognized and briefly reported on by physicians for decades (and may even have been recognized by the early Greek practitioners), George Huntington published the first comprehensive description of the disease in 1872, just 1 year after he graduated from Columbia Medical School. His monograph received widespread attention and the eponym stuck. In the ensuing 140 years many physicians and scientists have studied and written about HD. It was one of the first disorders to be studied as a problem in population genetics. From early on there were many questions. Where did it arise? Why was it late onset? Were new mutations rare or common? Why, given its lethality, was it so (comparatively) common?

For years some scientists posited that all patients were descended from a single English family and that English seaman, who spent much of their prime decades circumnavigating the globe on whaling ships, spread the mutated gene. Recently, molecular studies have proven that not to be true, but the fact remains that *new* mutations probably account for at most 5%–10% of cases. These apparently sporadic cases (i.e., in families without a history of the disorder) arise in a gene that is predisposed to error. Given its lethality, HD is (to a geneticist, at least) surprisingly common. In the United States there are about 30,000 patients and probably another 150,000 persons (siblings and children of affected persons) who have a one in two risk of developing the disease. The at-risk population is thought to be so large for two reasons: The disease may not manifest until middle adulthood, and it is possible (but unproven) that those at risk have a higher fertility rate than the average person.

In 1979, then still the dark ages of gene hunting, Nancy Wexler, a clinical psychologist at Columbia, who was herself at one in two risk for developing HD, and other colleagues, instituted an unprecedented gene mapping project by enlisting the cooperation of a huge extended family around Lake Maracaibo, Venezuela. By diagnosing the disorder and obtaining hundreds of blood samples from both affected and unaffected family members, over several years the research team was able to use a growing number of DNA markers to determine with which of them the mutation was consistently associated. In 1983 (nearly two decades before the human genome was sequenced), Dr. James Gusella and his team at the Massachusetts

General Hospital showed that the HD gene resided on the short arm of chromosome 4. By 1989 they had refined the location to a much smaller region of the chromosome, and in 1993 the gene was isolated and cloned. HD was the first autosomal dominant disease gene to be so mapped.

To everyone's surprise, further analysis of the mutated gene determined that it arose because of the *unusual expansion* of a string of three base pairs (CAG) of DNA, a copying error that beyond a certain length disabled the protein (called Huntington) for which it coded and eventually caused great damage to certain areas of the brain. Today we know that all humans have a string of CAG repeats in each of their two copies of the HD gene. Almost everyone has fewer than 27 copies. In rare instances during meiosis—the process by which DNA is separated, copied, and distributed to the newly formed gametes—the CAG chain is extended. The larger the expansion, the greater is the risk of developing HD. Careful study of persons with the disease has shown that an embryo conceived with an egg or sperm that contains 40 or more CAG repeats will develop HD. Longer repeat lengths are associated with earlier onset of disease. HD is just one of about 20 single-gene disorders (almost all dominantly inherited) that arise because of an expansion of a section of a gene that normally has a relatively small number of triplet repeats.

As has been so often the case in medical genetics, the discovery of the causative gene has not yet led to a therapy. For the last two decades those at risk have faced the tough situation of knowing there is a test to tell them whether at conception they won or lost the genetic coin toss, but for whom bad news does not come with much hope. It is not so surprising that only ~10%–15% of those at risk for HD seek to be tested presymptomatically (i.e., to find out perhaps 10–20 years in advance whether they will develop the disorder). The arrival of the DNA-based predictive test created a new ethical dilemma. Theoretically, if every at-risk person was tested and if those who were positive did not have biological offspring, the incidence of disease could be reduced by 90% in about a generation. The few rare publications endorsing this option have not been well received. Although it was possible to do so, during the period 1990–2005 few persons at risk for HD chose prenatal diagnosis to avoid the birth of children with the same disorder, but with the development of PGD behaviors are changing.

In 2014 I met Andrew and Amanda Stricos, a newlywed couple in their early 30s who live north of Boston. Andrew was about 13 when he learned that his father was beginning to show signs of HD and that he and his brother faced a one in two risk of becoming afflicted. Amanda knew nothing of the disease until she started dating Andrew who had not been tested. As they were becoming more deeply involved in their relationship, Amanda began to notice that Andrew was developing subtle early signs of the disease—slightly slurred speech, deterioration of his handwriting, and occasional falling. When they decided that they would eventually marry, they faced the question of having children head on. Both agreed that they did not want to bring children into the world with a one in two risk for developing HD.

Serendipitously, it was during a visit with a doctor for endometriosis that Amanda learned about PGD, which offered the couple precisely the avenue they sought. To get the insurance company to pay for PGD, Andrew had to undergo a DNA test to confirm what he well knew—that he was affected. When she learned of the positive DNA test, Amanda was not dissuaded. She knew both families would be there for them, and in her mind there was never a doubt that HD would drive her away from Andrew. In 2012 the couple started the long and uncomfortable process of having IVF and PGD. The first cycle of egg harvests did not go well. Of the nine eggs that doctors retrieved from Amanda and fertilized with Andrew's sperm, *eight* had inherited the HD mutation. The doctors transferred the single unaffected embryo, but Amanda did not become pregnant. Fortunately, the next cycle of hormonal treatments and egg harvesting led to the creation of four unaffected embryos. The doctors transferred two, but, once again, Amanda did not become pregnant. So the doctors retrieved the remaining two unaffected embryos from their frozen storage and implanted them. This time, everything clicked. In May of 2014, Amanda and Andrew welcomed twins, Jack and Scarlett. Amanda was all smiles when she described the happy outcome, and quite pleased to say there would be no more pregnancies!

There is no database to which one can turn to find precisely how many HD couples are using PGD to avoid the birth of any affected child. But most couples who face similar risks for severe monogenic disorders indicate that they find PGD ethically much more acceptable than abortion. Of course, even if PGD substantially reduces the births of infants destined to develop HD, there are several hundred thousand people alive today in the world who will develop HD, and who will need a breakthrough treatment!

5

Stem Cells

Creating Human Mosaics

History

THE ORIGINS OF CELL BIOLOGY LIE DEEP IN the 17th century. In 1665, Robert Hooke, an English botanist, coined the word cell, an imperfect choice as the word implies a walled *empty* space (but, to be fair, correctly captures the look of many plant cells under low-power microscopy). During the 17th and 18th centuries, the few scientists who focused on understanding cells were mainly concerned with whether new life arose primarily from the sperm (the spermatists) or the egg (the ovists), a debate that was finally settled in 1759 when Caspar Friedrich Wolff proved that a fertilized egg does not at the beginning contain any *preformed* structures, but that they arise gradually. Although knowledge accretes over time and all scientists are indebted to the work of their predecessors, the origins of modern cell theory are forever linked with the genius of two German biologists, Matthias Schleiden and Theodor Schwann, who in 1839–1840 articulated the core tenet of cell biology: that the intricate organization of all species arises from an orderly program of cell division and differentiation.

Schleiden (1804–1881) was born into a wealthy family in Hamburg. He reluctantly studied law at Heidelberg University and then returned home to start a practice. But he did not enjoy legal work, and he spent more and more time pursuing his love of botany. After suffering a serious depression, he quit the legal profession, returned to school, and studied biology and medicine. He eventually earned an appointment at the University of Jena where he became a pioneer in using microscopy for purposes beyond mere classification. By 1837 his intensive study of plant development led him to conclude that all cells derived from other cells. Essentially the father of plant cytology, Schleiden wrote a botany text that dominated teaching for two decades.

Schwann (1810–1882) was born in Neuss and studied in Cologne and Bonn where he assisted another great early microscopist, Johannes Peter Müller, in preparing an influential textbook. Of his many notable discoveries, the most famous is that of a specific type of cell that wraps around and supports nerve fibers—a cell that will forever bear his name. Beginning in 1837, after a conversation with Schleiden about structures in plant cells that we today know as nuclei, Schwann, who immediately realized that he had seen similar structures in nerve cells, extended Schleiden's idea that the nucleus somehow controlled cell division to include animal cells.

In 1855, the great German pathologist, Rudolf Virchow, captured the principle that Schleiden and Schwann had discovered in an elegant Latinism: "omnis cellula e cellula"—all cells arise from cells. As microscopic study of organisms expanded, long-held beliefs concerning spontaneous generation (also known as *vitalism*) that had held sway for centuries were finally laid to rest. Incontrovertibly, the first cell from which an organism developed must contain *all* the instructions needed to form every specialized cell in the fully developed being. Leading biologists began to grasp (however tentatively and timidly, as it challenged the dominant Christian worldview) that just as all individuals of all species begin as a single cell, all of life in its many wondrous forms must have unfolded over an immensely long and dimly reckoned continuum from a single, primordial cell. The publication of *On the Origin of Species* (1859) by a self-taught naturalist named Charles Darwin forever changed our perception of life. Except among the Christian literalists, the biblical timescale quickly crumbled, giving way for a dim past of unknown dimensions.

A few years later (1866) the equally great German biologist, Ernst Haeckel, notoriously married ideas about embryonic development and evolution with the aphorism "ontogeny recapitulates phylogeny," in essence arguing that as humans proceed through gestation the embryo remolds itself along the memory of evolution. For example, the gill slits we have in early embryonic life recall the evolutionary rise of the fishes.

Virchow's magisterial aphorism left a crucial question unanswered. How could a dividing cell in an early embryo initiate the lineages destined to become the brain, the heart, and other organs *and* at the same time maintain the characteristics of the cell from which it derived? Scientists have been pursuing this mystery for nearly two centuries. Although we have learned a great deal (including enough to generate several Nobel prizes), there are still many missing pieces in the puzzle. Of this we can be sure, the concept of "*stemness*"—the property of a cell that allows it to differentiate as it divides so that one of its daughters takes on a new destiny while the other maintains the parental features—was familiar to cell biologists during the last third of the 19th century. Haeckel may have been the first (1868) to use the term in roughly the manner we do today. In the 1896 edition of *The Cell in Development and Heredity* (the first comprehensive text on the subject, written by the Columbia University biologist, Edmund Wilson), one can find a full-page diagram positing the existence and function of "stem cells." In 1901, Claude Regaud, a French biologist, used the term to describe spermatogenesis, the process by which the testis produces millions of sperm cells each day. He posited that these specialized cells must derive from a self-renewing ancestral cell.

During the first half of the 20th century, understanding of stem cells unfolded slowly. Several research groups showed that the various cells in the blood—the red cells, the various white cells, and platelets—arose from precursor cells in the bone marrow. However, the first *proof* of the self-renewing power of certain cells in the bone marrow was published in 1960 when two Canadian scientists, Ernest McCulloch and James Till, showed that if they injected bone marrow cells into irradiated mice (which lacked the immune system to reject foreign cells), nodules of cells formed into the animals' spleens in direct proportion to the number of cells injected. Three years later, they and a graduate student, Andrew Becker, used studies of cell surface markers to prove that these nodules were *clonal* (each cell mass was derived from a single cell). Over the next decade, McCulloch and Till proved the existence of the "hematopoetic stem cell," one that had the capacity to differentiate into any of the various cells in the blood. Their extraordinary work, for which they won a Lasker Award in 2005, provided the intellectual basis for bone marrow transplants as a lifesaving therapy. McCulloch died in 2005, so he will not win the Nobel Prize (which is not awarded posthumously) that he richly deserves. Today, the two men are regarded as the "fathers of stem cell science."

The extraordinary advances made in the scientific understanding of how the progenitor "stem cells" in our marrow make billions on billions of blood cells every hour, which occurred in the late 1950s and 1960s, were greatly influenced by the Cold War. As the USSR developed and stockpiled atomic bombs, quickly ending our nuclear hegemony, the United States Department of Defense grew interested in funding research to treat radiation sickness. It was clear from the horrific suffering of patients who survived the bombing of Hiroshima and Nagasaki that persons exposed to a certain level of radiation soon died (largely of infection or anemia) because of bone marrow failure. Interest in the possibility of reconstituting a patient's bone marrow through donor transplant exploded.

Organ transplantation, which is now performed tens of thousands of times each year in the world, began at the Peter Bent Brigham Hospital in Boston in 1954. Dr. Joseph Murray, a 35-year-old surgeon, had honed his skills doing trauma surgery during World War II. Desperate to help his patients, Murray had removed skin from dead soldiers and grafted it to badly burned soldiers. Although the skin grafts lasted only about 10 days, even that brief success encouraged him to believe that it might be possible to develop transplant medicine. In the late 1940s, Murray conducted research on kidney transplantation in dogs, but few colleagues encouraged his dream of developing the operation for humans. Most had deep ethical concerns about subjecting healthy persons (donors) to the risk of major surgery and the removal of an organ that they might someday need to stay alive.

In the fall of 1954, Dr. Murray was asked to see a 23-year-old man named Richard Herrick who was dying of a chronic kidney disease and who had a healthy identical twin. After consulting with clergy and lawyers and spending weeks practicing kidney transplants on cadavers, Murray decided to perform the surgery. On December 23, 1954, he led the team and became the first person in the world to see blood flow into a transplanted human kidney. The patient lived 8 more years, but succumbed to a recurrence of his disease. His donor brother lived until 2010. Murray went on to perform about 20 kidney transplants between identical twins, almost single-handedly creating a new surgical specialty. He later devoted most of his career to developing the field of reconstructive surgery to help patients born with severe malformations or who were badly burned.

In the mid-1950s, Murray met Dr. E. Donnall Thomas, a Texas-born physician who received his medical degree from Harvard in 1946 and who was taking his medical residency at the Peter Bent Brigham Hospital where Murray worked. The two became lifelong colleagues. After his residency, Thomas, who had become interested in leukemia in medical school, began doing research on bone marrow, first with Dr. Sidney Farber (the father of cancer chemotherapy) and later with a team at Massachusetts Institute of Technology (MIT). In 1955, Thomas, who had accepted a faculty appointment at Columbia University, moved to its affiliate hospital in Cooperstown, New York to join Dr. Joseph Ferrebee who was conducting experimental research in bone marrow transplant (BMT) on dogs. By chance, later that year they were asked what could be done for a child dying of leukemia and who had an identical twin. They reasoned that total body radiation (to wipe out the leukemic cells), followed by an infusion of bone marrow from an identical twin, might avert otherwise certain death. It did. Thus, perhaps the first successful bone marrow transplantation in history was performed in a community hospital in upstate New York. Thomas spent most of his extraordinary career in Seattle where he built the world's preeminent BMT center. In 1990, he and Joseph Murray shared the Nobel Prize in Physiology or Medicine.

In April of 1958, Dr. Murray was asked to treat a young woman named Gladys Lowman who was near death from renal failure. After giving a massive dose of radiation to her, he became the first surgeon to transplant a kidney donated by someone other than an identical donor into a patient. The grafted kidney functioned well, but the woman died a month later from an infection that she could not fight because her immune system was so weak.

In the late 1950s, BMT could only be used with identical twins, a relatively rare situation (one in 300 births). No one yet knew how to transplant healthy cells from nonidentical donors. A critically important advance occurred in 1958 when Jean Dausset, a French scientist, discovered that the distinction between self and nonself was largely governed by markers on cell surfaces called *histocompatibility* antigens. Essentially, Dausset discovered that certain proteins that are located on the surface of cells play a central defensive role by defining what is self, and recognizing proteins that derive from alien cells. Being able to "type" (create a profile of) these so-called human leukocyte antigens (HLAs), a process that researchers improved over the next 10 years, enabled the rapid growth of transplant medicine. By screening populations, one could find persons who, because they had the same or nearly the same HLA profile as a particular patient, could act as donors *even though they were unrelated.*

Scientists and clinicians at the University of Minnesota, especially Dr. Robert Good (1922–2003), played a particularly important role in pushing clinical use of the research. Good, who obtained his undergraduate education, his medical degree, and his PhD from the University of Minnesota, joined its medical school faculty in 1950. His first major achievement came in 1962 when he showed that the little-studied thymus gland played a key role in establishing the immune system. In 1965 he showed that the much-maligned tonsils also play a role in developing the body's immune system (a finding that spared untold thousands of children unnecessary tonsillectomies). During the 1960s, he focused his attention on studying the HLA system, in part to determine how much of a HLA mismatch a patient could tolerate without rejecting the grafted tissue.

BMTs

About 1968, Good made the daring decision to lead a clinical team that performed perhaps the world's first successful human BMT from a nonidentical donor. Because virtually all BMT research to that point had focused on trying to cure an identical twin with leukemia, it is surprising that the first person to receive a stem cell transplant from a nonidentical donor was a baby with a rare genetic disease. The patient was a 4-month-old boy with a disorder called severe combined immune deficiency syndrome (popularly known as "bubble boy syndrome" because the only chance of survival for affected children was to live in a germ-free environment) that had killed *11* of his male relatives. His 8-year-old sister provided the tissue. The BMT cured the little boy who, now in his late 40s, is in good health. The idea of using stem cell transplants to treat or cure rare *genetic* disorders was born.

It was Good's younger colleague, William Krivit (who had served on the first transplant team), who would lead the development of BMT as a therapy for rare genetic disorders. Even in college, it was clear that William Krivit was a driven man. Chyrrel, his wife of more than five decades, once recalled that as an undergraduate at Duke, Krivit used to open a window on the first floor of the library so he could sneak back in to study after hours. After graduating from Tulane University School of Medicine in 1948 and completing his residency in pediatrics at the University of Utah, Krivit joined the faculty of the University of Minnesota Medical School in 1952, and began working with Dr. Good. For the next two decades they often collaborated, and in 1962 they were both promoted to full professors.

During the 1970s, the team at the University of Minnesota, as well as others, began to aggressively treat children with severe, uniformly fatal leukemias with BMTs, usually with marrow from sibling donors. Because the procedure requires the complete destruction of the patient's bone marrow (hopefully, including all the cancer cells), it carries great risks. The immediate risks were that the child would die of an infection (because his natural defenses were compromised) or that the donor cells would fail to engraft (to take hold and build a cancer-free system), requiring a second donation (and, in those days, often causing death). Even if the child survived the operation, he or she faced about a 30% risk over the ensuing months of developing graft versus host disease (GVHD), which is often fatal. In GVHD some of the donor's immune cells attack various types of the cells in the recipient, including the oral mucosa, the skin, the liver, and the kidney. In performing these early operations, when so little was understood about the immune system, child, family, and the medical team were undertaking a heroic journey that often ended badly, but sometimes resulted in a complete cure of what had been a uniformly fatal disorder.

As success with treating childhood leukemia and lymphoma grew, Krivit began to think more and more about whether BMT could be used to treat the many rare, fatal genetic disorders, especially those that were caused by the toxic buildup in cells of unprocessed chemicals due to a mutation in a gene responsible for making the enzyme that normally degraded that material. When Krivit first proposed that the team at Minnesota attempt to treat children with several of these "storage" disorders, he met stiff resistance from his colleagues. In treating leukemia, physicians used drugs to destroy cancer cells in the blood and replaced them with donor marrow containing stem cells that could quickly create a normal population of new white cells. In the case of patients with storage disorders, however, they would be trying to use BMT to treat many different types of damaged cells in many different organ systems—liver, spleen, muscle, and brain, to name a few. There was no convincing experimental evidence to show that enzymes made by cells derived from donated bone marrow would ever reach these targets.

"People thought he was nuts when he first started," recalls Dr. Elsa Shapiro, a national expert in evaluating the cognitive function of children with genetic brain diseases, who worked with Dr. Krivit for more than two decades. Behind his back, they called him "Wild Bill," she remembered when she was interviewed about him in 2005. "He was always willing to take risks that other physicians would not." Despite the understandable resistance of his colleagues, Krivit persisted; he viewed many of the orphan storage disorders as being just as terrible as the childhood leukemias, even though the cancers often killed children much more quickly.

In 1982, Krivit and his colleagues performed perhaps the first BMT to treat a single-gene storage disorder. The patient was a little girl with an orphan condition called Maroteaux–Lamy syndrome (MLS), who was dying of heart failure caused by a massive buildup of waste material in her cells. MLS, which was first formally described by two French physicians with those names in 1963, is a recessive disease that affects only about one in 100,000 children. Patients are born with a defect in an enzyme called arylsulfatase B, which leads to the accumulation of a chemical called dermatan sulfate in their tissues. In those days, most patients died of heart or lung failure in their teenage years.

Krivit and his team performed the BMT on the 13-year-old girl, using cells donated by her unaffected sister. Then they crossed their fingers and waited. Because the donor was immunologically well matched to the patient, Krivit was confident that the girl would survive the transplant. But he had no idea if the cells derived from the transplanted bone marrow would eventually make enough arylsulfatase B (an enzyme that cells secrete, making it possible that it could enter *other* cells and partially repair them) to reverse the damage done over 13 years. To everyone's delight, within months there were positive clinical signs. After 2 years of follow-up, the transplant team reported the patient's dramatic improvement in *The New England Journal of Medicine*. Over that period, the activity of the critical enzyme had increased from 2% to 16% of normal, there had been a decrease in the amount of storage material that could be detected in the girl's urine, and, most dramatically, her heart function had much improved. Thirty years later the patient is still alive, albeit with compromised lung function.

The successful use of BMT in the teenager with MLS marked the start of an extraordinary 20-year period in which Krivit moved aggressively to treat children with a wide variety of severe, otherwise untreatable, orphan storage disorders. Not infrequently, physicians and bioethicists at other universities criticized the effort as too risky, arguing that if the Food and Drug Administration (FDA) had jurisdiction over new uses of BMT, these interventions would not have been permitted. As BMT is essentially a form of surgery in which new cells are placed in a patient, it is not regulated in the rigorous way that drug development is monitored. Whether it was sufficiently regulated or not, during the 1980s and 1990s Dr. Krivit and his team almost single-handedly developed this new approach to orphan genetic disorders. Transplant centers at Duke University and in Paris and London also entered the field.

When an individual successfully undergoes bone marrow transplantation, he or she becomes a *chimera*. I do not mean the goat-headed, lion-bodied, snake-tailed, fire-breathing monster of Greek legend. But forever after, he or she lives with cell lines ultimately derived from *two* different sources: the fertilized egg contributed by his parents and the individual who acted as the bone marrow donor. Those cells make his blood, fight off his infections, and migrate to his brain and other organs. Decades after the transplant, DNA tests can discern which cells are descendants of the bone marrow donation that the patient received.

In the early 1980s, it was difficult to find a donor who could provide the HLA-matched cells needed to perform a relatively safe BMT procedure. Although there were often many willing donors, the odds of finding an acceptable match declined greatly if the person in need did not have a sibling who had inherited the same (or similar) HLA gene pattern from the parents. This requirement that there be a very close match between certain proteins on the surface of the donor's cells and those of the recipient is a consequence of the evolution of our immune system over many millions of years. To survive, we must be able to recognize foreign proteins and quickly destroy them. Although humans have many HLA genes, transplant doctors seek matches for only four or five. Each one has many variants, so the odds of finding a "perfect 10" (each parent contributes five variants to a child) from a nonrelative is much less than one in 10,000.

There was only one way to overcome the challenge posed by the immense genetic diversity of the human immune system—build a national bone marrow donor registry of immense size. To do so, one would have to create an HLA profile of each volunteer and put it into a computer database to which transplant teams around the country could have rapid access. The creation and growth of what today is called the National Bone Marrow Registry dates to 1979 when physicians at the Fred Hutchinson Cancer Research Center in Seattle, to which Dr. E. Donnall Thomas had moved in 1975, struggled to find a suitable donor for a woman named Laura Graves who was dying of leukemia. By searching records in their blood bank they found that a member of their own staff had a suitable match. The BMT performed with his cells drove the leukemia into remission for 2 years. Although Laura Graves succumbed to her disease, her family was determined to help build a donor registry. During the early 1980s, the Graves family members started bone marrow registries in Milwaukee, Saint Paul, Seattle, and Iowa City, in each case by studying records in blood banks and approaching blood donors. But because HLA typing was expensive, it soon became obvious that only a well-funded national effort could grow the donor registry to a size that could offer hope to thousands of dying patients.

Responding to a sustained lobbying effort by the Graves family and many others, in 1984 Congress passed the National Organ Transplant Act. Among the key advocates was Admiral E.R. Zumwalt whose son had survived lymphoma after receiving bone marrow from his brother. The National Bone Marrow Donor Registry was launched in July 1986 at the Saint Paul Regional Blood Services Center in Minnesota. In less than a year, 39 donor centers in 22 states had opened satellite centers. In 1988 the organization renamed itself as the

National Marrow Donor Program (NMDP). In 1990 it became an independent nonprofit organization (taking oversight from the American Red Cross) and began the hard work of growing the organization, coordinating the registries, and seeking donors among underrepresented minority groups. They enjoyed immense success. By 1996 NMDP had enrolled two million donors who had provided lifesaving stem cells to 5000 patients. Today, the NMDP is international with cooperating registries operating in more than 35 countries. The volunteer list exceeds 6,000,000 names. More than 25,000 volunteers have donated bone marrow; each month that number grows by nearly 300.

The pioneering work at the University of Minnesota initiated a new era in the treatment of genetic disease. Several other pediatric transplant centers, especially Duke University in North Carolina and the Necker Hospital in Paris, expanded their original focus on treating children with leukemia to include orphan genetic diseases. In addition, centers in Europe, notably in Italy, began to move more aggressively to offer BMT to children with β-thalassemia.

The universe of severe genetic disorders is vast, each disease is rare, and even the few physicians who are truly expert about a particular disease readily acknowledge the limits of their understanding. Thus, the use of BMT to treat severe, often rapidly progressive, genetic diseases was shrouded in uncertainty. Because the disorders were individually rare, it would take many years to assess the impact of successful BMT. This harsh reality made it difficult to obtain a truly informed consent from the parents. But desperate situations called for desperate measures. A summary of the 30-year history of BMT as a new therapy for orphan genetic disorders, which one may fairly describe as one long, uncontrolled experiment, is best told by reviewing the experience in two categories: immune deficiency disorders and lysosomal storage disorders.

As is apparent from the name, genetic immune deficiency disorders arise from any mutation that interrupts a key element in the development of either of the two arms of our immunological defense system: the T cells (which directly attack foreign invaders) and the B cells (the source of protective antibodies). In some cases the genetic disorder affects both arms and is called a "combined" immune deficiency disorder. Depending on whether one is a lumper or a splitter, the term includes more than a score of disorders. From the perspective of transplant physicians, the many different single-gene forms of this class share a common origin—a defect in cells that are derived from the bone marrow. Thus, BMT directly targets the crucial compartment in the body from which these diseases arise.

One of the first of these disorders that transplant physicians attacked was Wiskott–Aldrich syndrome (WAS), a rare X-linked disorder characterized by low platelet counts, eczema, and recurrent infections that usually killed the boys by mid-childhood. Between 1980 and 1995, physicians in Zurich used BMT to treat 26 patients. Among 10 who received cells from HLA-identical donors, eight became long-term survivors. Among those who were treated with cells derived from partially matched donors, six patients survived and seven died of infections. Between 1990 and 2005 a team at the University of Brescia treated 23 boys with this disorder. Of that number, 18 survived, of whom 16 had either full or partial engraftment of donor cells and were effectively cured. The deaths were more likely to occur among those who received cells from a donor who only partially matched with their own. Of those who received well-matched marrow from relatives, *all* were cured. By 2009 researchers were able to compile data on the long-term outcome of stem cell transplants on 194 WAS patients. The overall survival rate was 84%, and among those who received stem cell therapy since 2000, the long-term survival rate was 89%. Of the survivors, 72% had a full donor chimerism and were cured; of the remainder who had mixed chimerism (they still had some of their own stem cells) all reconstituted an imperfect immune system, but were clearly improved. By 2010 it was fair to assert that the once fatal WAS had become a curable genetic disease—*if* the patient had timely access to a fully matched HLA donor. But even with the rapid growth of bone marrow registries, ~20% of boys still had to undergo BMT with mismatched cells, and therefore faced about a 10%–15% risk of death. During this era, many research teams around the world tirelessly attacked the problem of how to improve the safety and efficacy of stem cell transplants from imperfectly matched donors. There has been progress, but even in the most experienced centers the procedure-related risk of death approaches 10%.

Perhaps the most devastating and rapidly progressive form of immune deficiency is known as severe combined immunodeficiency disease (SCID). Regardless of the underlying defect, the infants are born essentially defenseless, having few T cells, relatively few B cells, low levels of immunoglobulins, and no specific antibodies. Until the advent of BMT, the infants typically died within months of birth. In the first decade (1968–1977) of using BMT to treat orphan genetic disorders, SCID was one of the disorders about which transplant teams were most hopeful. But, in retrospect, a review of the 80 SCID patients in the world who were treated with BMT is sobering. A review published in 1979 reported that *only* 18 of the 80 were still alive. Fifteen of the 18 had received BMT from HLA-matched individuals. Only three among 62 patients who received cells from mismatched donors survived. One interesting finding was that among the 18 survivors, three had a particular form of the syndrome known as adenosine deaminase deficiency (ADA), which suggested they were responding well to this novel therapy.

Desperate for donors, researchers tried many approaches. One was to use cells harvested from an HLA-haploidentical parent (which constituted a "half-match") coupled with drugs that killed the donor T cells lurking in the marrow (which they reasoned would reduce the risk of GVHD). In 1985, a team in Europe reported heartening news. Of 15 infants that they had treated, 13 developed donor-dependent T-cell functions. Eleven of the patients were long-term survivors, and nine were at home living with normal T cells and evidence of slowly reemerging B-cell function. In 1989, the team, led by Alain Fischer at Necker Hospital in Paris, reported on 183 patients with SCID that it had treated since 1968. The 70 children who had received HLA-identical bone marrow had a 76% probability of survival at a median follow-up of a little more than 6 years (probability of survival depends on time from treatment, so in successful therapies, it increases with time). Reflecting a steady improvement in procedural and postoperative care, of the 32 treated since 1983, 31 had been cured! Among the 100 infants who received nonidentical, T-cell-depleted marrow, the survival rate was 52% at 4 years. Although much lower than with identically matched donors, it still represented a 50% cure rate for a uniformly fatal disease—far greater than most cure rates in cancer! In 1991, the Minnesota team reported that the long-term survival among 18 patients with SCID or other lethal immune disorders who received donor marrow from *unrelated* persons was 68%.

But inadequate access to HLA-identical donors remained a reality, so physicians had no choice but to work with mismatched donors. In 1998, Fischer and his colleagues at Necker Hospital reported on the follow-up of 193 SCID patients who had been treated with nonidentical T-cell-depleted marrow (which they hoped would reduce the severity of GVHD) at 18 European centers. In that group, 77 had died within 6 months of transplant, usually from infection, graft failure, and/or acute GVHD. Another 24 patients died after 6 months, usually from chronic GVHD (a slower form of the phenomenon that was poorly understood). But, despite the grisly numbers, doctors were curing about one-half of the children, all of whom would have otherwise died. In 1999, physicians in Italy reported that nation's experience among 28 centers that had treated SCID children with partially matched marrow from unrelated donors. At 5 years the survival rate was 62%. Progress was being made.

During this period one major research focus was on containing GVHD. The strategy was to use a cocktail of powerful drugs to weaken the attack of the donor cells, thus hastening the time to stem cell engraftment and elevation of the white cells count, which narrows the window of exposure to the risk of life-threatening infections. In 2000 (an era in which the bone marrow registries were able to find matches for only ~40% of white children), a team in Toronto reported that by using the drugs methylprednisone, cyclosporine A, and methotrexate to attack donor T cells, they had achieved a long-term survival (at 4 years) rate of 75% among 16 patients with genetic immune deficiency disorders who received marrow from unmatched donors.

Emboldened by success in treating uniformly fatal immunodeficiency diseases, during the 1980s transplant physicians expanded the number of rare genetic disorders they would consider for treatment. Among this vast list of possible targets, they focused particularly on lysosomal storage disorders (LSDs). All of the more than 40 known LSDs are caused by mutations in genes that code for enzymes that play a role in breaking down chemicals that accumulate during normal cellular processes. Failure to metabolize those chemicals leads to a buildup of the material in the lysosomes (organelles in cells that operate as detoxification units). Although all LSDs are rare, collectively they affect about one in 6000 children. BMT emerged as a therapy at about the same time that scientists were making great progress in identifying the individual genes and enzymes that were faulty in each of the LSDs. Because there was nothing beyond "supportive" care for affected children, and because many with severe forms died in childhood, attempting to use BMT to reconstitute the marrow with cells that would make the normal form of the defective enzyme was logical. Even though the new cells would not adequately populate most organs, the hope was that BMT could supply enough of the normal enzyme to ameliorate the disorder.

Because treatment teams were willing to treat many different LSDs, because each patient with a LSD has his or her own distinct genetic background, because transplant science and approaches to treating GVHD were steadily evolving, and because it took 12–18 months to see early signals of a therapeutic effect and many more years to compile convincing evidence concerning long-term safety and efficacy, the case for using BMT for a particular LSD was not made quickly. Most of the early papers are case reports involving a varied mix of patients treated with various different sources of bone marrow and followed for various periods of time. One of the first large reports from Europe in 1995 reviewed the use of bone marrow to treat 63 patients with some form of LSD. They confirmed that BMT with HLA-identical marrow carried a "low" procedure-related mortality rate of 10%, whereas procedures with mismatched donors carried a frighteningly high 20%–25% mortality rate. Imagine the dilemma faced by parents when they were presented with such information! They *had to choose* between subjecting their child to a near-term risk of ~20% of a painful death in the hope of arresting a terrible disorder that took many years to kill *or* of waiting for a safer therapy that might not emerge in time to help their child.

Of the patients who had been followed long enough to yield some insight into the impact of BMT on disease progression, it appeared that in the 11 patients with serious skeletal defects caused by storage disorders the disease stabilized, but that the symptoms had not much improved. Somewhat better was the news that as enzymes derived from donor cells did their work (eliminating the chemicals stored in the lysosomes), the grossly oversized organs started to shrink to normal size! The most disappointing finding was that in patients with diseases that were characterized by *brain* dysfunction, BMT seemed to offer little benefit. The best news of all was that all of five patients with a rare disorder called nonneuronopathic Gaucher disease (type I) experienced virtually complete regression of symptoms.

By 1999 more than 400 patients with a variety of LSDs had undergone BMT. Overall, the picture remained the same. Transplants from unrelated donors carried a high risk of death or chronic GVHD. But among those who survived the intervention, disease symptoms tended to stabilize, and life expectancy increased. By 2000 researchers were beginning to get a clearer picture of the value of BMT in particular disorders. The experience with a storage disorder called Hurler syndrome is illustrative.

In 1981 British researchers reported on a 1-year-old boy with this severe disorder who received a haploidentical transplant from his mother (in effect, a half-match) in an attempt to provide an enzyme called α-L-iduronidase. In addition to being able to control the inevitable GVHD, they reported that about 4 months after transplant they found evidence that the donated enzyme was working. His enlarged liver and spleen had returned to normal size, the clouding in his corneas had disappeared, and his development looked normal. In 1988, the team led by Bill Krivit at Minnesota provided a similar, encouraging case report.

Still, the overarching challenge was the difficulty of finding suitable donors. The first major study on the outcome of *unrelated* BMTs in rare genetic disorders appeared in 1996 when physicians at the University of Iowa reported on outcomes among 40 children with Hurler syndrome. One-half of the children were living 2 years after being transplanted (sufficiently long so that no one was likely to die of GVHD). Hurler syndrome causes intellectual disability, so it was especially important to study the impact of the therapy on cognitive development. Of six children who had a mental development score in the low-normal or better range at the time of treatment, *none* showed a decline in their abilities, and four were developing normally. This was miraculous news, and it got better. In 1998, the Storage Disease Collaborative Study Group reported outcomes on 54 children with Hurler syndrome who had been treated with BMT with marrow either from HLA genetically identical siblings or haploidentical unrelated donors. In 39 (72%) of the patients, donated cells had engrafted with the first BMT, and the probability of survival at 5 years was 64%. Most impressive were the data that showed that the earlier the transplant, the better the intellectual status of the patient. Nine of 14 patients treated before 24 months of age had normal or near-normal cognitive development; among those who received BMT when they were older than 2 years, only three of 12 showed normal or near-normal development. The researchers concluded that BMT offered the best therapy for children with Hurler syndrome who were younger than 2 years of age and whose cognitive status at the time of treatment was still in the normal range.

The team at the University of Minnesota Medical School remains one of the most active centers in the world in the use of stem cells to treat orphan disorders. Over the last 15 years it has been especially involved in using stem cells to treat children with an otherwise rapidly fatal disorder known as childhood cerebral adrenoleukodystrophy (CCALD) (discussed in Chapter 7). This X-linked disorder, which affects about one in 20,000 boys, is caused by mutation in a gene (*ABCD1*) that makes a protein that is crucial to the breakdown and elimination of very-long-chain fatty acids (VLCFAs) in a cellular organelle called the peroxisome. Before the decision to use BMT in an effort to arrest the inexorable progress of this disorder, affected children, who typically developed symptoms in early childhood, died slowly over about 5 years.

The first BMT to treat a boy with CCALD was performed about 1982. Over the ensuing 17 years the treatment was attempted in about 125 other patients. The number might have been considerably higher, but in more than a few cases either a delay in diagnosis of this mysterious disease or very rapid disease progression meant that the disorder was too far advanced to ethically justify the therapy. In 2004 the Minnesota group reported on 94 of the 126 boys for whom full data could be obtained. Among the leading cause of death was rapid disease progression. This is so because it takes about 12–18 months for the descendants of the donated cells to populate the brain in sufficient quantity to supply enough protein to begin to eliminate the overabundance of toxic VLCFAs. Still, among this group, 5- and 8-year survival was 56%, higher than survival among untreated patients. More importantly, among children who were diagnosed and treated early, the 5-year survival was 92%!

In 2010 the University of Minnesota team reported on its experiences over a decade in treating 60 CCALD boys with stem cells (from both bone marrow and cord blood, which I discuss below). This large study confirmed the 2004 report. If affected children were diagnosed at an early stage, if disease progression was not unusually rapid, if they were fortunate enough to receive well-matched cells, and if they survived the transplant, there was a about a 2 out of 3 chance that the disease would be arrested in a phase that would allow the children to lead a fairly normal life. Sadly, that is a disturbing concatenation of ifs.

During the 1990s, stem cell therapy was still hobbled by lack of access to donor cells. Despite the growing number of persons in the United States and Europe who were registering to be bone marrow donors, the odds against quickly finding a person whose HLA genetic profile closely matched that of a needy patient remained quite low. This harsh reality accelerated research efforts that had begun in the mid-1980s to investigate the use of cord blood—cells derived from the umbilical cords of newborn babies—as an alternative source of lifesaving stem cells.

Cord Blood

The observation that mammalian cord blood contains cells with characteristics that appear to be the same as the stem cells in bone marrow derives from experiments conducted in mice in the 1970s. At that time, Edward A. Boyse (1924–2007), a British-born scientist (who became an American citizen), was using highly inbred strains of mice to advance our knowledge of the critically important role of T cells in protecting animals from infection. In 1975, Boyse and Dr. Harvey Cantor published a paper proving the existence of helper T cells and killer T cells, one of the most famous discoveries in the field of immunology. In recognition of his scientific contributions, Boyse became the first person in history to be elected to the Royal

Society, the National Academy of Sciences, and the American Academy of Arts and Sciences.

In 1982 (a time when the mortality risk of transplant between nonidentical individuals was >20%), Boyse proposed to his colleague and collaborator, Hal Broxmeyer, a stem cell biologist who has since 1983 been a professor at the Indiana University School of Medicine, that human cord blood might provide an alternative to bone marrow as a source of these lifesaving cells. Broxmeyer expanded his research on the properties of stem cells to include studies of cord blood. In 1989 his team at Indiana and Boyse published a study of 100 samples of human cord blood. They were able to show that both before and after freezing for long-term storage (followed by thawing) the aliquots of human cord blood contained a similar number of multipotent progenitor (stem) cells as were typically found in an aliquot of bone marrow used for transplantation. Their observations about the quantity and quality of stem cells that remained in a cord blood sample after freezing and thawing showed that a well typed sample could be *stored* in a tissue bank for transport to any hospital in the world where a patient needed cells with that particular HLA profile.

In 1988, Eliane Gluckman and her team at the Necker Hospital in Paris performed the *world's* first therapeutic cord blood transfer in an attempt to treat a 6-year-old boy who was dying of an exceedingly rare genetic blood disorder called Fanconi anemia. The disease, which affects about one in 300,000 infants, is named for Guido Fanconi (1892–1979), a Swiss pediatrician who spent 45 years on the staff of the Children's Hospital at the University of Zurich. In 1927 he described the very unusual case of a young child with small stature, a failure to make the various white and red cells in sufficient quantity, and a pigment disorder. The donor source for Dr. Gluckman's pioneering therapy was the patient's newly born HLA-identical baby brother. During the mother's pregnancy, doctors had developed a method to test for Fanconi anemia on cells derived from amniotic fluid and had determined the baby would not have the disorder. In addition, he would be a perfect match. It was as if the family had won two bets (a three in four bet that the new baby would not be affected and a one in four bet that he would be the right HLA type) in a row! The transplant was successful and the child (now in his early 30s and healthy) was *cured* of his disease. A new chapter in the treatment of rare genetic disorders opened.

In its earliest days, therapeutic cord blood transplant raised an ethical issue that for a time captured the attention of medical ethicists and the general public (even becoming a cover story for *Time*). When doctors were unable to find well-matched donors for young children with severe genetic disorders and time permitted, despite the one in four risk of having a second affected child, parents started another pregnancy, hoping to give birth to a child who would be unaffected *and* an HLA-matched donor. As early as possible, they used prenatal diagnosis to determine whether the fetus was affected. In a few cases, they terminated pregnancies when they learned that the fetus was not affected, but would not be able to provide the matching cells. They then tried promptly to have another pregnancy. Many people were critical of the manner in which the fetus was being "commodified" (treated as a donor), but more seemed to recognize the right of the parents to engage in this highly unusual last-ditch effort to save an existing child. With the rapid growth of cord blood registries over the last 20 years, such unusual cases have all but disappeared.

Impressed by rapidly accelerating clinical research, in 1992 the National Institutes of Health (NIH) funded the first public cord blood bank at the New York Blood Center (NYBC), a step toward making sure that all persons in need of stem cell transplants would have timely access to appropriately matched donor tissue. The NYBC launched an educational campaign urging obstetricians in New York to inform women about the program and to invite them to bank blood taken from the umbilical cord when they delivered. The first private (for profit) cord blood bank in the United States (the Cord Blood Registry) opened in 1995. Since then, both private (for profit) banking companies and public (state-based) banking systems have developed in the United States in parallel.

During the 1990s, most clinical experimentation with stem cells derived from cord blood focused on young children with potentially fatal leukemia, none of whose family members had closely enough matched HLA tissue profiles. By the early 1990s, many hematologists and transplant surgeons were convinced that if they could develop large-scale cord blood repositories, they could overcome the donor shortage problem that in the 1970s and 1980s so frequently had denied lifesaving therapy to persons not lucky enough to have a well-matched relative. The need was particularly acute for African-Americans who were statistically only about half as likely as whites to find an HLA-matched donor in a time of crisis.

The growth of public cord blood banks—those to which women donate their cord blood to be cryopreserved and on call for *anyone* with a need—was accelerated by the enactment of new laws in many states that required or encouraged physicians to request the donation. By 2005 public cord blood banks held over 150,000 typed samples, but many women were choosing (at significant expense) instead to store their cord blood for private use—as insurance for their children (or other close relatives) against the about one in 1500 lifetime chance that they would need the cells. Over the last 5–10 years, donations by women to public cord blood repositories has increased dramatically, as has the use of such donated material in stem cell transplantation, but private cord blood banking continues to outpace it. According to the results of one survey, in 2011 more than 400,000 units of cord blood were stored in more than 100 quality-controlled public cord blood banks and more than 780,000 units were stored in more than 130 private repositories. By 2012 the number of stem cell transplants performed in the world that used cord blood had passed 30,000, the vast majority to treat children with leukemia or failures of bone marrow called aplastic anemias.

The use of stem cell transplants—derived from both bone marrow and cord blood—to treat children with rare genetic disorders is increasing slowly but steadily. Reporting for the decade 2000–2009, the NMDP provided 5-year survival data on 42 patients with Hurler syndrome and 30 patients with either CCALD or metachromatic leukodystrophy (MLD), which are both fatal brain degeneration disorders. About 30% of the children were dead within 1 year of the transplant, but over the following 4 years there were only three deaths.

For more than a decade in the United States, the majority of stem cell transplants using donated bone marrow or matched cord blood to treat children with *rare single-gene disorders* (as opposed to leukemias in which the practice is widespread) have been performed at just two centers in the United States: the University of Minnesota Medical School and Duke University Medical Center in North Carolina. Although it started its program more recently, Duke University School of Medicine has become one of the largest pediatric transplant centers in the world, having performed more than 1700 procedures since it started in 1990. Both centers currently offer stem cell transplant (with bone marrow or cord blood) for more than 20 orphan genetic diseases. But the experience in treating children with monogenic disorders is far less than in treating childhood cancers. For many orphan disorders, our scientific and clinical knowledge of the disease remains sufficiently limited that a team needs to treat many patients with the disorder to gain a good understanding of the challenges it faces.

With the exception of the hemoglobin disorders, all the single-gene disorders in question are individually *less* common than childhood cancers. Further, the risk of the procedure (even for transplant between fully HLA-matched siblings) remains substantial. Overall, ~10% of the patients die within 1 year of transplant, usually either from graft failure and subsequent infection or from acute or chronic GVHD (in which immune cells in the transplant tissue attack the patient's foreign cells). In addition, another 10%–15% survive with chronic GVHD, a difficult, lifelong illness. Another confounding factor is that for many rare fatal genetic disorders the disease may be so advanced at treatment or progressing too rapidly to halt it with transplant.

Despite the still limited success with treating *storage* disorders, one might believe that by now the use of stem cell transplants to treat the hemoglobin disorders would be widespread. Why have transplant centers not moved more aggressively to use stem cells to cure β-thalassemia and sickle cell anemia? The answer is complicated. In both cases existing therapies have significantly improved the lives of affected persons, turning fatal childhood disorders into chronic diseases. But in both cases current therapy is unquestionably suboptimal, and cost and compliance issues become problematic as patients become adults.

Over the last decade the use of stem cell transplants to treat persons with β-thalassemia had grown steadily, now numbering more than 1000. The report of the transplant program in Ottawa is typical of the evolving picture. In 2011 the team described its experience in treating 179 children with cells derived from *HLA-matched siblings* (the best possible source). The median age at treatment was 7, and there was an average follow-up of 6 years. Eleven patients died from events directly related to the procedure (graft failures and infections), and 13% of the total suffered from *chronic* GVHD. But among those patients classified as having a medium level of disease (known as Pesaro class II), the 5-year probability of disease-free survival was 88%! Overall, the data strongly suggested that if one treated children under age 7 and *before* they had reached Pesaro class III status, that they had a 90% chance of being cured. But this good news is tempered by the harsh fact that if they choose such therapy for their children, parents must accept that they are gambling against about a 10% chance that their child will die from the intervention at a time when he or she is fairly healthy. This is, understandably, the major reason why many thousands more stem cell transplants have not been performed for β-thalassemia.

In 1998 the transplant team at Emory University School of Medicine became the first in the United States to use cord blood to cure sickle cell disease in an affected child. The little boy remains alive and well, but 15 years later, he is one of only about 250 such children with equally happy outcomes in the United States. There are many reasons for this: the availability of alternative, albeit noncurative, therapies, inadequate numbers of matched samples derived from African–Americans, and the reluctance among African–Americans to undergo therapies perceived as research.

The slowly growing experience with BMT for patients with sickle cell disease suggests that marrow donations from well-matched sources achieve a 90% chance of disease-free survival 5 years after transplant. The picture is not so good when the cells are derived from cord blood. In one recent trial, eight children with severe sickle cell disease received cord blood cells from unrelated donors with at least a match at 5/6 of the major genetic loci. At a little over a year of follow-up, one child had died of chronic GVHD and respiratory failure, and only 3 of 8 were alive with neither graft failure nor disease recurrence. The outcome was so disappointing that enrollment in the unrelated cord blood arm of the study was suspended. It may be that in sickle cell disease engraftment is less successful than it is in other disorders.

Today, BMT using cells from well-matched donors offers about a 95% chance of survival to patients who are healthy enough to undergo this debilitating procedure. Unfortunately, despite great advances in understanding GVHD and the availability of several new immunosuppressive drugs to avert GVHD, there does not appear to be any transformative intervention on the horizon to reduce the 5%–10% risk of death associated with the intervention. Although one can expect steady progress over the next 20 years to improve rates of engraftment and decrease rates of GVHD, stem cells derived from unrelated individuals for transplant therapy will continue to pose a considerable risk to the patient. This is one reason for the great interest in developing gene therapies for orphan diseases (discussed in Chapter 7). For many such disorders, one could remove some bone marrow stem cells from the patient, insert a normal copy of the gene of interest into those cells, and then return them to the patient, where they and their millions of descendants could provide a normal version of a needed protein—hopefully enough to ameliorate the basic molecular defect.

6

Enzyme Replacement Therapy

Genetically Engineered Drugs

Gaucher Disease

IN 1882, A DILIGENT FRENCH MEDICAL STUDENT named Philippe Charles Ernest Gaucher encountered a very ill woman in his clinic. When he examined her, he found that her liver and her spleen (which can sometimes be felt on the left side of the body below the rib cage) were massively enlarged, and that her legs bones were misshapen. She soon died, and Gaucher, who like all European medical students of his day had to prepare a thesis to graduate, decided to focus on her mysterious condition. At autopsy, he found that the enlarged liver, spleen, and bone marrow compartments could be explained by the fact that within each of these organs the *cells* were greatly enlarged as well. He carefully wrote up his findings, and his thesis earned him an eponym in medicine. The cells he first saw are today called Gaucher cells, and the disorder he described (i.e., a disease caused by the accumulation of material in cells that injures them) is called Gaucher disease.

As is often the case, the first published report of this disorder opened the eyes of clinicians, and other case reports soon followed. Physicians described patients with the same disorder, but who also had lung and kidney problems, bone problems, chronic joint pain, easy bruising, nosebleeds, and anemia. In several cases, they found that some patients had severe neurological problems. All agreed that the swollen cells were most likely stuffed with some fatty substance that was disrupting function. But biochemistry was in its infancy, and 50 years elapsed before a French chemist named Albert Aghion characterized the substance as a glucocerebroside (a fatty acid with sugar side chains).

Aghion's discovery stimulated much speculation. Did people with Gaucher disease overproduce the substance or were they unable to break it down? Perhaps the cells did process it, but for some reason they could not get rid of it. These were hard questions, and biochemists and cell biologists did not have the tools to answer them. The answer, which would not come for nearly 30 years after Aghion's discovery, required major advances in protein chemistry and the development of the electron microscope.

In 1949, a team led by Christian de Duve, then Chairman of the Laboratory of Physiological Chemistry at the Catholic University of Louvain, was frustrated by its inability to understand the function of an enzyme that they were studying. By chance, they found that this was because the enzyme was "stuck" to some cellular membrane that no one had yet described. Using centrifugation techniques (a way to fractionate cell components by weight) to pursue this curious finding, they eventually determined that cells contained little sac-like structures that contained enzymes that digested certain chemicals. Dr. de Duve named these organelles "lysosomes." Just a few years later his laboratory was able to use the electron microscope to define the physical structure of the lysosome. He won the Nobel Prize for this work in 1974.

We now know that lysosomes are a critical component of virtually all cells and that in each they function as the key waste processor. Since de Duve's work, researchers have discovered and described about 50 orphan lysosomal storage diseases that arise because of genetic errors that disable enzymes assigned to work inside this organelle. By serendipity and hard work, Dr. Gaucher discovered and described the first of them. By serendipity and hard work, Dr. de Duve provided a key discovery that would eventually lead to the creation of a new form of lifesaving treatment called *enzyme replacement therapy* (ERT) that has revolutionized the care of people with Gaucher disease and has helped patients with a steadily growing number of other *lysosomal storage disorders* (LSDs).

The first steps on the long road to developing therapies for genetic LSDs were taken by an extraordinary biochemist named Roscoe Brady. His lifelong interest in understanding the role of enzymes in disease began during World War II when as a Harvard medical student rotating on the wards of Boston City Hospital, he saw patients with extraordinarily high levels of lipids (fats) in their blood die young of heart disease. After an internship at the Hospital of the University of Pennsylvania, he opted for a career in research. In 1954, after a 2-year stint in the U.S. Navy (which put him in charge of its clinical chemistry laboratory in Bethesda), Dr. Brady joined the staff of the Institute of Neurological Disease and Stroke at the National Institutes of Health (NIH), where he was to spend the rest of his professional life. Few careers have been more focused or more fruitful.

After his appointment to the neurological institute, he redirected his research efforts to galactocerebroside, which scientists had shown was the most common fatty acid in *brain* tissue. This soon led him to become interested in Gaucher disease and the discovery by Aghion that had linked it to excess glucocerebroside in cells. Brady decided to carry Aghion's work further.

He initially surmised that patients were ill because their cells substituted glucose for galactose, but that turned out not to be the case. He next guessed that the key metabolic error might be an inability to make enough cerebroside, but that too turned out not to be the problem. This led him to hypothesize that in Gaucher disease the metabolic error was the inability to degrade glucocerebrosides, that is, to break them down for reprocessing.

Brady set out on a journey that over three decades resulted in many remarkable advances. In the mid-1960s he showed that all mammals have an enzyme that has the job of cutting glucose away from glucocerebrosides. In 1964 he showed that Gaucher patients lacked adequate amounts of this enzyme. In 1965 he showed that of the three different posited forms of the disease (based on careful study of the patients over time), the severity of each correlated with how much residual enzyme activity the patients had (many genetic diseases cause a reduction in enzyme activity rather than a complete elimination thereof). His work on the cell biology of Gaucher disease led to the intriguing thesis that there were many other rare storage diseases that arose for similar reasons. Over a period of 3 years in the late 1960s, other researchers elucidated the enzymatic errors that were the cause of several of them, including Niemann–Pick disease, Fabry disease, and Tay–Sachs disease. Over the next 5 years Brady and others developed relatively simple blood tests to diagnose these diseases, and showed the tests could be adapted for prenatal diagnosis.

In 1966, Brady suggested the innovative idea that patients suffering from LSDs would benefit from being treated with highly purified forms of the missing enzymes. This seemed logical, but there were many challenges. The most difficult was manufacturing. At the time no one knew how to make such big molecules, so the only solution appeared to be to collect them from the tissues of other animals (much as was then being done to obtain insulin from pigs and cows). But Brady and others worried that a nonhuman form of the enzyme would cause serious immunological reactions. Where, he wondered, could he find the large quantity of the human enzyme that would be eventually needed to conduct clinical trials? Brady recalls that, "one evening, I thought of the placenta." The very next day he started studying enzymes in discarded human placentas, and he quickly showed that he could find good quantities of a similar enzyme that was missing in Fabry disease. Unfortunately, when harvested, purified, and given to volunteers, that enzyme disappeared from the body so quickly that it did not seem to offer any hope of therapeutic benefit. But Brady persisted.

A single human placenta contains only a minuscule quantity of glucocerebrosidase, but each year hospitals discard millions of placentas, so Brady concluded it might be possible to set up a large-scale collection effort. Most important, early experiments suggested that the enzyme is fairly stable, making it easier to work with in the laboratory than had been the case with the enzyme needed to treat Fabry disease. In 1977, Brady and his team reported on their multiyear effort to develop a system to support extraction of the enzyme from human placentas. Along the way they had also figured out how to avoid using a purification step that threatened to introduce a small amount of a potentially hazardous material (concanavalin A) into the final product. The first small clinical experiment was gratifying. Brady and his team injected tiny amounts of purified enzyme into two adult patients with comparatively mild manifestations of Gaucher disease. Both men showed a significant drop in blood levels of the glucocerebrosides.

Few quests are as seductive or as demanding as the search to cure a disease. Brady and his team had set out on what became a nearly 20-year journey. After a year's effort and many thousands of placentas, they were able to purify just 9 mg of the enzyme. Then—to their surprise and despair—when they injected some of the precious material into a third patient, it had no discernible effect. Her cells had apparently accumulated so much toxic substance that the enzyme did not make a dent. Further, another experiment determined that when it was injected, most of the precious enzyme was taken up by liver cells, which prevented it from reaching the rest of the body.

Brady and his team faced a big challenge. It was not enough to purify the enzyme; it was essential that they find a way to enable it to leave the circulation and enter the cells that so needed it. Fortunately, in the early 1980s, a colleague of Brady's named Dr. John Barranger suggested an innovative way to solve this problem. Much of the offending lipid accumulated in white cells called macrophages. On their surface these common cells had a molecule that had a high binding affinity for sugars called mannoses. Glucocerebrosidase had lots of mannoses on its side chains. The only problem was they were largely hidden from the surface of the macrophages. Brady's team ran with Barranger's idea, and figured out a way to strip away part of the side chains off the enzyme, leaving the mannoses in plain view to the macrophages. The team then showed that so modified the purified enzyme was *50 times* more effective at getting into the fat laden cells in which it could degrade the toxic molecules!

In 1983, Brady obtained the approval of authorities at the NIH to use the purified and deglycosylated enzyme to treat a severely ill child with Gaucher disease, a young boy who was the son of a person who also worked at the NIH. As is typical for children with untreated Gaucher disease, he suffered from severe anemia, very low platelets, a massively enlarged spleen, and bony abnormalities. After much reflection, the research team selected as their clinical target an improvement in the boy's chronic anemia that, if successful, might allow him to avoid an otherwise inevitable splenectomy.

The researchers administered the modified glucocerebrosidase to the boy by intravenous (IV) infusion weekly for 2 years. After about week 26 of treatment, the boy's blood counts began slowly and steadily to improve. Over the next year his hemoglobin level reached 10 g/dL, a level compatible with a normal life. The platelet count also significantly improved. Perhaps more importantly, in Brady's words, "... the patient appeared much more vigorous. The abdomen was less protuberant and splenic size appeared to decrease on physical examination ... ." In addition, serial X rays showed improved bone mineralization. When the researchers stopped giving the enzyme (from experimental weeks 105 to 130) all the clinical improvements abated. When they reinitiated therapy, the patient again improved.

This encouraging development led Brady and his team to plan and conduct a much more ambitious trial. They would carefully design a study using similar clinical end points, recruit and treat a dozen patients for a year, and compile objective evidence on response to therapy. It took several years to plan, obtain approval for, recruit the patients, and initiate and complete the study. But they succeeded. In the spring of 1991, Brady and his team reported the results of the clinical trial in which they treated 12 patients (four young adults and eight children) with Gaucher disease in *The New England Journal of Medicine*. After a year of therapy the hemoglobin concentration improved in all the patients, the plasma levels of glucocerebrosides decreased in nine patients, the spleen shrunk in size in all, whereas the liver size decreased in five patients. This single paper provided much of the foundational clinical knowledge to support the development of ERT to treat patients afflicted with other LSDs.

The work of Dr. Roscoe Brady and his colleagues provided the foundation for ERT, but it fell to Henri Termeer, the CEO of a then tiny company called Genzyme, to bring the drug to several thousand patients scattered around the globe. Born in 1946 in the village of Tilburg in The Netherlands, Henri was the fourth of six children in a devout Catholic family. During his teenage years he became so obsessed with chess that his school grades suffered badly, and his parents forbade him to play. Right after high school, at the age of 19, Henri chose to perform his required 1-year military service. He was assigned to Officer's School and then given a managerial job. After graduating with a degree in economics from Erasmus University in Rotterdam, he worked briefly for a shoe company and then attended the Darden School of Business at the University of Virginia. From there he took a job at Baxter Travenol and began his rapid rise in that large health products company. Within months he won the task of overseeing the production of blood protein products (including Factor VIII for hemophilia) in its Hyland division in California. In 1976 he was promoted to run Hyland's German division, which, because of its focus on hemophilia, deepened his interest in rare genetic diseases. The late 1970s saw the emergence of the first great biotech companies such as Genentech and the Genetics Institute, and Termeer took notice as several colleagues at Baxter departed to work for them.

Perhaps the most important telephone call in Termeer's life came in 1983 from a small group in Boston that was funding a start-up company called Genzyme. Founded in 1981 by Henry Blair, a biochemist at Tufts, George M. Whitesides, a chemist at Harvard, and an entrepreneur named Sheridan Snyder, one of Genzyme's first goals was to try to industrialize the production of glucocerebrosidase to enable Roscoe Brady's clinical trials in Gaucher disease. The company was looking for a full-time CEO who knew the medical products business. Despite the company's tiny size and uncertain future, Termeer took the job.

Coincidentally, it was in 1983 that Congress enacted the Orphan Drug Act (ODA), a law that eventually would help Genzyme. The ODA provided a number of incentives to companies to begin working on diseases that because of their apparently small market size were of little interest to large pharmaceutical companies. The legislation was the culmination of a lengthy, persistent, lobbying effort by many small patient groups that had coalesced around the need to help loved ones with orphan disorders. Recognizing that the pharmaceutical industry had little economic incentive to develop drugs for orphan disorders, the loose coalition of patient groups concluded that the most effective way to drive such research was to guarantee economic rewards to smaller companies that made the effort. The product of years of discussion between patient advocates and the pharmaceutical industry, the ODA did just that. The law, which was amended in 1984 to define an "orphan" drug as one affecting less than 200,000 Americans, provides 7 years of market exclusivity—essentially a monopoly for that period. In addition, it provides tax *credits* for the cost of clinical trials, and other financial incentives. One of the most interesting aspects of the ODA is that two senators who were ideologically as far apart as any two members of the Senate—Ted Kennedy and Orrin Hatch—worked closely to make it a reality. In 1999 the European Union adopted similar legislation.

The ODA altered the economic calculus concerning rare disorders. In 1984 a drug called Panhematin developed to treat patients with a liver disorder called acute intermittent porphyria (AIP) became the first to achieve "orphan" drug status. During the ensuing three decades, the ODA has enabled the approval of more than 300 new therapies (including many for uncommon cancers and other severe disorders that are not monogenetic). Today ~20% of the drugs approved each year by the Food and Drug Administration (FDA) have an "orphan" designation. Many analysts believe that the ODA provided the critical stimulus to the explosive growth of the biotechnology industry. In fact, so many extremely highly priced drugs have been marketed under the protections afforded by the ODA that a few in Congress have periodically suggested that it should be repealed. Such concerns have had little impact.

One happy result of the struggle to obtain passage of the ODA was the creation of the National Organization for Rare Disorders (NORD) under the leadership of Abbey Meyers. No book about orphan disorders should be written without pausing to recognize her role. In 1982, Abbey, who has a son with Tourette's syndrome (an uncommon neurological disease associated with tics and behavioral problems), was hopeful that he would benefit from participating in a clinical trial sponsored by Johnson & Johnson to see if a drug it was developing for another disease would also help people with his disorder. One day she learned from her son's doctor that the company planned to stop developing this drug because of concerns that the market was too small. Meyers decided to challenge the company and began a campaign to enlist the help of other families burdened with rare disorders. She eventually met with Johnson & Johnson, and it decided to continue work on the drug.

A few months later, heartened by the support of many small groups built around concern for individual rare diseases, Meyers founded NORD, which she would lead for 20 years. During that time, Meyers became a well-known and effective advocate on Capitol Hill. For example, in 2002 Congress enacted the Rare Diseases Act, which directed the NIH to increase its efforts in studying rare genetic disorders, a law that she did much to drive forward. Today, NORD, which is an umbrella group that includes scores of organizations focused on individual rare disorders among its members, operates with a budget of more than $10 million, and under the leadership of its new director, Peter Saltonstall, continues to grow in influence.

From the start, Termeer instilled a key principle in the Genzyme culture: The patient always comes first. In his first years he made frequent trips to the NIH to meet with Roscoe Brady and to get to know the little boy who would become the first child treated at length with ERT. Termeer knew that the time line for developing the drug would be long and that many people might not have the patience to work in a company that had no products and offered its investors only hope. In fact, some early employees did leave. But that did not bother him. Termeer wanted people on his team who had self-selected to join him on his long journey. He served as CEO of Genzyme for more than 25 years, the longest tenure in the biotechnology industry and among the longest in the history of American industry.

Now in his mid-60s, Henri Termeer is (and looks like) the consummate CEO. Still blessed with leonine head of hair and an engaging Dutch accent, Henri, an elegant dresser, has an embracing smile and the uncommon knack for remembering everyone's name. At a meeting that I attended in April of 2012 on the challenges in creating new companies to develop new therapies for orphan genetic disorders, Henri used his time to recall the early years at Genzyme, which became one of the world's most successful biotechnology companies. "There is only one question you have to determine," he opined. "Can you make a drug that changes the lives of the patients you hope to help? If you can, then all the things you worry about—Are there enough patients to treat? Can we make the drug cheaply enough? Through what hurdles will the FDA make us jump?—will be resolved."

At the NIH during the 1980s, Roscoe Brady, to whom Henri often refers to as "the real hero" in the Gaucher story, had been trying repeatedly—without success—to obtain enough replacement enzyme to show clinical benefit in the little boy whose mother worked there. But despite giving the child every bit that he could extract from placental tissue, Brady simply could not produce or acquire enough of the missing enzyme (glucocerebrosidase) to reverse or halt the inexorable disease. The child steadily worsened. The production process needed to be industrialized—made at a scale that is far beyond the reach of an academic laboratory.

This is where Genzyme stepped in. At that time in Europe scientists in a village in France were removing the fluids from hundreds of thousands of placentas that were sent to them from hospitals throughout Western Europe (~70% of all that were potentially available). They were using that fluid as the starting material to isolate and purify an important immunoglobulin called IGG, which played a crucial role in helping people with immune system disorders. They had no other interest in the placental tissue, which they routinely incinerated after the fluid was taken. Genzyme got access to the used placentas.

Genzyme built a processing facility and began to extract the glucocerebrosidase from the placentas. Initially, scientists calculated that they would have to process 22,000 placentas to purify enough of the enzyme to treat just *one* patient with Gaucher disease for *one* year! When Genzyme started to make what it would call Ceredase, it had absolutely no certainty that it could produce the product at a cost that was acceptable to society. But that was not the toughest challenge.

"Imagine," recalled Termeer, "the FDA's reaction when we said that we wanted to make a novel biological drug by pooling cells from tens of thousands of placentas." Genzyme proposed this plan at the height of the AIDS crisis, and it was possible that some of the placentas contained HIV. At first, officials at the FDA thought it would be impossible for Genzyme to purify the enzyme to a level at which they could convince the agency that Ceredase was safe. But Genzyme, despite many doubters, persevered. It built a facility in Albuquerque to which it sent material extracted in France for extensive purification. Less than a year later, authorities in Europe decided that the threat of HIV infection was too great to permit further harvesting IGG from placentas. But Genzyme prevailed on the regulatory authorities to let it continue isolating glucocerebrosidase and managed to convince them that they could produce an uncontaminated final product.

Of course, while visiting the NIH Henri met the boy whom Brady hoped to treat and his mother. As is so often true with efforts to cure rare disorders, a parent's love and dedication is an important part of this story. When she learned that her son had Gaucher disease, the boy's mother, Dr. Robin Berman, ended her clinical practice and went to work at the NIH to focus on helping her son and other children like him. In addition to working with Dr. Brady, she founded the National Gaucher Foundation. As time passed, the little boy was getting progressively weaker from a severe anemia caused by destruction of the bone marrow. Most days he lay in bed, pale and listless. While the needed enzyme was being purified in Arizona, Genzyme obtained the FDA's permission to conduct what is called a Phase I safety trial of Ceredase. Safety trials are not planned to show that the drug is efficacious. The dose given to the human subject is typically less than what the researchers believe will be therapeutically effective.

Because of the rarity of the disorder and the limited amount of available drug, the FDA took the unusual step of permitting Genzyme to bypass the usual requirement that the drug be put in adults first (a practice that demands the production of much more drug, which is usually dosed according to body weight). In 1983 Brian Berman became the first child to receive therapeutic doses of the purified enzyme. When Termeer, for whom it was a core principle that he would get to know the patients for whom Genzyme was developing drugs, visited the child a few months into trial, he could not believe his eyes. The little boy was scampering through the hospital wards. Although many regulatory hurdles would have to be met, Termeer was convinced. "I did not need another clinical trial to prove to me that Ceredase worked," he recalled with a big smile. He now knew that the challenges of manufacturing the enzyme, the studies of efficacy (required even if the small safety trials show evidence of therapeutic benefit), the dose ranging studies (to determine the correct dose to give), and obtaining regulatory approvals could with diligence all be met.

In the fall of 1990, about 1 year after Genzyme had filed a New Drug Application (NDA), the FDA's Endocrinological and Metabolic Drugs Advisory Committee unanimously advised the FDA to approve the modified enzyme for treating humans with Gaucher disease. In its review, the committee's major safety concern was the risk that a product derived from human placentas would be contaminated with *prions*, the tiny bodies that had been found to contaminate extracts of human pituitary glands that had provided a source of human growth hormone. Prions were known to cause a rare degenerative and fatal brain disorder called Creutzfeldt–Jakob disease. An exhaustive literature search found no evidence of any discovery of such transmission from placental tissue, so the committee gave the green light. The FDA approved Ceredase, as the drug was named, for human use in 1991.

It is difficult to convey just how astounding this success is. It took an extraordinary effort to purify the human glucocerebrosidase, but that was only one hurdle. The molecule is so large that few biochemists at the time thought it could be given to humans in a manner that would permit it to circulate and reach the necessary cells before it was chewed up by the body's defenses. In addition, many understandably feared that by presenting a protein to patients whose immune systems would recognize it as foreign carried a serious risk of a severe, perhaps life-threatening response. Even after the drug was approved, many physicians still had concerns about safety.

At that time, I was on the board of a small genetic testing company called Vivigen in Santa Fe, New Mexico that Genzyme was seeking to acquire as part of its effort to grow its genetic testing business. The plan was to acquire Vivigen by issuing Genzyme stock, and the future value of the stock depended almost completely on sales of Ceredase. If safety issues arose after the sale (which often happens), our board's decision to sell the company to Genzyme might turn out to be disastrous. I called Dr. Robert Desnick, one of the world's experts in Gaucher disease, who was a professor at Mount Sinai School of Medicine in New York City. More than 20 years later, I still remember our exchange. I asked, "How good is this drug?" Without missing a beat, he replied, "It's a miracle."

The approval of Ceredase to treat patients with Gaucher type 1 disease was a triumph, but extracting enzyme from tens of thousands of human placentas was a formidable and inefficient process. Even while they were developing Ceredase, scientists at Genzyme were also working hard to harness the new methods of recombinant DNA technology to make the drug from scratch, freeing them from the arduous task of harvesting and purifying the complex protein from human cells.

To accomplish this, they developed a highly sophisticated fermentation system. They isolated the DNA sequence that codes for the human protein, they spliced that sequence into a DNA circle called a plasmid, and then they infected a master cell line made from Chinese hamster ovary cells. The use of nonhuman cells as a fermentation system was attractive as they were far less likely to contain viruses that might be able to contaminate the production process and—down the line—harm humans. Once the master cell lines had been coaxed into making substantial amounts of the enzyme, the scientists scaled up the process, eventually growing them in 2500-L bioreactors in which the cells were constantly bathed in a highly supportive growth media. On a daily basis the media, into which the cells secreted the enzyme, was removed and subjected to a complicated series of purification steps of which the final product was the lyophilized (dried) enzyme.

Once the Genzyme scientists had perfected this much more efficient means of production, the clinicians had to determine if its safety and efficacy was comparable to that of Ceredase. They conducted a Phase III trial comparing the two drugs in 30 patients with Gaucher disease. It showed that imiglucerase (Cerezyme), the enzyme made in the bioreactors, was as effective as alglucerase (Ceredase), and that it elicited a much lower immunological response. Genzyme gained approval for Cerezyme in 1994 and use of Ceredase was phased out. Today, about 6000 patients scattered among

100 countries rely on Cerezyme to control the signs and symptoms of Gaucher disease. The drug generates about $1 billion in revenue, and it is still the gold standard for treating the disease, despite the recent emergence of competitive products. Cerezyme is among the most expensive drugs in the world, but there is no question that it saves lives.

The development of Ceredase and Cerezyme created a new field in the pharmaceutical industry. Theoretically, the approach and processes used to develop those drugs could also be used to develop breakthrough therapies for many of the three dozen or so LSDs. During the 1990s, Genzyme rapidly expanded its research and development effort to do just that, and a few other biotechnology companies followed suit.

Fabry Disease

In 1897, two astute European dermatologists, William Anderson and Johannes Fabry, independently recognized some unusual clinical signs in patients that initiated the long quest to understand a second, and perhaps more mysterious, lysosomal disorder. At the time, Fabry, then 37, was already the chair of dermatology at the city hospital in Dortmund, a post he would hold until his death in 1930. One day in clinic he examined a 13-year-old boy who told him that over the last 4 years he had developed small, distinct, dark-red skin lesions in an unusual distribution limited largely to his thighs and buttocks. Although Fabry had no idea as to the cause, he dutifully published his findings, describing them as "purpura haemorraghica nodularis" (a typical Latinism of the day that is, thankfully, no longer in use) and speculated that they arose because of some kind of congenital vascular defect. Within weeks of when Fabry made his observations, William Anderson, a British dermatologist working at St. Thomas Hospital in London, recorded similar findings in a young patient, calling them "angiokeratoma corporis diffusum universale" and published it as "A Case of Angio-keratoma," without comment as to cause. The term, often shortened to "ACD," remained in clinical parlance for six decades. Unlike Anderson, Fabry retained a lifelong interest in the disorder, followed the course in this and a few other patients, and published autopsy findings on the first patient in 1930. Thus, the fact that today we no longer call this disorder Anderson–Fabry disease is defensible.

During the first four decades of the 20th century, physicians sporadically published case reports of patients with these distinctive skin findings, extending the description of the disease to include a rapidly widening array of signs and symptoms involving other organs, not the least of which was a chronic pain syndrome. During the 1940s, two Swedish physicians took a special interest in the disorder. Among several other papers, in 1947 they published detailed autopsy results in two patients under a title that awarded Fabry the eponymic ownership of the syndrome. Their work also helped advance the understanding that the tendency over time of the disease to affect the kidney, the heart, and the eye was due to the widespread deposition of a fatty substance in the cells.

Because it can manifest in so many different ways, with a wide spectrum in age of onset, organ systems that are most affected, and severity, it is not surprising that many years would elapse before some astute physician recognized that the variety of signs and symptoms others had reported could be best understood as the variable manifestations of a single rare monogenic disorder. Some patients with Fabry disease might manifest only skin lesions and choose not to consult a physician about them. Others only learn they have Fabry disease later in life when they develop signs of heart disease. Still others learn their diagnosis when they develop signs of kidney failure. In the past, the chronic pain that some patients feel in their joints or in the abdomen misled physicians to suspect psychiatric illness.

Relatively recently, researchers who analyzed data generated from a natural history study of 318 affected men and 337 affected women, have been able to show that the vast majority of patients have signs and symptoms in *childhood* that should permit doctors to reach the correct diagnosis. For example, by age 9, 40% of boys and 23% of girls had some skin lesions; by age 12, 27% of boys and 27% of girls had tinnitus (ringing in the ears); and by age 7 about 35% of boys and 24% of girls had recurring bouts of abdominal pain after eating (signifying involvement of the autonomic nervous system). In retrospect, it is difficult to understand how, until about a decade or so ago, even experts discounted disease manifestations in carrier girls and women.

A major advance in understanding the genetics of Fabry disease came in 1965 when a team led by a remarkable clinician named John Opitz, then a professor of pediatrics at the University of Wisconsin, used biochemical techniques to prove that the disorder was caused by a mutation in a gene on the X chromosome. During the 1970s and 1980s, Opitz, an unusually tall man who at age 15 emigrated from Germany to the United States, regularly dazzled me as he reported his latest studies at scientific meetings. At this writing he is an emeritus professor at the University of Utah, and he is certainly among the most erudite clinical geneticists in the world.

The confirmation that Fabry disease was X-linked explained why many women in Fabry disease families seemed to have a milder version of the disorder. In one-half of their cells the X chromosome with the mutation is active, but in the other half the cells operate with a normal copy of the gene. The history of our misunderstanding of the impact of the mutation on women represents a sustained underestimate and obvious delay (lasting some 30 years) of the impact of the disorder on women who are carriers. It was not until the launch of a long-term study called the Fabry Outcome Study (FOS) that followed the course of the disorder in more than 300 affected women that it became clear that many have more than a "mild" case of the disorder. Thanks to that study and others, we know that on average, women who are Fabry carriers have a life expectancy that is more than a decade less than women in the general population and that they are far more likely to die of heart disease in their 50s or develop renal failure requiring dialysis in their 60s. We still do not understand why, but a possible explanation is that—unlike the case with many other X-linked disorders—human cells need >50% of the normal enzyme to function properly.

As with Gaucher disease, the first hint that ERT might greatly improve the lives of patients with Fabry disease came in the late 1960s. In 1967 the team led by Roscoe Brady discovered the enzymatic defect. Between 1967 and 1970 they and others conducted further experiments to advance our understanding of the function of the enzyme called α-galactosyl hydrolase. But it turned out that this enzyme was particularly difficult to purify from human tissues and impossible to make in quantity. Another decade elapsed until bioengineers were able—using the same approach that had been used to manufacture Cerezyme—to manufacture the material in quantities that might be enough to ameliorate the disease.

Another major breakthrough came in the mid-1990s when scientists, again including Brady, developed a "knockout" mouse that lacked the ability to make the murine version of the human gene, and which showed signs of disease that were quite similar to those seen in humans. From there the pace of the advance toward therapy quickened. A team led by Dr. Robert Desnick showed that ERT reversed signs of Fabry disease in mice with the disease. Working with Genzyme, which was scaling up manufacture of the enzyme, he and others ran a Phase I/II (safety and dose ranging trial) of ERT in Fabry patients.

In 2001, they published their encouraging findings in *The New England Journal of Medicine.* They reported on a randomized, placebo-controlled, double blind (the treating doctors did not know which patients were receiving the ERT and which were not) of 58 men with Fabry disease. The key clinical end point chosen to support a claim of drug efficacy was to use biopsies of kidney tissue to measure clearance of the globotriaosylceramide (GL-3) (the harmful accumulated material) after treatment. Twenty of the 29 men who received ERT had no GL-3 in their biopsies, whereas all 29 members of the control group still had the deposits in their cells. In an "extension" study in which all 58 persons received ERT for 6 months, the GL-3 disappeared in all but two patients. Other less measurable factors, such as reduction in pain scores and improvement in overall quality of life, also improved markedly. In both Europe and the United States the regulatory bodies worked closely with Genzyme to streamline the approval process. The European Medical Agency approved the drug late in 2001 and the FDA (after Genzyme readily agreed to follow the patients in a long-term registry) approved it for human use early in 2003.

Pompe Disease

In the late 1990s, Genzyme also decided to attempt to develop ERT for an extremely debilitating storage disorder called Pompe disease. The disease is named after Joannes C. Pompe, a Dutch pathologist who first described the associated constellation of abnormalities in 1932. Pompe studied medicine at the University of Utrecht. On December 27, 1930, while still in training, he conducted an autopsy on a 7-mo-old girl who died of pneumonia. He immediately noticed that her heart was abnormally enlarged. Through the microscope he found that muscle cells were highly distorted into a mesh-like system. Because other storage disorders had recently been described, Pompe deduced that the infant girl died from a disease in which some accumulated chemical distorted cell function. Soon thereafter, he and his colleagues showed that the child had a vast excess of glycogen (a starch-like compound) in her cells. Pompe continued to study this disorder and his doctoral thesis was based on his analysis of the pathological findings in the heart. In 1939, after working for a time at a hospital in Nijmegen, Pompe was appointed the chief of pathology at a hospital in Amsterdam.

We do not often think of academic physicians as patriotic heroes, but Joannes Pompe deserves that recognition. In the spring of 1940, after the Nazis invaded the Netherlands, he joined the Dutch army until its surrender on May 15. He then took part in the Dutch resistance, at first using back rooms in the Onze Lieve Vrouwe Gasthuis (OLVG) Hospital to hide Jews. In the spring of 1943 he refused to enroll in the Nazi Medical Association and tendered his resignation to the hospital authorities, but they chose to ignore it, and he continued working there. Because his laboratory was somewhat isolated, in November 1944 he agreed to let the Resistance hide a radio transmitter in the animal laboratory and use it to transmit information to the Allied Forces in the United Kingdom. On February 25, 1945, Nazi soldiers discovered the transmitter and shot the radio operator in the hospital courtyard. They found Pompe coming home from church, and immediately imprisoned him. On April 14, 1945, the Dutch resistance blew up a railroad bridge, destroying a German train in the process. In reprisal, the Nazis took Pompe and 19 other prisoners to a meadow near the bridge, where they were summarily executed and buried in a mass grave. Today, a plaque commemorating the lives of Pompe and the other hospital employees who died with him, hangs near the entrance to the hospital in which some of the early studies in Pompe disease were conducted. I wonder how much more quickly research into this disorder might have progressed had Dr. Pompe not died at the hands of the Nazis.

Pompe disease is a glycogen storage disorder caused by defects in an enzyme called α-glucosidase (or acid maltase). Typically, depending on where in the gene the mutation is located, children either develop a profoundly severe disorder in infancy or a milder form in childhood or later. In profoundly affected infants, the muscles are so weak that they cannot breathe on their own; within months most are dependent on breathing machines to survive. When Genzyme embarked on this development effort, >90% of the children born with the "infantile" form died within the first year of life.

Genzyme's decision to take on Pompe disease was in part owing to convincing early data from naturally arising animal models of the disease. In 1998 a team from Tokyo published a study of the effects of dosing Japanese quail with the avian version of Pompe disease (arising from mutations in the same gene) with recombinant acid maltase. Normally, the affected birds suffer a severe myopathy, are never able to fly, and die of profound muscle failure. After just 2 weeks of therapy (only a few injections of the drug) the four treated birds became able to flap their wings and one was even able to fly a few feet.

I have heard Henri Termeer recall that in 1998, when a researcher from Duke University School of Medicine approached Genzyme with a proposal to develop a therapy for the condition in humans, that he did not ask the physician how many patients there were. It was sufficient that there was evidence in an animal model that by supplying replacement enzyme one could improve muscles and keep animals alive. In fact, when Genzyme initially tried to determine how many patients there were, it could barely find any. Because the children lived so briefly no one really knew the incidence (annual number of births per country) or the prevalence (total living patients) of the disease. This triggered debate within the company, but Termeer did not worry. His view was that if Genzyme made an effective drug, there would be enough patients to justify having done so.

Infantile Pompe is an ultraorphan disorder. It is so rare and so lethal that clinical experts at Genzyme thought that it would be unethical to study it in a randomized clinical trial, one that compared a group of treated with a group of untreated patients. Instead they undertook an extensive natural history study. At great expense, they found and studied the records from around the world of 150 patients who had been diagnosed with infantile Pompe disease. From these records they compiled the world's largest database on the progression of the disease. Eventually, they were able to convince the FDA to let them attempt to show efficacy by comparing the results of the clinical trials under which they would give replacement enzyme (which the company called Myozyme) to the clinical course of the untreated babies in the recent past. The results were striking. Whereas essentially all of the untreated babies had died within their first year, all of the 16 infants that Genzyme enrolled in the clinical trial were alive at 1 year, and none required a respirator.

Termeer was right; the patients were out there. If a disorder is almost uniformly fatal and there is no meaningful treatment for it scientists have little incentive to measure its incidence and prevalence, but that does not mean there are no patients. When Genzyme announced to the world in 2006 that the FDA had approved Myozyme for use in the treatment of infantile Pompe disease, within days it was inundated with calls about access to the new drug. Clinical demand outstripped the company's ability to supply the drug. Executives scrambled to improve production. This was an unanticipated problem for a company in which the majority view had been that there were not enough patients for which to develop the therapy! In 2012 Austrian researchers reported that a genetic study of 34,736 consecutive newborns had found four infants with Pompe disease—an incidence of 1:8684, much higher than the literature suggested. Most of the newly discovered children have missense mutations, those that usually result in the less severe, later-onset form of the disease.

As he had with the families of children with Gaucher disease, Termeer reached out to the families of children with Pompe disease. Staff at

Genzyme helped them organize, and the company supported the creation of a patient registry, essentially a database that compiles information about the disorder. Termeer knew that registries are worth their weight in gold, for they provide quick access to finding patients who might be willing to participate in future clinical trials, and they help build the market for the drug.

Physicians have now been treating children with infantile Pompe disease with Myozyme for about 8 years—enough time to acquire some sense of its efficacy. As anticipated, the evidence that it extends survival, delays time to use of a ventilator, and improves muscle function in severely affected infants is overall unassailable, but the drug falls far short of a cure. Patients often have strong immunological responses to the protein (that their bodies recognize as foreign), the enzyme does not return muscle function to normal, and some children still die of respiratory or cardiac failure, albeit at an older age. As the average child with the infantile form of the disease lives longer, a new phenotype is emerging. For example, because glycogen accumulates in brain tissue, some of these children with early-onset disease may later manifest intellectual disabilities. Despite its limitations, Myozyme and its recently approved (2010) companion, Lumizyme (for treatment of persons with milder and later-onset versions of the disease) has saved or improved the lives of more than 1400 patients with Pompe disease.

Genzyme was neither the first biotech company nor the first genetic disease company, but its immense success in developing ERT for Gaucher disease and for other genetic storage disorders stimulated great interest in the new industry. By the mid-1990s others were launching companies in the same arena. Probably, the most important is BioMarin, a California-based company that has also focused on LSDs. In 1997, Elizabeth Neufeld, a professor at UCLA and one of the nation's leading biochemical geneticists, and Emil Kakkis, a pediatrician who was a research fellow in her laboratory, successfully developed a Chinese hamster ovary cell line to make α-iduronidase, the enzyme that is defective in a disease called Hurler syndrome or MPSI (mucopolysaccharidosis type I).

Initially called "gargoylism" because of the coarse features seen in affected children, this rare (less than 1:100,000 births) storage disease was first described by an English physician about 1900. But the eponym, Hurler syndrome, honors a young pediatrician who published a case report of a patient she evaluated in Munich in 1919. Although the eponym stuck, several decades ago clinical geneticists reclassified nine different MPS syndromes by number, and today most call this condition MPSI. By the age of 3 or 4 the impact of having cells in virtually all their organs stuffed with large molecules called glycosaminoglycans (GAGs) has devastated the children. They typically have swollen spleens and livers, misshapen spines, compromised breathing, corneal clouding, hearing loss, joint problems, and in some cases mental retardation. Until about 15 years ago, many died in late childhood.

The demonstration by Neufeld and Kakkis that it was possible to make a normal version of the enzyme that malfunctioned in MPSI led to the creation of BioMarin, which was started with a mere $1.5 million investment from a company called Glyko. Within a year the start-up had raised another $11 million from private investors. Remarkably, just a year later the tiny company entered into a joint venture with Genzyme to develop ERT for the disorder. Progress was—by drug development standards—fast. In 1998 BioMarin went public, raising $67 million in its initial public offering. By now Emil Kakkis was the Chief Medical Officer of the company. Over the ensuing 15 years BioMarin has steadily expanded its research and development agenda. The FDA approved Aldurazyme, as the drug to treat MPS1 is known, in 2003. It approved a second drug, called Naglazyme, to treat a related disease called MPS VI in 2005. Since then BioMarin has won approvals for drugs to treat MPS IVA (known also as Morquio syndrome) and a drug to treat patients with phenylketonuria (PKU). In mid-2014 BioMarin had a market capitalization of $8.5 billion.

What might Henri Termeer advise those who would like to conquer a rare disease? Develop as much preclinical evidence as possible. Acknowledge that one cannot resolve all safety and immunogenicity problems. Understand that the FDA will require that treated patients be followed for years to assess the safety risks (even though ERT—giving a human protein to a human in which it is not present or not working—probably poses less safety risk than does treating with a newly developed small molecule that may have many "off-target" effects). Do not be deterred about the number of treatable patients. Time and again, Genzyme and other companies found that once a drug was approved, the number of known patients doubled or tripled. Remember, the regulatory agencies do understand how hard it is to conduct clinical trials for rare diseases, and they will be flexible and will work with you to find a reasonable path.

During 1990–2010, Genzyme took on one lysosomal storage disease after another, ultimately winning FDA approval for three new orphan drugs: Fabrazyme (α-galactosidase) to treat Fabry disease (2003), Aldurazyme (laronidase) to treat Hurler disease (2003), and Myozyme (alglucosidase α) to treat Pompe disease (2006). During this era it was the undisputed leader in the biotech world in developing drugs for orphan monogenic disorders, and along the way it delivered impressive value to its shareholders. When the European pharmaceutical giant Sanofi acquired it in 2011, Genzyme was valued at $20 billion, and Termeer was among the nation's longest serving CEOs.

I have spent some time speculating how many lives have been saved because of access to drugs that Genzyme developed. My *guess* is that by 2014 its drugs had saved more than 10,000 lives. Thus, one unconventional way of looking at the Sanofi transaction is that Sanofi valued Genzyme's drugs at much more than the $1 million dollar per life it had saved. Because most of the patients who are treated with ERT began treatment as children, it will cost many millions to treat them throughout life. However, ERT also eliminates the need for other expensive care and (certainly in the case of Gaucher disease and Fabry diseases) it permits many patients to live productive lives. It would probably be impossible to conduct a conclusive cost–benefit analysis, but thus far society is willing to tolerate these expensive drugs. Indeed, it is almost certain that the cost of ERT elicits far better economic return to society than does a cancer drug that costs $100,000 to extend lives for several months.

By 2013 the FDA had approved five man-made enzymes for the treatment of LSDs and several more for other monogenic enzyme diseases. Of these, Cerezyme, the first to be approved (for treatment of Gaucher disease), has had the most beneficial effects. Each of the other enzyme replacement strategies has ameliorated some aspects of the genetic LSDs for which they were designed. But, in those instances in which the disease harms the brain, their therapeutic value is poor. These large molecules cannot cross the blood–brain barrier (BBB).

During embryonic life and early childhood the capillary endothelial cells that line the smallest blood vessels in our brain develop exceedingly tight junctions that allow only a few small molecules to enter brain cells. Unlike other cells, those that make up the capillaries that support the brain are embraced by astrocytes. In addition, brain capillaries have a thicker, less penetrant, basement membrane. Each brain cell demands a relatively high level of energy to do its work, so the capillary network has to be immense to meet its task of providing nutrients to every cell. Some have estimated that laid end to end the capillaries in our brains would measure more than 300 *miles*.

In effect, because they cannot reach the brain cells, ERT drugs offer essentially no value in slowing the inexorable brain damage caused by some genetic storage disorders. Fortunately, the most common form of Gaucher disease spares the brain. Unfortunately, several others—for example, Hunter syndrome (also called MPS I) and Hurler syndrome (also called MPS II)—do cause neurological damage over time, even when children are placed early on maximal ERT.

Over the last few years, a few research groups have attempted to overcome this critically important therapeutic problem by devising ways to deliver the manufactured enzyme by infusion into the spinal fluid that bathes the brain. This is far from an optimal method of delivery as treatment would almost certainly have to be periodic and lifelong. Getting and maintaining access to the spinal fluid is inconvenient and carries some risk. But evidence is emerging that intrathecal delivery of ERT may confer significant benefit. Since about 2009, scientists have been injecting enzymes into the spinal fluid of mice that have been genetically engineered to have the murine version of Hunter and Hurler syndromes. The subsequent reduction of harmful storage material in brains suggests that the animals have benefited from the treatments, and that humans might. In 2014 a study in using intrathecal ERT in dogs with Batten disease offered further encouragement.

The first clinical trials in humans with these two disorders have started. Joseph Muenzer, a professor at the University of North Carolina, is the PI (principal investigator) on a trial sponsored by Shire Pharmaceuticals, the maker of Elaprase (idursulfase), to deliver a slightly reformulated version of it intrathecally (to the spinal fluid) to children with Hunter syndrome. The study enrolled 16 affected children (3–18) who were divided into groups of four and either treated with placebo or given one of three different doses of the drug (a so-called dose ranging study). Results have not yet been reported. At the University of Minnesota, Dr. Paul Orchard is leading a team that will conduct a clinical trial in which children with Hurler syndrome will be treated with bone marrow transplant and periodic intrathecal delivery of Aldurazyme (laronidase), the enzyme missing in that disorder. That study is just starting to recruit children, and the first safety readout will likely not be made until 2016.

The regular intrathecal delivery of enzyme faces several hurdles, the most difficult of which is the challenge of delivering sufficient enzyme throughout the brain (for storage disorders affect essentially every cell in the brain). Although news from the experiments on mice is positive, it may not be possible to deliver enzyme broadly enough to a human brain to arrest the disease. One positive note is that in 2007 a group led by Dr. Emil Kakkis (then at BioMarin) showed that the delivery of high doses of laronidase to the spinal fluid of dogs with the canine equivalent of the disease resulted in long-term correction of the levels of GAGs, a superabundance of which characterizes the disease.

Overall, in 2014, ERT drugs were given to ameliorate the severe burden of LSDs in about 10,000 patients, and in aggregate generated about $3 billion in revenue. Genzyme was responsible for about one-half of this, most related to the treatment of patients with non-neuronopathic form of Gaucher disease. In 2014 the International Gaucher Registry was following the clinical course of more than 6000 consented patients to better understand the long-term risks and benefits associated with ERT.

Substrate Reduction

Good as it is, ERT is ameliorative, not curative. In addition, the extremely high cost of ERT and the immunological problems the drugs cause for many patients have spurred researchers to develop other treatments. One plausible approach to the treatment of LSDs would be to reduce the amount of substrate—that chemical that accumulates to toxic levels in cells because a patient's mutated enzyme is unable to break it down. In a sense the low phenylalanine diet (discussed in Chapter 1) is a form of substrate reduction therapy. In LSDs (which are caused by a failure to break down large molecules made by cells) dietary manipulations do not work. Nevertheless, it might be possible to develop molecules that reduce the concentration of stored material in one of several ways: by developing a drug to slow the accumulation of the toxic substrate to a level at which a partially functional mutant protein could metabolize enough of it to reduce disease severity, by improving the efficacy of some alternative metabolic pathway, or by reducing the stability of the stored material. In 1979, a group at Johns Hopkins proposed that one could improve the health of children with disorders of the urea cycle (nitrogen metabolism) by using a small drug to eliminate toxic products along a secondary biochemical pathway. Since about 2000, scientists have attempted to develop substrate reduction therapy (first proposed in the early 1980s) for several disorders with, thus far, only modest success.

About 15 years ago researchers showed that chemicals called iminosugars could slow the rate at which toxic gangliosides build up in the cells of children with Tay–Sachs disease (and the closely related Sandhoff disease). Unfortunately, subsequent research has not shown a significant clinical benefit of treating affected children in this way. About a decade ago, after one research group showed that a small molecule called cysteamine

(Cystagon) reduced the amount of stored material in cells from patients with a severe juvenile onset disorder known as Batten disease, others launched a small clinical trial of that drug. After several years of follow-up here too there was no clear evidence of benefit. Researchers have seen the most encouraging results from attempting to use substrate reduction therapy (SRT) to treat Gaucher disease. During the late 1990s and early 2000s a biotech company named Actelion developed miglustat (now marketed under the name Zavesca), to inhibit the production of the first enzyme in the pathway by which cells make glycosphingolipids, the compounds that (when further processed) become the glucocerebrosides that accumulate to harm patients with that disease. In 2003 the FDA approved miglustat for treating Gaucher disease in patients who cannot use ERT. SRT should be much less expensive than ERT. For some patients, physicians are able to lower the dose of the expensive recombinant enzyme by using miglustat as an adjunctive therapy. Perhaps the most recent hopeful news is that in the spring of 2015 Sanofi/Genzyme announced that it would conduct a phase 2a study of a novel oral substrate reducing agent in patients with Fabry disease. The drug (GZ/SAR402671) inhibits an enzyme to block the formation of glucosylceramide (GL-1), a key intermediate in the pathway that leads to the toxic buildup in Fabry disease.

One exciting aspect of SRT is that the same compound may be efficacious for ameliorating *several* different orphan disorders, each of which arises from an enzymatic defect at different steps in the *same* synthetic pathway. Miglustat has also been shown to be of modest clinical value in a related, but more severe and even rarer storage disorder called Niemann–Pick C (NP-C) disease. Actelion's efforts to gain regulatory approval in the European Union, Russia, and elsewhere to use miglustat to treat children with Niemann–Pick C disease succeeded. However, in 2010 the FDA, reviewing essentially the same data package, refused to approve miglustat for expanded use. In a "Complete Response Letter," the agency essentially asked to see more clinical evidence of efficacy. This confused many parents of children with NP-C, and outraged some. How could a drug be approved in Europe and then fail to pass muster in the United States? In fairness to the FDA, by any measure, the data package showed quite modest evidence of clinical improvement and little evidence on which to predict long-term benefit.

Still, there is no other therapy for affected children. In the United States access to miglustat for kids with NP-C depends on whether insurers are willing to pay for off-label use of an expensive drug. My guess is that currently about one-half of affected children somehow obtain access to it. Over the last few years several small clinical trials have generated results that miglustat treatment does stabilize some aspects of NP-C disease, especially eye movements and the ability to swallow properly.

Many of the mucopolysaccharidosis (MPS) syndromes arise because of the toxic effects of molecules called GAGs. There is evidence that a small molecule called genistein inhibits the synthesis of GAGs and clinical trials are under way to determine if it can reduce the burden of disease. In 2013 Genzyme reported the results of a Phase III trial of a next-generation substrate reduction drug called eliglustat. In a trial of 28 patients with Gaucher disease who had not yet been treated with other drugs, eliglustat met its goal of significantly reducing spleen volume, reducing liver size, and improving platelet counts. In the summer of 2014, the FDA approved the new drug (now marketed as Cerdelga) for first-line oral therapy for certain Gaucher I patients.

Although progress with SRT has been slow, it remains an active area of research in regard to LSDs, and several disease-modifying drugs should be approved in the coming decade. By its nature, however, SRT is a strategy to achieve disease amelioration; it will not provide a cure. There is little reason at this time to anticipate that SRT drugs will reverse the course of serious orphan disorders.

7

Gene Therapy

Using Viruses to Deliver Normal Genes

Recombinant DNA

IN ONE SECTION OF THE MARVELOUS DISNEY MOVIE, *Fantasia*, Mickey Mouse, cast as a sorcerer's apprentice, disobeys his master, plays with powers forbidden to him, and quickly loses control, causing a devastating flood. The great conductor, Leopold Stokowski, directed the movie's musical score, a collection of eight classical pieces (including works by Beethoven, Shubert, Dukas, and Mussorgsky) set to animation, at the Philadelphia Academy of Music in 1939. When I think of the origins of gene therapy, I sometimes recall the flood sequence in *The Sorcerer's Apprentice*. Why? Once molecular biologists realized in the early 1970s that they had discovered tools that would enable them to manipulate DNA, they were eager to explore a dizzying array of scientific questions. But some among them raised concerns that the exercise of their new powers might unintentionally cause great harm. One fear was that someone might use recombinant DNA technology to modify a strain of bacteria that would, if it escaped the confines of the laboratory, devastate crops or kill people. The dawn of molecular biology generated unprecedented ethical reflection among biologists. I will return to the debate that it generated, but I will first recall some early events that laid the groundwork for gene therapy. Some of this précis relies on a fine book called *The Coming of the Golden Age*, written by Gunther Stent, one of the founders of molecular biology.

New as it sounds, the term "molecular biology" was coined in the 1930s by a British scientist named W.T. Astbury who specialized in X-ray crystallography. At that time the use of X rays to study protein structure was a rapidly developing field that would, among many other outcomes, permit Watson and Crick to deduce the helical structure of DNA in 1953. The neologism was slow to catch on. If there is a single person who reified the term, it was Max Delbrück, a German physicist who had studied with Bohr, but who in the later 1930s redirected his powerful intellect to exploring the *physical* nature of the gene. Although Mendel's notion of a gene as a discrete entity that could faithfully transmit information when cells reproduced had been dogma for four decades, when Delbrück began to study the gene, scientists still did not know anything about its composition. Indeed, because of the far greater complexity of proteins compared to the apparent simplicity of DNA molecules, most scientists thought protein was the hereditary material.

In the early 1940s, Delbrück started collaborating with two other scientists, Salvador Luria and Alfred Hershey, who shared similar interests. These three were the core founders of what became known as the American Phage Group (phage was the word give to tiny viral-like particles that had been discovered two decades earlier and that were known to have the capacity to infect bacteria and produce multiple copies of themselves). The now legendary phage group, which throughout the 1940s met each summer for several-week sessions at Cold Spring Harbor on Long Island, made some of the fundamental discoveries that gave meaning to the term, molecular biology. I still recall the trepidation I felt when in 1975 I delivered the first invited seminar of my life to the Department of Biology at Caltech, and realized that Max Delbrück was in the audience.

The greatest leap in advancing understanding of the gene was made in 1944 in the laboratory of Oswald Avery at the Rockefeller Institution in Manhattan. In experiments with differing strains of bacteria, Oswald showed that the DNA from the "donor" could change the *heredity* information in the recipient. Therefore, DNA must be the hereditary material, and the DNA molecule must be structured in a manner that permitted it to carry and faithfully transmit vast amounts of information.

James Watson, a student of Luria's at Indiana University, was the youngest member of the phage group, which numbered no more than 50 people throughout the 1940s. The story of his postdoctoral adventures and his collaboration with Francis Crick in 1952–1953 at the University of Cambridge in England, which culminated in the discovery of the double helix, has been retold many times. Yet, surprisingly, during the 1950s there was not an immense flowering of discoveries built on this mighty finding (for which Watson and Crick became among the youngest of Nobel Laureates in 1962). Scientists, who did not yet have the chemical tools they needed to interrogate DNA, focused mostly on classical and bacterial genetics to slowly tease out the structure and function of the gene. Among the most important contributions were those of Seymour Benzer who painstakingly elucidated the fine structure of bacterial genes (and, in so doing, confirmed and extended the thesis that "one gene codes for one enzyme" that had been developed in the 1930s). Benzer's work also set the stage for Jacob and Monod, two French scientists, who made important early contributions to understanding gene regulation.

The origins of gene therapy are tied to the work I have just so briefly summarized and to the work of virologists who were deeply interested in how simple viruses entered the cells of the complex organisms and then commandeered their enzymatic machinery. One important event in the prehistory of gene therapy was a discovery made by Renato Dulbecco, an Italian physician-scientist who had fought in the resistance against Nazi Germany. After the war, he emigrated to the United States to work with his countryman, Salvador Luria, at the University of Indiana. A few years later he moved to Caltech to join Max Delbrück. In 1964 Dulbecco showed that a tumor virus called SV40 transformed normal mammalian cells into cancer cells by transmitting its DNA into those cells in a way that allowed them to alter the expression of genes in the mammalian cell genome. The conversion of a normal cell to a cancer cell is not good for an organism, but the experiment hinted that it might someday be possible to harness viruses to carry normal genes into cells to correct genetic disorders.

This idea gained more credence when Dr. Stanfield Rogers later showed that another virus called TMV (tobacco mosaic virus), which naturally infects some plant cells, caused them to make a protein that they, the cells, were *not* normally able to make (a discovery in line with Avery's work of 1944). About 1971, scientists in Dulbecco's group, including Dr. Theodore Friedman, began to explore if it might be possible to use naturally occurring versions of the papovavirus to transport desirable DNA sequences into cells.

The discovery about 1970 that many bacteria make enzymes (dubbed "restriction enzymes") that cut DNA sequences at precise locations soon led to an array of new tools that permitted scientists to isolate genes, and to insert them inside viral capsids. During the 1970s, scientists perfected techniques to combine isolated genes with viral vectors to create such payloads. In the same era, two of Dulbecco's students, Howard Temin and David Baltimore, independently discovered an enzyme (called reverse transcriptase) that permitted cells to read the *RNA* code of viruses and to make a corresponding DNA sequence. In 1975, Dulbecco and his two students shared a Nobel Prize for this and related work.

Another provocative development occurred in 1972 when Rogers, who was working with the Shope papillomavirus (SPV), discovered that some of his colleagues in the laboratory had low serum levels of an amino acid called arginine and reasoned that it was owing to exposure to the virus. Rogers knew that there was a very rare human disease called argininemia (today called arginase deficiency) in which patients apparently lack an enzyme to metabolize arginine, causing blood levels that are toxic. In an experiment that would not receive the approval of human studies committees today, Rogers treated three German girls with this rare disorder by infecting them with the Shope virus with the hope that it would infect their cells, transfer its gene for arginase, and correct their metabolic abnormality. The girls survived, but the experiment (which evoked ethical concerns) did not result in any clinical benefit.

About 1972, Paul Berg, a biochemist at Stanford who would become a Nobel laureate in 1980, showed that he could use the now readily available "restriction enzymes" to cut up the DNA of SV40 and of a bacteriophage called lambda, and then *splice* (recombine) *sequences from the two distinct life forms together.* This extraordinary experiment, and others like it, gave birth to the field of genetic engineering and set the stage for a true technological revolution. To this day, I remember the standing room only crowd and the palpable excitement in a large lecture hall at the Baylor College of Medicine in Houston when Berg lectured on his discovery. But the early experiments in recombination also raised the specter of environmental catastrophes, cancer epidemics, and frightening new forms of biological terrorism. What, for example, might be the consequences of accidentally infecting *Escherichia coli,* a bacterium that lives in every human gut, with a cancer-causing gene? Berg and others moved quickly to address these concerns.

In 1975 they organized a meeting at the Asilomar Conference Center in Monterey, California at which about 140 leading biologists and a few lawyers and ethicists met to discuss the risks of genetic engineering, and to consider how to evaluate and contain them. As one outcome of this extraordinary meeting, the Asilomar group drafted a set of voluntary guidelines that focused on the requirement that any organism selected for experiments in genetic manipulation must be biologically crippled so that it could not survive outside a special environment in the research laboratory in which it was studied. The knowledge and experience of microbiologists who worked in the esoteric field of bacterial taxonomy was particularly influential, as they could point out ways to disable the bacteria in question.

During the late 1970s, there was much public discussion about the risks associated with recombinant DNA experiments, and federal rules were drafted requiring different levels of laboratory containment for different kinds of recombinant DNA experiments. Some local political leaders, most famously Alfred Vellucci, then mayor of Cambridge, Massachusetts, were deeply wary of the new technology, and enacted ordinances to forbid or strictly curtail the research. Over the next few years a substantial number of scientific papers reported on efforts to explore the environmental risks associated with moving DNA sequences between species, in general providing a large measure of reassurance. Genetically hobbled laboratory organisms just could not survive outside their protected environment.

By the late 1970s, scientists had made significant progress in assembling DNA payloads and inserting them into the genomes of viral vectors that were naturally able to infect human cells. Much of this work was performed with the human β-globin gene (hundreds of thousands of people suffer from sickle cell anemia and β-thalassemia because of mutations in it). One major interest of these early genetic engineers was to harvest progenitor (stem) cells from patients with genetic blood diseases, transduce those cells with viral vectors that carried the normal genes, and then return them to the patient. The plausible biological reasoning was that the transduced stem cells would proliferate and that their progeny would produce enough normal hemoglobin to ameliorate the disease. But no one yet knew how to transduce enough stem cells to have any confidence that once they were returned to the patient and made their way to their natural home in the bone marrow, that they would produce a therapeutic level of hemoglobin.

The dream of attaining this goal led to one of the more infamous ethical misdeeds in the annals of modern medicine. Knowing he could not (owing to safety concerns) obtain regulatory approval for experiments in gene therapy to attempt to cure these often fatal blood disorders, Dr. Martin Cline, a hematologist at UCLA, who had had some success in 1980 in transferring virally transduced bone marrow cells into mice, left the United States, and attempted to treat two little girls with β-thalassemia—one in Italy and one in Israel. The experimental therapy did not have a beneficial clinical effect. Although Cline did not break any laws in those countries, he clearly violated research rules in the United States. After a full review, the National Institutes of Health (NIH) censured his work and ruled that his future applications for further funding would undergo special review.

In 1981, several groups reported that they had been able to alter retroviruses to make them less dangerous to humans and to enable them to carry and deliver a gene payload. Early evidence that one could insert a normal gene into a modified retrovirus, transduce (deliver them into human) cells, and observe a detectable improvement came in 1983. For example, a team led by Theodore Friedmann used a retroviral vector to partially correct the metabolic defect in cells taken from a child with a rare orphan disorder (characterized by severe developmental delay and self-injurious behavior) called Lesch–Nyhan syndrome. The period 1985–1995 saw rapid growth of gene delivery systems, with research groups working on both viral and nonviral systems such as liposomes (tiny globules of fats into which a long stretch of DNA could be inserted, and that have the ability to fuse with normal cells and deliver their payload to the cell's interior). The proliferation of delivery systems pulled hundreds of young scientists into what had been an esoteric area, and during the 1990s the number of publications addressing gene therapy soared.

As good a date as any for the birth of clinical gene therapy is September 14, 1990, the day that W. French Anderson, a NIH researcher, treated a little girl with an orphan genetic disorder called adenosine deaminase deficiency (lack of this enzyme greatly reduces the ability of immune cells to fight infection). Using a simple intravenous infusion, he returned millions of white cells that had been collected from her bone marrow and genetically engineered with an adenovirus that carried a normal copy of a gene that coded for adenosine deaminase (ADA). It was a smart choice to take on this disorder because one did not have to correct all the cells in the body, only those bone marrow cells that made white cells. It would take years of following many such patients to make sure that their immune systems grew strong and stayed strong, but the success in treating ADA suggested that gene therapy was coming of age. Sadly, Anderson's career ended ignominiously some years later when he was incarcerated for child molestation.

By 1995, federal regulatory authorities in the United States and several European countries had approved more than 200 clinical gene therapy trials, and more than 1000 patients were enrolled as subjects. Optimism was high, but it was obvious that many technical problems (especially in regard to safely delivering enough of the virus carrying the gene) needed to be solved before gene therapy could enter regular clinical care. In the fall of 1995, the results of three different gene therapy trials—to treat cystic fibrosis, ADA deficiency, and a genetic form of very high cholesterol—all concluded with disappointing clinical results. In response, two NIH advisory committees expressed skepticism about the efforts, and recommended more rigorous oversight.

Still, optimism continued, and interest in the biotech industry burgeoned. For example, about 1993, Genetix Pharmaceuticals, a Cambridge-based start-up that planned to develop vectors to deliver the gene for β-hemoglobin, became among the first in the nation to garner private investments devoted to commercializing gene therapy. In 1995, Harold Varmus, then head of the NIH, apparently impressed with the growing evidence that gene therapy trials could be conducted safely, reduced the regulatory power of the federal Recombinant DNA Advisory Committee (RAC), cut its membership from 25 to 15, and ruled that its views were advisory only. His message was clear: Gene therapy did not need the extra level of review that had been in place for the first few years. Writing a brief history of gene therapy in 1998, Theodore Friedmann was hopeful. He noted that the Food and Drug Administration (FDA) had approved the launch of some 200 trials and that more than 2500 human subjects had been enrolled.

Ironically, just after Friedmann's book was published, prospects for human gene therapy imploded. In the fall of 1998, Jesse Gelsinger, an 18-year-old young man from Arizona who had been diagnosed at age 2 with an orphan X-linked genetic disorder called ornithine transcarbamylase deficiency (OTC), agreed to enter a gene therapy trial that had been organized by Dr. James Wilson at the University of Pennsylvania. In its most severe form, OTC, a liver disorder, renders affected children unable to metabolize nitrogen compounds, rapidly causing a massive buildup of ammonia in the blood, which can severely damage the brain in a few days. In that era some children slipped into a coma and died within a few days of birth, often diagnosed too late to permit doctors to get their severe metabolic abnormalities under control.

The form of OTC deficiency that Gelsinger had been born with was unusual; he was a "mosaic." He had not inherited a mutated X chromosome from his mother; rather a new mutation had arisen spontaneously on his X chromosome shortly after conception, so some of his cells were normal and others were not. Thus, at birth he appeared well, and was not diagnosed until he was 2 years old. Doctors could control his disease by keeping him on a very low-protein diet and giving him a chemical called sodium benzoate that sponged up excess nitrogen.

By careful attention throughout his life to adhering to a low-protein diet, Gelsinger, who had a mild form of OTC, had been able to maintain good health and lead the life of a normal teenager. He was not so ill as to need to take a chance with a novel new therapy with unquantifiable risks. But the local Institutional Review Board (IRB) at the University of Pennsylvania had made the reasonable decision that for safety reasons the researchers should treat adults with mild OTC deficiency before they could enroll more severely ill infants in the clinical trial. The IRB was also concerned that anxious parents could not really give truly informed consent to place a very ill child in the experimental protocol. So Wilson and his clinical colleague, Dr. Mark Batshaw, proceeded to invite older patients with milder forms of OTC to enter into a "safety trial" (one focused on assessing the risks rather than the benefits of a new therapy).

Of course, the researchers—in preparation for treating humans—had already performed extensive animal studies in mice, rhesus monkeys, and baboons. Some of the monkeys had developed signs of liver inflammation, but they had resolved without incident. To try to reduce the unknown risk of liver damage to human subjects, the plan was to inject the adenoviral vector carrying the normal OTC gene directly into just one part of the liver (the right lobe). No serious adverse effects developed in the first 17 human subjects. Jesse was the 18th person to be treated.

On September 17, 1990, a physician injected about 30 milliliters of fluid containing billions of copies of the genetically engineered adenovirus into the right lobe of Jesse's liver without incident. He soon spiked a temperature, which was not particularly unusual, but less than 48 hours after treatment he began to show signs of jaundice—a warning sign that the liver was in trouble. Despite heroic medical measures, Jesse quickly sunk into liver failure. In essence, the adenovirus had elicited an overwhelming immune response that severely damaged the very organ that the research team hoped to protect. Within days his kidneys and other organs were failing. On September 29, 1990, in the presence of his parents, Jesse was declared brain dead, and his respirator was turned off. His family scattered his ashes from the top of Mount Wrightson (near Tucson) that he had so loved to climb.

Federal authorities quickly launched a detailed review of the clinical trial. They concluded that Wilson had not properly disclosed that he held an equity interest in a company that planned someday to commercialize gene therapy, thus raising allegations of conflict of interest. This triggered an extensive investigation by the United States Department of Justice, which after protracted legal proceedings, led Wilson to agree (without having to make a formal admission of wrongdoing) to a 5-year suspension of his right to be the principal investigator of certain NIH-funded research. The Gelsinger family sued the University of Pennsylvania; the case was eventually settled. Having complied with the NIH settlement, Wilson continued his important work on gene therapy. Over the last 15 years, his large research team has made many fine contributions to the field. One of the most important is that Wilson has played a major scientific role in the discovery and characterization of a number of adeno-associated viruses (AAVs), which are today considered the safest agents to use in gene therapy.

The death of Jesse Gelsinger had a deep impact on gene therapy in the United States. Regulatory authorities declared a brief moratorium on gene therapy trials, one that would last until the death could be fully evaluated. Many of the most ardent proponents of gene therapy were shaken and entered into a sustained period of doubt. But extensive scientific review of Gelsinger's death found no serious weaknesses in the approved protocol, no evidence from animal studies that the human trial should not have been initiated, and no deviation from the treatment plan. The reviewing team concluded that Jesse had died from a massive, but unpredictable, inflammatory response to the virus, something that had not happened to any other patient in the trial.

Gene Therapy Comes of Age

The circumstances surrounding Gelsinger's death had a somewhat smaller impact in Europe than in the United States. In France, especially, there was much enthusiasm to drive gene therapy forward to treat otherwise incurable monogenic disorders of the hematopoietic (blood-forming cells) system. In the late 1990s, building on work performed by other scientists with mouse models, medical researchers at Necker Hospital in Paris led by the noted transplant specialist, Dr. Alain Fischer, initiated a clinical trial to attempt to cure a rare genetic disorder of the immune system called X-linked severe combined immune deficiency (X-SCID). The disorder is caused by a mutation in the gene that codes for a protein called the *γc subunit* of certain cytokine receptors. This in turns leads to the failure to make cell surface receptors that are needed to interact with various chemical messengers called interleukins (specifically numbers 2, 4, 7, 9, and 15). These chemical messengers are crucial to the formation of T cells and "natural killer" cells, which protect us from the world of bacteria and viruses in which we live. Boys born with this genetic disorder cannot defend themselves from infection, making X-SCID a fatal disorder of early childhood. In the late 1990s the only hope for cure was a bone marrow transplant (BMT), a heroic intervention that itself carried a 15% risk of death.

In 1998, Fischer and his team, notably a remarkable Italian hematologist named Marina Cavazzana-Calvo and a talented pharmacologist named Salima Hacein-Bey-Abina, reported the results of a clinical trial in which they removed stem cells from boys with X-SCID, transduced those cells with a

Moloney retrovirus carrying the normal DNA sequence for the γc subunit, and returned them to the childrens' bodies where it was hoped that they would propagate to reconstitute normally functioning immune systems. In 2000 they published a paper in *Science* that provided convincing evidence that in their first two subjects, the experimental intervention had been without significant side effects, and that *the genetically engineered cells were making the protein needed to correct the immunological defects.* In 2002 the team reported the results to date from treating nine patients, demonstrating the presence of large, healthy populations of T and natural killer cells just 4 months after treatment. The numbers of normally functioning B cells, although lower than desired, were adequate to protect the children from infection. Worldwide, the popular press heralded a breakthrough in gene therapy. And, so it was. But a major problem soon surfaced.

In 2003 the team reported that of the 10 boys with X-SCID that it had treated with gene therapy, the bone marrow of nine had been sufficiently repopulated to develop a protective immune system—essentially a cure. Of great concern, however, was that the two youngest boys had been diagnosed with leukemia. Molecular analysis revealed that in both cases the viral vector had inserted itself into a stretch of DNA that coded for the promoter of an oncogene (cancer risk gene) called *LMO2*. For some unknown reason, the vector preferentially inserted DNA in that spot and derepressed control of a gene known to cause cancer. Unfortunately, scientists cannot control where viral vectors insert into the host cell DNA. In effect, they were at that time relying on the sheer vastness of the human genome, most of which does not carry coding information, hoping that *insertional mutagenesis,* as it is called, would be very rare. Now, it did not appear that such an assumption was valid. About the same time that the group in Paris was treating boys with X-SCID, a group in London was running a similar study, and reaching similar results. Among 10 boys that they treated with gene therapy, all benefited from a reconstituted immune system, but one developed acute T cell lymphoblastic leukemia. Fortunately, aggressive chemotherapy induced a long-term remission in his disease.

Fifteen years have elapsed since the first two trials were initiated. Has gene therapy for X-SCID succeeded or failed? All but one child benefited from a genetically reconstituted T cell defense system. Most children developed low, but adequate numbers of functioning B cells, and about half no longer need periodic injections of intravenous γ-globulin. But five of the 20 boys treated with gene therapy developed leukemia, in each case owing to the disruption in regulatory controls that the vector caused when it preferentially inserted in or near the *LMO2* gene. Of those five, four responded well to treatment of their leukemia and remain in long-term remission. One boy died. Nineteen of the 20 boys, all of whom were born with an *inevitably* fatal genetic disorder for whom the only choice had been a risky bone marrow donation from a well-matched donor, are alive and well, and do not have to deal with the lifelong care that comes with BMT. These trials delivered wonderful scientific news—that it is possible to use gene therapy to achieve long-term reconstitution of immune function in children with X-SCID. But they also taught that it was crucially important to develop safer vectors.

Over the last decade, this has been a major scientific quest, on which considerable progress has been made. Perhaps most important has been the rapid expansion in use of AAV, a virus that easily infects human cells, but has *never* been known to cause an illness (most of us have been exposed to one or more forms of AAV, evidenced by the existence of antibodies to them in our blood). In the late 1990s and early 2000s scientists working in Wilson's laboratory in Philadelphia, notably Guangping Gao, discovered many new forms of AAV, most of which have not yet been completely studied, but such work is accelerating. It is extremely rare for AAV to insert its DNA into human DNA. It typically resides as an independent episome (DNA circle) in the cytoplasm of the cell, so the risk that it might cause cancer is much lower than for viruses that hone to the human DNA. AAV-based gene therapy is a young field, and there are many challenges still to be overcome. In addition to the need to discover or develop more nonintegrating vectors, we need to better understand how our immune systems respond to these viruses, design vectors that will only infect particular cell types (this is called tropism), and (the toughest challenge) find ways to control expression of the new gene after it has been delivered.

Despite the terrible problems caused by the leukemias that arose directly from the gene therapy given in the X-SCID trials, from about 2003 forward researchers working to develop gene therapy to treat human disease have become more optimistic. One measure of this was that in 2000 the gene therapy community was sufficiently large to support the creation of the American Society of Gene and Cell Therapy (ASGCT), which publishes a key journal and holds annual meetings now attended by more than 1000 scientists. In addition to the growing evidence that gene therapy could restore immune function in children with SCID, progress in other laboratories buoyed the whole field.

Of particular note is the research conducted by Dr. Katherine High, a hematologist, who works at Children's Hospital in Philadelphia. Although trained as a clinician, High maintained a strong interest in chemistry and early in her career almost chose to complete a PhD rather than an MD. In 1984, as High was completing her hematology fellowship at Yale, scientists cloned the genes responsible for the proteins that caused two of the major forms of hemophilia. This was instrumental in leading her to undertake research on gene therapy for hemophilia B, a goal she has tenaciously pursued for three decades. In 2002, High and her research team reported a successful effort to treat *dogs* with severe hemophilia B (caused by a "null" mutation, meaning the dogs made no Factor IX).

To do this work her team first designed and manufactured an AAV vector that carried as its payload a copy of the canine Factor IX gene. They delivered a dose of 10 billion vector genomes per kilogram of dog weight by catheter to the livers of four affected dogs, and then carefully monitored the animals for 18 months. They showed that after a *single* treatment, three of the dogs were making between 5% and 12% of the levels of Factor IX found in normal dogs. Many studies in humans have shown that people who have just 3% of the normal amount of protein are essentially disease free. In agreement with the level of proteins, the three dogs had normal clotting times and were in good health. The fourth dog showed only transient benefit because he had neutralizing antibodies that prevented the protein from working. Since then, Dr. High and her team have persisted in their study of the immunological challenges to gene therapy, including those posed by neutralizing antibodies.

High's work with dogs raised hope for human trials. In 2011, in what was undoubtedly a critically important paper for the expansion of human gene therapy, a multi-institutional team that included many of the leading names in the field, reported the results of the experimental treatment of six humans with hemophilia B. The researchers delivered a single dose of a vector derived from AAV2 that carried the normal version of the gene by intravenous infusion into men who were known to produce <1% of the normal amount of Factor IX, and then followed them for 6–16 months. The study participants were enrolled sequentially in groups of two to receive escalating doses of the vector. After treatment, four of the six men were able to *stop* receiving the regular injections of Factor IX protein that had sustained them; the other two were able to remain healthy on a substantially reduced level of factor supplement. Overall, the transgene delivered by the vector led to production of 2%–11% of normal levels. The one important caution was that both of the men who received the highest dose showed some transient evidence of liver inflammation. High's 20-year dedication to pursuing gene therapy in humans seemed within reach. When the work, the first author of which is Amit Nathwani, a scientist at University College London, appeared in *The New England Journal of Medicine* in December of 2011, it ignited a firestorm of interest.

In the spring of 2014, there were several Phase I/II clinical trials of AAV vectors under way to treat persons with severe hemophilia. Reflecting renewed commercial interest in gene therapy, two of the three were sponsored by recently funded AAV-based biotechnology companies. Asklepios BioPharmaceuticals, Inc. was founded by gene therapy researchers at the University of North Carolina, including Jude Samulski, one of the most prominent scientists in the gene therapy field. In 2013 the Children's Hospital of Philadelphia (CHOP), recognizing the greatly talented team assembled there, decided to commit $50 million to create Spark Therapeutics. Dr. High is currently the President and Chief Scientific Officer of Spark. Other founders include Fraser Wright, one of the world's top experts in manufacturing AAV vectors, and Jean Bennett, an ophthalmologist who has been a vital force in driving gene therapy for severe genetic eye disorders, a story that I recount below. In the spring of 2014, Spark raised an additional $73 million from a consortium of venture capital groups, making it one of the most well financed gene therapy companies in the United States.

Leber Congenital Amaurosis

In 1869, Theodor Leber, a German ophthalmologist, became the first physician to characterize in detail the congenital forms of the several retinal diseases that now bear his name. Like the other great 19th century physicians mentioned in this book, Leber—often referred to as the father of experimental ophthalmology—merits a few words. Born in 1840 in Karlsruhe, Germany to a well-educated family, Leber became deeply interested in science as a child, and attended medical school in Heidelberg. He made his first major contribution to his chosen field at the age of 24 when he used chemical dyes to delineate the complex blood circulation within and about the eye. A century later, textbooks were still using circulation maps that he drew. In his early career, Leber worked at hospitals in Paris and Berlin. It was in Berlin that he first encountered and closely examined a family in which three young children had the degenerative eye disease that now bears his name. Leber noted that the disorder is characterized by severely restricted vision from birth owing to the malfunction of cells in the retina.

Despite careful clinical study, understanding of the physiology of vision, hampered by the lack of research tools, made little progress until the 1950s. Perhaps the first great molecular advance came from the laboratory of George Wald of Harvard University, who in 1967 won a Nobel Prize for elucidating the key role of vitamin A in the biochemistry of vision. Nevertheless, in the 1970s ophthalmologists could do little more for patients with genetic disorders of the retina than use some of their new clinical tools to extend the descriptive work that Leber had initiated. For example, electrophysiological studies could be used to document that affected eyes showed little response to light. In the 1990s, scientists began using DNA analysis to study families burdened with rare genetic eye diseases. In 1993, Dr. T. Michael Redmond at the National Eye Institute in Bethesda cloned the gene for RPE65, a protein that played a key role in processing vitamin A in the retinal cells. Four years later it was shown that mutations in *RPE65* caused one type of retinal disease called Leber congenital amaurosis (LCA) in some children. Over the last two decades, molecular biologists have teamed up with ophthalmologists to parse LCA into at least 19 distinct forms, each owing to mutations in a different gene coding for some protein of key importance to the function of the rod or cone cells in the eye. The majority of these disorders are autosomal recessive.

In 1999, Swedish scientists reported that a canine eye disease called congenital stationary night blindness was caused by mutations in the canine gene that coded for RPE65. The discovery of a spontaneously arising large animal model of LCA came just a year after the creation of the first "knockout" mouse model of the disease, and offered a critically important system for studying gene therapy for the disorder. Soon thereafter, a team of researchers from Cornell and the University of Pennsylvania launched a project to use gene therapy to restore vision to these animals.

After the molecular biologists made a gene vector in which an adeno-associated virus type 2 (AAV2) was used to carry the normal version of *RPE65* (AAV-*RPE65*), the retinal surgeons injected the vectors into the eyes of four Briard dogs that by 3–4 months of age had lost essentially all functional vision. One dog was not injected and used as a control; one dog was treated with subretinal injection in one eye and not injected in the other, and two were treated with subretinal injections in one eye and intravitreal injections (into the front chamber of the eye) in the other. Four months later the dogs were subjected to extensive visual testing and observed by experts as they were challenged to negotiate an obstacle course under normal and low light. The physiological tests showed that in the eyes that were treated with subretinal injection, there was a substantial *restoration* of function. Five of five observers rated these dogs as having functionally normal vision. The results provided powerful support for an effort to attempt to treat children with the human version of the disease that afflicted the dogs. From 2001 forward, several research teams extended the dog studies and continued to report both an excellent safety profile and long-term restoration of vision. It appeared that the scientists had *permanently* ameliorated blindness with a *single injection* of a vector carrying the normal version of the gene!

The idea that gene therapy should be tried as a transformative treatment to prevent, or even reverse, loss of vision is not as brash as it might at first seem. Targeting eye disease has two solid safety advantages. First, it is an immune-privileged organ; that is, it is largely isolated from the rest of the body. Thus, the delivery of a vector designed to carry a therapeutic gene into the eye would not likely result in that vector making its way throughout the rest of the body. Second, in any clinical trial, one could treat one eye at a time so that if a serious adverse event occurred, the other eye would be preserved.

Taking on genetic eye disease is also attractive for practical reasons. Since scientists began seriously to consider human gene therapy in the late 1980s, the challenge of making the therapeutic vector at reasonable cost has remained formidable. The challenge of making sufficient quantities of vector to target a few tens of thousands of cells in the retina is much less daunting than trying to make enough to deliver to a much larger organ (such as the brain) or to deliver it systemically. Many retinal surgeons are highly skilled at directly injecting substances into the back of the eye, so they can deliver small amounts of the vector very close to the key target cells.

The pace of clinical research is always slower than patients or scientists would like, and this is especially true when one is developing a novel therapy for use in children. Slowly, but steadily, the team at CHOP, and others at the University of Florida and in London, went through the time-consuming, but necessary, steps (including animal safety studies, production of high-quality vectors, and review and approval by human studies committees) to lay the groundwork for experimentation in human beings. In 2003 the group at CHOP was given approval to enroll 12 patients (ages 12–44) with LCA2 in a Phase I gene therapy safety trial. But the group did not immediately move to enroll patients.

The team led by Jean Bennett, a Harvard-trained physician who had been the senior author on the 2001 dog paper, wanted first to deepen its understanding of how best to deliver gene therapy to a human eye. Working with a mouse model of the disease (that may more closely mimic the human disorder than does the disease in Briard dogs), in 2005 Bennett's team showed that there was a better response to therapy if the vectors were delivered to locations on the retina that retained a comparatively intact layer of photoreceptor cells. Also in 2005, the team reported encouraging long-term follow-up on the Briard dogs, demonstrating that the animals retained stable rod and cone cell function. Before the FDA will permit researchers to give a new drug to humans, scientists are required to conduct toxicology studies in animals. In 2006 the group at CHOP and others reported reassuring data about the response of the dogs to high doses of the vector.

Results of the first human clinical results were published in May of 2008 in back-to-back papers in *The New England Journal of Medicine*. The CHOP team and another led by Dr. J.W. Bainbridge and Dr. R.R. Ali at the Institute of Ophthalmology at University College London each reported early results of AAV gene therapy in three young adults with severe vision loss owing to mutations in *RPE65*. In the London trial, one of the three patients showed a significant improvement in seeing light and shadows, a finding that was confirmed by specialized ophthalmological examinations. In Philadelphia, the three adults, who all underwent injection of the therapeutic vector into their right eyes, experienced modest improvements in visual function on subjective tests of visual acuity. One patient experienced a small retinal tear, almost certainly the result of the surgery itself. The report from CHOP was remarkably similar to contemporaneous reports from the two other groups (London and Florida) that had pursued similar efforts. Going forward, the teams regularly updated their evaluations. Improvements in visual function seemed to persist. But no one could be sure for how long they would last. The duration of positive therapeutic effect is one of the major unanswered questions in gene therapy, one that will not be fully answered for 10–20 years.

In 2009 the CHOP group published the not-unexpected finding that among the group of 12 patients with *RPE65* deficiency, the greatest improvements in various measures of visual function were recorded in children, a finding that might have important implications for eligibility for treatment. A year later, another group reported follow-up on 15 children and adults, but did not find an age-dependent response to treatment. In 2013, follow-up of five patients treated in Italy found that their early gains in vision after gene therapy were maintained over time. This reassuring finding was undercut, however, in 2013 when a team including the veterinarian who had led the first treatment of the dogs, reported that, despite gene therapy, in the dogs degeneration of photoreceptors continued.

Because the patients in the study were not completely blind (e.g., they could sense change when a hand was waved in front of them), it was critically important to control for any placebo effect of the surgery. One key objective measure was change in the pupillary light reflex. When one shines light into one pupil, both constrict (this is called a consensual response and is mediated by the brain). When Dr. Bennett and her team shined light into the treated eye, they elicited a much more robust pupillary response in both eyes than when they did it in untreated control patients. But when they shined light into the untreated left eye, they did not elicit a strong response. The patients also had an improvement in nystagmus, an abnormal eye movement often seen in low vision. The most impressive results were from the low light obstacle course. Before therapy, patients 1 and 2 could not negotiate the maze, bumping into most of the large objects. A few weeks after treatment, patient 2 could walk the course by reading the directing arrows, and did not bump into objects.

In May of 2010, I attended the annual meeting of the American Society of Cell and Gene Therapy in Washington, D.C., in large part to hear Jean Bennett speak. After presenting her scientific data, in an unusual, but quite delightful, move, Dr. Bennett invited a boy with LCA who had undergone gene therapy to join her at the podium. There, in an extraordinary session, she showed the video that compared his ability to negotiate an obstacle course before and after gene therapy. Dr. Bennett then interviewed him about the impact of the treatment. Understandably shy before an audience of about 1000 persons, he nevertheless made a huge impact on the multitude when he reported that the best part of the therapy was that he could now catch a baseball! His remarks drew a standing ovation.

Extended studies of AAV gene therapy for RPE65 patients are ongoing in several centers in the United States and Europe. Of course early studies have to focus on safety. When a new drug is introduced into the body there is always great concern about causing an immunological response. One way to test this risk is in follow-up treatment. About 2 years after he treated the right eyes of the first three patients, Dr. Maguire injected the AAV vector carrying the *RPE65* transgene into the subretinal space of their left eyes. The patients did not generate a worrisome immune response, and their vision further improved.

In 2015 there were about nine open clinical trials dedicated to treating RPE65 with AAV gene therapy. The team led by Bennett in Philadelphia is conducting a Phase III trial that hopes to treat 24 young children with the LCA2. If this trial meets its clinical end points, the AAV2 vector and the transgene it carries could become the first gene therapy drug approved for clinical use in the United States. Groups in France and Israel are working to replicate and extend the work by teams in Philadelphia and London. In the spring of 2015 two research groups reported on the long-term follow-up of a small cohort of patients. Unfortunately, it appeared that improvements in vision that had been detectable 1 year after treatment were fading, a finding unlike those in the experimental dogs, which seemed to maintain visual gains. This may mean that human patients need a higher dose of the "drug." Because RPE65 is but one of a large number of retinal degenerative disorders that might be treated with a similar approach (subretinal injection of a viral vector carrying the normal version of the dysfunctional gene), physicians and patient groups are eager to see the early success extended to other orphan disorders.

Although much profoundly important research that provides the foundation for the development of breakthrough therapies is performed in universities and medical schools, these institutions are not structured in a way that permits them to undertake the extraordinarily expensive and time-consuming task of carrying new drug development efforts all the way to approval by the FDA. As with all drugs, the road to gain approval of a gene therapy product will require tens of millions of dollars, and is fraught with the risk of failure. The common wisdom is that only about one in 10 drug efforts that reaches safety trials in humans is ever approved. No company has yet developed a gene therapy drug to the point that the FDA has approved it for human use. Further, for diseases like *RPE65* deficiency, which are ultrarare disorders, it is extremely difficult to fashion a commercially successful strategy on which to justify investing millions of dollars in a drug development program. There are just too few patients to justify the costs that would be incurred.

We need new economic models for developing drugs for ultraorphan disorders. It will be tragic if great academic scientists and clinicians repeatedly show proof that a therapy works in children with such disorders, but then are unable to pass the baton for commercialization of the drug. The decision in 2013 by CHOP to commit about $50 million to launch Spark Therapeutics, Inc. to accelerate and translate the work of High and her colleagues was bold and innovative.

Childhood X-Linked Adrenoleukodystrophy

I first met Dr. Hugo Moser in the early 1990s when he visited The Eunice Kennedy Shriver Center for Mental Retardation in Waltham, Massachusetts where I was then the Executive Director. Hugo had worked as a research neurologist at Harvard Medical School and the Shriver Center for a few years in the mid-1970s, before he moved to what is now the Kennedy-Krieger Institute, which is associated with The Johns Hopkins School of Medicine. Born in Switzerland in 1925, Hugo spent his childhood in Berlin until 1933 when his parents fled the emerging Nazi regime. They settled in New York, where Hugo did his undergraduate degree at Columbia before attending Harvard Medical School. After a stint in the army during the Korean War, he returned to Harvard to train under the legendary neurologist Raymond D. Adams. He went on to become one of the nation's first pediatric neurologists. Ever curious, both a talented researcher and a compassionate physician, Hugo later became fascinated by a mysterious disease called X-linked adrenoleukodystrophy (or XLALD). Teaming up with his wife, Anne, also a physician and a talented biochemist, he devoted much of his later life to understanding this orphan disorder.

Some readers will remember the portrayal of Hugo as the skeptical physician (played by Peter Ustinov) in *Lorenzo's Oil*, a Hollywood film about a family's effort to find a cure for their affected son. In fact, Hugo, a big man with a big heart and a reserved and gentle manner, did become deeply interested in the idea advanced by some parents that a diet heavily supplemented with certain fatty acids might ameliorate the childhood form of XLALD called CCALD. He organized clinical trials of Lorenzo's oil, and he included the young man's parents as coauthors on papers he published.

To convey a sense for this frightening disorder, I will tell a story about a family that struggled with it (I have changed their names). Kathy Griffin's nightmare began one afternoon with a phone call from her son's second grade school teacher, asking her to come in for a conference. Naturally, Kathy was a bit anxious. Peter had been acting out a bit lately at home. Was he also misbehaving at school? To her surprise, when Kathy met with the teacher the next afternoon, she confronted a more serious problem. "It is not that he is misbehaving; Peter just seems disconnected. He used to be so eager to learn. Now he just can't seem to focus. He has also started crying twice for no apparent reason." The teacher was worried that there might be a serious problem at home (another child in the class was suffering through his parents' divorce). That was not the reason; Peter lived in a happy family with two older sisters and doting parents.

Kathy and her husband, John, wasted no time booking an appointment with their pediatrician. Sensing the parents' alarm, the doctor, who had seen children with sudden changes in behavior many times in her career, was particularly thorough. On physical examination everything seemed fine, except that Peter seemed to lack energy. The pediatrician decided to order some blood tests, but as the family left, she told them that she had found nothing wrong.

A few days later, the doctor called Kathy. To her surprise, the laboratory tests had uncovered abnormalities; Peter's blood had high potassium and low sodium, findings that warned of Addison's disease—the failure of the adrenal glands (which sit on top of the kidneys) to make enough cortisol. Within days a more sophisticated test confirmed the diagnosis. Fortunately, Addison's disease is treatable—by supplementing the body's supply of cortisol. But the bigger question was *why* had Peter developed this problem? The search to find out became a diagnostic odyssey that lasted months and ended badly.

Over the rest of the school year, Peter's performance (even though medicines corrected his low serum cortisol levels) continued slowly but steadily to deteriorate. By April it was clear to the teacher that rather than advancing, this formerly bright little boy was losing his math skills. Over 3 months, three more trips to the pediatrician had uncovered no obvious new physical problems, but both the parents and the doctor were now really worried. At the end of the third visit, Kathy pushed to have Peter evaluated by a pediatric neurologist at a medical school.

Two weeks later, Kathy and John waited anxiously as the neurologist did a painstakingly thorough examination, one that was quite different from the examinations performed by their pediatrician. Although they watched everything that the neurologist did, they did not notice anything too abnormal. They were right. At the end of the examination the doctor said that he had not found anything seriously off, but there were a couple of subtle findings that worried him, especially the fact that Peter seemed not to be able to see the doctor's fingers when he wiggled them near the side of his face. "I think we really need to do a brain MRI, just to make sure nothing is wrong."

It was at the next visit a week later that things turned much worse. The MRI showed lesions in two areas of Peter's brain: at the back in the area that controls vision and in the middle in an area that influences how thoughts are processed. In addition, a special dye used to look for inflammation had lit up around the edge of the lesions. Peter had an inflammatory process in his brain. Although it would require a special blood test—for chemicals called very long chain fatty acids—to confirm it, the neurologist told the parents that it was highly likely that Peter had a rare genetic disease called childhood cerebral adrenoleukodystrophy (CCALD). In a moment the Griffin's wonderfully happy world crumbled. A tsunami of confusion and terror swept over them. The first question was, "How could it be a genetic disease? Both of us are healthy." The neurologist explained that because there was no family history, in Peter's case CCALD probably arose because of a new mutation that had occurred for unfathomable reasons in the DNA of the egg that had been fertilized by John's sperm.

The next question was, "You can't change genes. Can anything be done for Peter?" In 2002, the year in which this story unfolded, the only hope for treating progressive CCALD was a stem cell transplant. Pioneering work in the 1990s by Dr. William Krivit, at the University of Minnesota and Dr. Patrick Aubourg, a pediatric neurologist at the Hôpital Saint-Vincent-de-Paul in Paris, had shown that if you treated boys with CCALD with BMT early in their illness—and, if they survived this risky therapy—they could have a relatively normal life. If one could quickly identify an appropriate bone marrow donor or cord blood sample, it might be possible to stabilize the disease. Peter would have to undergo full destruction of his immune system and be transplanted with stem cells that would divide and grow in his bone marrow. Some of them would become macrophages and travel to his brain where they would cross the blood–brain barrier and further differentiate into *microglial* cells that would have the capacity to chew up the very long chain fatty acids (VLCFA), the accumulation of which is a hallmark of CCALD.

Beginning that day, the Griffin family fell down the rabbit hole into the strange and often forbidding world of esoteric medicine. Because Peter's CCALD was obviously progressing (in ~5%–10% of boys with CCALD, for unknown reasons the disorder does *not* progress), time was of the essence. It took 2 weeks to determine that no one in the nuclear or immediate family had a bone marrow that was sufficiently similar (human leukocyte antigen [HLA]-matched) to Peter's to act as a donor. After another 3 weeks, the hospital was able to locate sufficiently (but imperfectly) matched cord blood cells to provide the needed stem cells.

Once they knew that Peter could be transplanted, the Griffins had to face one of the most difficult decisions that parents could ever face. At that time, in about 10% of the cases in which this therapy was attempted, the child died (usually within a year) of complications of the procedure—either early from a massive infection or later from severe acute graft versus host disease. Another 20% of those who survived would be burdened for life with chronic graft versus host disease. By deciding to go ahead, the parents would put Peter through a terribly difficult time with weeks of isolation in hospital and—almost certainly—many painful side effects. To give their son the chance of stopping the relentless progression of CCALD, they had to force him to take the risk that he would die as a result of the therapy.

But the situation was worse even than that. As is often the case with CCALD, at the time of the stem cell transplant, Peter's disease had progressed to the point in which even those most experienced with this disorder could not predict whether a successful transplant would leave him with a level of brain function that would allow a relatively normal life or whether he would continue to decline. The two most important factors in deciding on the status of the disease are a neurological examination that focuses on assessing the areas in which CCALD patients usually lose the most function (vision, hearing, speech, and motor skills) and the application of a special scoring system to the MRI scans. The neurological examination—sometimes called the Raymond scale for Dr. Gerald Raymond, a doctor then at Johns Hopkins who developed it—scored Peter as a 2. This essentially meant that he had obvious deficits, but that if his CCALD could be controlled quickly, he might do fairly well. But his MRI score was 10, which suggested that during the time that it would take for the therapy to work his disease could advance to the point where he would survive, but have severe intellectual disabilities. Experience with only about 300 cases in the *world* suggested that the child was approaching the end of the period in which transplant offered much hope. Finally, the decision was complicated by the fact that even in the best of circumstances it took about 9–12 months for the transplant to stabilize the disease. After a successful transplant, Peter's disease would almost certainly get worse before it got better.

But what options did Peter's parents have? If they did not decide to have Peter treated, he would almost certainly continue to lose function, and some years later (maybe 3, maybe 7) wind up in a chronic care facility unable to speak or communicate, unable to walk, unable to eat, suffering from recurring pneumonias until he died. The Griffins opted for the transplant. During the first 10 days after the treatment Peter did well, but on the 11th day he developed an infection in his blood that soon spread to his brain. He died 3 days later.

I could have shared a less painful story. Some boys who undergo allogeneic (meaning cells from nonself) stem cell transplant for CCALD do quite well. They survive transplant, the disease stabilizes, and they have a much higher level of cognitive function than they would have if they had not been treated. Unfortunately, there is little information on exactly how well they do. Since the first stem cell transplant was offered to a CCALD boy in 1986 through 2012, only about 300 procedures have been performed. During that period only a few papers have been published about the outcomes, and most of them have only discussed survival, reporting little about the functional status of the children. In May of 2011, the transplant team at the University of Minnesota published a follow-up study of 60 CCALD boys who had undergone transplant there over the prior decade—the largest number ever transplanted at single teaching hospital. But even that paper provided little of the outcome details that would be most important to parents.

As recently as 2012, despite heroic efforts by transplant teams to use bone marrow or cord blood stem cell transplants to save the lives of boys with CCALD, unless the child had the good luck to have a brother or sister who was a "matched sib" (meaning the two were identical for all the major antigens that determined if the body would accept or reject donated stem cells), the child's risk of dying within 1 year of the operation (from either the procedure or the progression of the disease) was still about 15%. Of course, this mortality risk depends on how willing a center is to offer BMT to a boy with advanced disease. If centers refuse to treat children with advanced disease, the mortality rate associated with intervention is lower.

bluebird bio

About a year after I started working at Third Rock Ventures, I became deeply involved in an effort to use gene therapy to treat boys with CCALD. In February of 2009, I sent an email to Dr. Alfred Slanetz, the CEO of a company named Genetix Pharmaceuticals, in Cambridge, Massachusetts. I had learned that it was trying to raise money, and I wanted to assess whether its scientific programs might represent an attractive project. Like the proverbial phoenix, Genetix was trying to rise from the ashes of self-immolation. It had come into being in the early 1990s, created by a team of scientific founders who were among the world's experts on how cells produce hemoglobin, the key oxygen-carrying molecule. In the simplest sense the billions of red blood cells in one's blood are little cargoes of hemoglobin, ferrying oxygen to the body's many tissues, picking up carbon dioxide, and returning to the lungs to dump carbon dioxide and reload with oxygen.

The original team had raised substantial funding from European venture groups, which over the years they devoted to solving the immensely challenging task of developing gene therapy to cure sickle cell anemia and β-thalassemia—two of the world's most common single-gene disorders (although still orphan diseases in the United States and Europe). Their first major success came in 2001 when a team led by Philippe Leboulch (an eminent molecular geneticist who would later direct a major research group in France and who was a key consultant to the company) published a paper in *Science* that offered convincing proof that his team had cured sickle cell anemia in a mouse model of the disease. To do this, the team constructed a lentiviral vector to use as a biological missile to carry a payload of the normal version of the gene coding for hemoglobin to stem cells that they collected from the sickle cell mice. When the researchers infused the virally transformed cells back into the mice, these stem cells traveled to the bone marrow, began to divide rapidly, and made enough normal red blood cells to reverse the burden of the disease. Most of the abnormally shaped sickle cells disappeared, the animals' spleens did not swell to gigantic proportions, a key aspect of kidney function remained normal, and unlike the untreated control animals, the treated mice did not die.

This was a significant scientific advance, but the company still had to undertake much exceedingly expensive preclinical and clinical research, and it needed to raise more money just as it was becoming much more difficult to do so in a shaky economy. Unable to raise a sufficient amount of new money, the board of directors of Genetix powered down, hoping for a better financial climate in the future. Unfortunately, the report in 2004 that several of the children with X-SCID in France who had undergone gene therapy had developed leukemia made many doctors and investors understandably cautious. Slanetz faced the unenviable task of raising money from new investors without capitulating to a steep decline in the valuation of the company (which was still afloat only because the original investors were loaning it funds on a monthly basis).

Two things about Genetix were exciting. First, it was sponsoring a small Phase I/II clinical trial in France that sought to evaluate the safety of a genetically engineered lentiviral vector (that was unlikely to cause the problem that had arisen with the vector used in the X-SCID trial) for treating people with a form of β-thalassemia. In 2006 the company had obtained permission from the French equivalent of the FDA to treat the first human subject. I learned that 3 years after undergoing gene therapy, the young Cambodian man who had been treated was doing exceedingly well. The autologous stem cells that had been taken from his bone marrow, transduced with the viral vector, and returned to him were now making about one-third of his body's hemoglobin. I was astonished to learn that he no longer required the periodic blood transfusions that he had needed to sustain life. Second, I learned that Dr. Slanetz had recently in-licensed rights to a Paris-based gene therapy trial to treat boys with CCALD. Under a confidentiality agreement, he told me that scientists in Paris had treated two boys who thus far were doing well. Regulatory authorities had given the company permission to treat two more boys.

Among the myriad biotech start-ups that are built on the dream of creating new therapies, most never make it to clinical trials. Here was a company that had not one, but two, investigational "drugs" being given to humans in approved clinical trials! If the clinical results from the CCALD trial continued to be positive, that fact combined with the results from treating the single young man with β-thalassemia would potentially be of great clinical and commercial value. In the summer and fall of 2009 some of my colleagues and I spent much time digesting the scientific and clinical information that Genetix provided. We gradually became convinced that investing in Genetix so that it could move its clinical trials forward offered potentially dramatic benefits to patients.

I was fairly familiar with the challenges of using gene therapy to treat genetic hemoglobin disorders, so I devoted more scientific due diligence to understanding the plan to treat CCALD. This required reading many of the clinical and all of the relevant scientific papers, interviewing experts, and meeting the clinical scientists who were conducting the clinical trial. That summer I learned that the Paris team, led by two academic physicians, Patrick Aubourg and Nathalie Cartier, were submitting a paper to *Science* that would provide details on the clinical status of the two boys with CCALD who had been treated with the Genetix gene vector.

The report was encouraging. There was no sign of any adverse event related to the insertion of the lentiviral vector into the DNA of the stem cells, and there was no sign of a preleukemic syndrome that doctors feared owing to the experience with the children with the genetic immune disorder. A significant fraction of the circulating white cells were descendants of the virally transduced stem cells, and in both boys the brain damage that is the hallmark of CCALD had not progressed during a full year after therapy. In concluding the paper, Aubourg, who had in earlier years been involved in using BMTs to treat 35 boys with CCALD, argued that given a mortality risk that in that disease approached 40%, lentiviral modified autologous cell therapy "could become the best therapeutic option." Of course, two patients are not very many, but having learned the devastating nature of CCALD and the great risks associated with BMT, our team was encouraged.

By any measure, autologous therapy in which genes were inserted into one's *own* cells would be much safer than allogeneic therapy (the use of *donor* cells). Although it only reported on two children, the report showed that thus far gene therapy was as at least as effective as BMT. If a biotech company could develop a drug that avoided the risk of death associated with BMT, it would likely become the treatment of choice.

Early on in my investigation, I had the good fortune to work with Nick Leschly, a partner at Third Rock. A Princeton grad with limitless energy, the father of five daughters, and passionate in his desire to help patients, Nick was leading the commercial aspects of our due diligence process. For the rest of the summer and throughout the fall and early winter, Nick and I (along with consultants who were expert in particular areas, such as evaluating intellectual property) focused on making as sure as we possibly could that the two clinical programs had a reasonably good chance of success. We reviewed several academically based gene therapy trials, partly to confirm that the entire field was making good progress and partly to get feedback on the CCALD trial. After November 6, 2009, the day that *Science* published the detailed report on the CCALD patients, the feedback from the "KOLs" (key opinion leaders) whom we interviewed was strongly positive. The major concern was one that we could not address: What were the long-term risks of undergoing gene therapy? However, because CCALD is such a terrible disease and because all the clinical experts stressed that allogeneic bone marrow therapy carries great risks, opinions were remarkably uniform: The early results suggested that virally modified autologous cell therapy, if safe, might constitute a major advance in treatment. The excitement over the possible use of gene therapy to cure hemoglobin disorders was even higher.

By January 2010 the diligence was complete, and our team was ready to present the case in favor of investing a substantial sum in a new version of Genetix. During his due diligence work, Nick had become so excited about the prospects for a rejuvenating the company that he told his partners that he wanted to devote all his considerable energy to driving the new company forward. The partners agreed to invest, and Nick became the new CEO and I became the interim Chief Medical Officer. In March of 2010 the two of us and the new Chief Scientific Officer, Mitch Finer, one of the world's experts in gene therapy, moved into 840 Memorial Drive in Cambridge, an aging, but celebrated biotech building, and began the exciting task of rebuilding the company. Nick renamed the new entity bluebird bio, writing like the poet e.e. cummings, who used no capital letters.

Initially, all of bluebird bio's clinical work was undertaken in Paris, the site of both clinical trials. I traveled there every few weeks, in part to get to know Drs. Aubourg and Cartier, and their colleagues. Tall, lean, nearly always wearing a dark turtleneck jersey and (when not in the laboratory or seeing patients) smoking a pipe, Patrick Aubourg epitomizes the old world image of the senior professor. In 1986 he organized the first BMT for a child with CCALD. Since then he had been involved in treating nearly 40 other boys, a huge number for such a rare disease. When a physician working almost anywhere in Western Europe diagnoses CCALD in a child, he usually refers the boy to Patrick. His colleague of many years, a pediatrician named Nathalie Cartier, a warm, stylish Parisian, was the lead author on the *Science* paper.

During the ensuing years bluebird bio, now well capitalized, grew rapidly, focusing intently on two goals: (1) improving the overall ability to deliver corrective genes into human stem cells and (2) advancing the clinical trials for CCALD and β-thalassemia and opening similar ones in the United States. bluebird bio planned to enroll 12 boys with CCALD at a relatively early stage of the disease. After obtaining stem cells from the boys, the company would transduce those cells with the vector carrying the corrective gene. When it had enough transduced cells that met various rigid safety standards, it would give the signal that the boys should be hospitalized. Doctors would then treat them for several days with powerful drugs that would destroy marrow cells. Next, they would infuse the genetically engineered stem cells into them. If all went well, these would "engraft" and proliferate in the bone marrow. Some of the new white cells would differentiate into cells, which would travel to the brain to microglial cells. These cells were the key: They would have the capacity to break down the excess fatty acids that affected boys were not able to metabolize.

bluebird bio had a strong argument that its therapy was safer than the only alternative. But the latest data from Europe on the only four boys in the world who had been treated with a similar vector suggested that going forward, it would be essential to treat as early as possible. None had suffered adverse side effects. However, although the first two boys (who had been treated in 2006 and 2007) showed ongoing evidence of disease stabilization, the third boy (treated in 2008) did not. The first two boys were in school, walking well, and leading relatively normal lives. The third boy had active brain lesions and was losing the ability to speak. The likely reason was that an adequate number of genetically engineered microglial cells did not get to his brain. A year and a half after being treated, the fourth boy seemed to be doing quite well, but it was too early to be sure. Gene therapy had worked in two of three boys, about the same ratio as the experience with allogeneic BMT. Of course, the sample size was exceedingly small. In 2012 the FDA agreed that bluebird bio could expand the clinical trial in the United States.

Since its inception in 2010, bluebird bio has made steady progress. It has opened gene therapy trials in the United States to advance its effort to treat individuals burdened with β-thalassemia. By the late summer of 2014 its extended CCALD trial had enrolled and treated 12 boys. In the fall of 2014 it announced that, under an experimental protocol it had treated two patients with sickle cell disease. The company went public in 2013. Of course, as Nick Leschly would be the first to note, the ultimate metric of success—saving the lives of children with these diseases—still lies ahead.

While the team at bluebird bio was driving hard to make lentiviral gene therapy a reality, a growing number of other, mostly new, companies were entering the gene therapy field. One, which was originally known as Amsterdam Molecular and that also had, despite promising scientific work, suffered severe financial woes, was recapitalized in 2011, and acquired a new name—uniQure. At that time it was deep into the process of seeking regulatory approval in Europe for a gene therapy (using an adeno-associated virus known as AAV) drug to treat an ultraorphan disorder called *lipoprotein lipase deficiency,* in which patients as young as 2 have extremely high levels of cholesterol and suffer recurring, life-threatening bouts of pancreatitis. In July of 2012, after a prolonged and difficult regulatory process that required four reviews, the European Medicines Agency (EMA), the functional equivalent of the FDA, approved its drug Glybera for human use, making it the first gene therapy product in the western world available for human use. The hope is that it will prove superior to the standard treatment, which is to put the patient on a diet that is nearly devoid of fats (a very difficult diet to follow) and treat with statins to lower cholesterol.

Glybera is administered as a small series of injections into muscles in the legs. The results of three small clinical trials involving just 27 patients (the disease affects about one in 1,000,000 people so it is difficult to recruit many people) show that over a period of a few months the AAV gene therapy drug greatly reduces both the levels of triglycerides in the blood and the frequency of pancreatitis. The latter benefit occurs because some of the new enzyme is secreted into the bloodstream where it breaks down large lipid complexes called chylomicrons. Of course, uncertainty about safety concerns means that the company will have to follow treated patients for many years.

A treatment with Glybera is currently priced at about $1,000,000. Although this may seem like a mind-boggling sum, the lifetime medical expenses currently generated by patients may be much higher.

During 2013 and 2014, investment in gene therapy experienced explosive growth. In addition to uniQure and Spark, about a dozen more companies—most with academically based scientists as founders—were able to find the venture backing needed to launch comprehensive programs. Following the lead of GlaxoSmithKline, which started to partner programs in gene therapy in 2009, pharmaceutical giants, including Pfizer and Roche, put more resources behind teams that had been created expressly to evaluate opportunities in the space.

One simple metric with which to characterize the growth of gene therapy is to summarize the activity at the 17th Annual Meeting of the ASGCT in May of 2014. More than 1000 scientists contributed 788 abstracts of their work, the majority focused on therapeutic efforts. The larger sessions included a dozen special symposia covering topics including the development of new, more efficacious viral vectors, improved design of the gene payloads, gene editing, and cancer immunotherapy. In recent years one of the special events at this meeting is the special lecture given as part of an award for "lifetime achievement" in the field. In 2014 the society recognized Luigi Naldini among the world's most influential scientists in advancing gene therapy from dream to reality.

Dr. Naldini trained in the laboratory of Inder Verma (one of gene therapies earliest champions) in the Scripps Institute in California in the mid-1990s. After a short stint working at Cell Genesys, one of the first commercial efforts in gene therapy, Naldini returned to an academic career in his native Italy. For many years he has been the Director to the San Raffaele Institute in Milan, where he has built and leads an extraordinary team of virologists, molecular biologists, and neuroscientists. Perhaps more than any other person, Dr. Naldini deserves credit for the clinical development of lentiviral therapy. Taking advantage of the high efficacy with which HIV enters cells, he was among those who reengineered the virus, eliminating the features that make it pathogenic and enhancing the features that make it a successful vector.

Although a basic scientist, Naldini has throughout his career thought continuously about adapting his work to treat children with severe disease. Several years ago, his work attracted the attention of the pharmaceutical company, GlaxoSmithKline, which entered into an agreement with him to fund research into several rare fatal genetic disorders. One of the most challenging is metachromatic leukodystrophy (MLD), a rare storage disorder arising from mutations in a gene coding for an enzyme called arylsulfatase A. Affected children fail to develop motor skills, and typically die by age 10. After designing a lentiviral vector to deliver the normal copy of the gene to the patient's hematopoietic stem cells, Naldini and his team spent many years preparing for and eventually conducting clinical trials attempting to halt or ameliorate the disease. Because the disorder strikes early in life and is relentless in its advance, Naldini and his team decided to focus on helping younger sibs of affected children, children that DNA testing foretold would be soon struck down by the disease. In a remarkable clinical trial, they have shown that a single treatment given *before* onset of symptoms in children otherwise destined to die of the disorder can seem (thus far) to prevent it from arising. This success is probably based on the phenomenon of "cross correction." The transduced cells migrate to the brain where they secrete normal enzyme that the brain cells are able to take up and use. If enough treated cells reach the brain, their ability to secrete the correct version of enzyme offers exciting prospects for halting or even reversing some monogenic disorders.

While I attended the annual meeting of the ASGCT, I had the pleasure of hearing Dr. Naldini deliver an address in which he traced the highlights of his career. Over 40 minutes, he summarized a 20-year period that moved steadily closer to the creation of new therapies. The most moving moment came when he showed a video of children with the genetic defect that causes MLD, but who were behaving completely normally thanks to "preventive gene therapy" that may have permanently prevented the onset of MLD. This is a magnificent accomplishment. It was a pleasure to join 1000 scientists in the audience when they rose to give him a standing ovation.

The early success in using gene therapy to treat boys with MLD who are presymptomatic raises the question of how late in the course of an illness gene therapy could be used to significantly ameliorate or even reverse it. The field is still so young that we can offer only anecdotal information, but some of it is cause for great optimism. This is especially true for diseases of the hematopoietic (blood-forming) system such as some of the genetically caused immune disorders and, in particular, β-thalassemia and sickle cell anemia. In June of 2014 at the European Society of Hematology meeting in Milan, scientists working on a trial funded by bluebird bio announced highly encouraging results concerning two patients who had undergone gene therapy to correct β-thalessemia. Only *weeks* after undergoing the procedure, the patients were now making enough hemoglobin (much of it from their new genetically engineered stem cells) to no longer need the monthly blood transfusions that had kept them alive since infancy. Both the level of transduction of the stem cells and the amount of hemoglobin that their descendant cells were making far exceeded expectations.

Although the excitement about gene therapy in the biotech community and among patient groups that are bound together to help children with rare genetic disorders is palpable, there are many other exciting developments under way. In the next three chapters I explore some of them.

8

Overcoming Mutations

THE VAST ARMY OF BIOMEDICAL RESEARCHERS NOW working in the United States, Europe, Japan, and a growing number of other nations include hundreds in laboratories that are focused on developing even more novel methods to treat rare genetic disorders. The unhappy fact is that the development of specialized diets to ameliorate inborn errors of metabolism, the administration of Factor VIII to persons with hemophilia, the use of frequent blood transfusions and iron chelation to control β-thalassemia, the delivery of enzyme replacements to treat children with lysosomal storage disorders, the infusion of viral vectors for gene therapy, and the newly emerging methods to manufacture large molecules for protein replacement therapy address only a fraction of all potentially treatable, single-gene disorders.

Recently many writers have asserted that there are about 7000 monogenic disorders. The number is probably derived from the number of entries in the catalog known as Online Mendelian Inheritance in Man (OMIM). It is almost certainly an underestimate. That catalog lists slightly *more than* 7000 distinct phenotypes associated with genetic variants, but many of them are too mild to be considered as diseases or disorders. One might plausibly argue that there must be more than 20,000 single-gene disorders, largely because there are at least that many genes in the human genome. But there are probably many monogenic disorders that have such devastating effects on early fetal development that they are lethal before birth, and it is unlikely that we will ever characterize or count them.

The more important question is, how many "treatable" monogenic disorders are there? Of course, the answer depends on how one defines "treatable." Scientists who have considered this issue have explored the concept of the "druggable genome." That is, given our skill set in basic biology and drug development, for how many single-gene disorders is there any reasonable chance that we could develop a medicine that would have a beneficial impact? The answer will be a function not only of understanding the relevant biochemistry, but assessing whether there is actually a reasonable opportunity to intervene. No one actually knows the number of treatable orphan disorders, but we do know that drug developers have created therapies for fewer than 5% of the recognized monogenic disorders. One fact is clear: There is at least a century of hard work ahead.

The biochemical challenges presented by the universe of single-gene disorders may seem remarkably uniform. Each disease—from alkaptonuria to xeroderma pigmentosum—arises because a mutation in a gene leads to a failure to produce a particular protein in its proper form, thus disrupting the cellular function in one or many tissues. Why not then treat every serious genetic disorder by supplying the normal copy of the gene or the protein for which it codes to the cells that need them? This is the goal of gene therapy and protein replacement therapy, in particular. But there are many disorders for which we do not yet know how to do this in a safe and effective manner. Some disorders constitute an extremely difficult challenge for drug delivery. For example, research in Duchenne muscular dystrophy suggests that if one delivered genes coding for the protein dystrophin to enough muscles, one might be able to improve the lives of affected children. But imagine the technical challenge of delivering (whether by viral vectors or some other vehicle) dystrophin to hundreds of muscles! There are also a number of disorders, such as osteogenesis imperfecta (OI), in which the problems arise because of the cell's failure to make a *structural* protein (one responsible for tissue integrity) as opposed to an enzyme that drives a particular *chemical* reaction. No one has yet created a therapy to ameliorate such problems (although there are some active research programs that I discuss in Chapter 9).

In this chapter I describe several lines of research that might greatly expand the number of disorders for which we could develop new therapies. The list is by no means exhaustive, but the vignettes provide a sense of how molecular biologists, medicinal chemists, and their colleagues develop new therapies. I will briefly discuss how one enterprising physician-scientist and his colleagues have tried to use a low-cost existing drug to treat Marfan syndrome; how scientists have been trying to treat persons with Friedreich's ataxia with small molecules to up-regulate the production of the protein (called frataxin), reduced levels of which cause the disease; how one determined group tried to develop drugs to inhibit or dial down the action of certain brain molecules to improve the behavioral problems that are a common feature of Fragile X syndrome; and how some researchers are trying to increase the expression of one gene to offset the loss of a neighboring gene as a way to treat children with a fatal disorder called spinal muscular atrophy type 1.

Marfan Syndrome

At six feet four inches, Abraham Lincoln was our tallest president. In the photographs of him visiting the diminutive George McClellan, Commander of the Army of the Potomac, Lincoln towers over the general and his staff officers. Other photos reveal that his coat sleeves are too short for his arms. Caught in full view, his hands seem to hang markedly low against his thighs. Among the more recondite hobbies of physicians is to speculate—usually based on scant evidence—whether certain historical figures had various peculiar diseases or disorders. Did mad King George III lose the American colonies because he suffered from acute intermittent porphyria? Did short, bowlegged Toulouse-Lautrec (born to parents who were first cousins) suffer from a monogenic form of dwarfism? Did Abraham Lincoln have Marfan syndrome, the hallmarks of which are unusual height and very long arms?

Nearly 20 years ago, Victor McKusick, then the Osler Professor of Medicine at Johns Hopkins School of Medicine, invited me to join a committee to deliberate if we *should* and, if so, *could* we determine if Lincoln did have Marfan syndrome. Based on photographs and biographical minutiae, McKusick offered odds at one in two that Lincoln was affected. Given that the chance that any person selected at random has Marfan syndrome is less than one in 5000, he had a high level of suspicion.

Besides the curiosity that naturally arises about great historical figures, why might a posthumous diagnosis matter? McKusick offered two answers: If we could prove that Lincoln had Marfan syndrome, it would suggest that even if John Wilkes Booth had not put a bullet in his brain, Lincoln might have had only a short time left to live and that the South might still not have benefitted from what would almost certainly have been his more compassionate approach to Reconstruction. Second, given Lincoln's extraordinary role in preserving the nation, it would generate a powerful message—that persons with disabilities are capable of achieving greatness—quite important in the unending battle to improve the lives of those citizens for whom prejudice is often a greater barrier than is their medical burden.

How could we find out if Lincoln had Marfan syndrome? Because the disorder is dominantly inherited, the easiest way would be if we could find living direct descendants who were affected with the disorder. Our genealogical research found no evidence of any living direct descendants. But McKusick knew that more than 100 strands of hair and about 15 g of bone fragments from Lincoln's skull repose in the Museum of the Armed Forces Institute of Pathology in Washington, D.C. One of the physicians who attended to Lincoln after he was shot had placed them there hours after his death, so their provenance is beyond question. Two key questions our committee faced were (1) would it be ethical to extract DNA from this material to look for mutations that are now known to cause Marfan syndrome, and (2) would it be technically possible to do so? The committee debated the ethical questions at great length, and concluded it would be permissible to subject Lincoln's tissue to testing, but we never acted on this conclusion. At that time (1998), we did not yet have the technical means of extracting the DNA from small bits of tissue that had been stored at room temperature in a museum box for more than a century. Even if we could extract some DNA, it would be so fragmented that it would not be possible to read the whole sequence of one gene. Today we probably could successfully analyze Lincoln's DNA, but no one is suggesting that we should.

Antoine Bernard-Jean Marfan (1858–1942) was born in southern France into the family of a small town country doctor. He started his medical studies in Toulouse, but 2 years later in 1877 he moved to Paris, where over the next 30 years he became one of the nation's leading pediatricians. In 1901 he was appointed as director of the Department of Diphtheria at the Hospital for Sick Children, and in 1910 he became a full professor (an even greater challenge then, than today). As would be expected in the late 19th century, Dr. Marfan's major academic interests included infectious disease and nutrition. He became an expert in diphtheria and tuberculosis. In 1897 he published his magnum opus, a *Treatise of Diseases of Children,* which earned him membership in the French Academy of Sciences. Although "Marfan" is an eponym for several findings in medicine (Marfan's law is based on an observation that children with certain forms of tuberculosis are at very low risk for recurrence, and a red triangle on the tip of the tongue, said to be a sign of typhoid fever, is known as Marfan's sign), it was a chance encounter with a little girl, named Gabrielle, that would make his name known to every physician.

When Dr. Marfan examined Gabrielle, whose mother was worried about her failure to thrive, he noticed that she had unusually long, spidery fingers (arachnodactyly), very long arm and leg bones (dolichostenomelia), and abnormalities of the chest wall (pectus excavatum) and spine. Perhaps because he also examined the child's mother, he guessed that these features were caused by a genetic condition. As was typical of the era, his case report stimulated other physicians to look for similar signs in patients, and over the years similar reports appeared in the literature. Marfan's patient died (apparently of tuberculosis) not long after their encounter, and he does not seem to have contributed much more to understanding the disorder that would take his name.

In an era before research became the cornerstone of medicine, efforts to delineate this condition were desultory. In 1914 a physician noted that dislocation of the lens of the eye was frequently seen in patients with the skeletal oddities that Marfan had noted. In 1931 a Belgian physician posited that patients with this steadily growing collection of physical signs might have a disorder of the connective tissue, and (perhaps for the first time) coined the term Marfan syndrome. In 1943 two physicians published case reports noting that life-threatening, aortic aneurisms were often found in young adults with this disorder. This suggested the logical inference that the structure of the aortic wall might become abnormally weak as the affected person aged. Recall that every second throughout one's life, the left ventricle contracts, sending a great jet of blood through the aortic valve to begin its journey along the vascular tree, ferrying untold billions of red blood cells with their life-sustaining oxygen molecules to the brain, the kidneys, and the muscles. The action of the heart maintains a high pressure inside the vessels, and each jet of blood strikes the walls of the aorta. In the aortas of children with Marfan syndrome, the walls are congenitally weaker. Slowly but steadily, year in and year out, pounded by a wave of blood, their aortas dilate.

Great strides were made in the 1950s, especially thanks to Victor McKusick, then a young cardiologist at Johns Hopkins who was metamorphosing into a clinical geneticist. In 1956, he published a treatise on heritable disorders of connective tissue with a focus on Marfan syndrome. More than 50 years had elapsed between Marfan's case report and the compilation of evidence that the most severe aspect of the syndrome was the risk in adulthood of sudden death from a burst thoracic aortic aneurysm.

About 1960 Marfan syndrome became a surgical disease. With the development of special polymers that could be used to make grafts to patch the aorta, it became possible to follow the growth of the aneurysms and repair them before they became inoperable. Before 1960 most persons with classical Marfan syndrome died of aortic rupture, often in young adulthood. Between 1976 and 1997 surgeons at The Johns Hopkins Hospital replaced the aortic roots of 231 patients who were at high risk for such ruptures. Five years after the operation, 88% of the patients were alive; 20 years after surgery, 75% were alive—an astounding success. Over the last three decades advances in medical therapy and cardiothoracic surgery have increased the median life expectancy of people with Marfan syndrome from the mid-40s to the early 70s—that is, to nearly normal!

During the same era, there were important advances in understanding the molecular basis for the disease, and in thinking through new methods of treatment. In 1990 (the Paleolithic age of gene mapping) a group of Finnish researchers established that the gene mutations that cause Marfan syndrome were localized to chromosome 15. About that same year, a young pediatric cardiologist at Johns Hopkins named Hal Dietz approached some of the geneticists there, and told them he wanted to retool as a research scientist focusing on Marfan syndrome. Over the next 15 years, Dietz would become the third professor (after McKusick and his student, Dr. Reed Pyeritz) at Johns Hopkins to advance our understanding of this disorder.

About 2 years after he commenced his work, Dietz was the lead author on a paper showing that the culprit gene coded for a protein called fibrillin, which was known to be a major component of fibrils in the extracellular matrix. It now seemed almost certain that the signs of Marfan syndrome arose owing to a structural defect in various tissues. At first, this was disheartening because it offered no obvious path to prevent the dilation of the aorta. If children with the disorder were born without enough of a structural protein, was not the aortic disease inevitable? But this view of causation did not well explain many other features of Marfan syndrome. Why, for example, should patients often also have dislocation of the lens in the eye, bony overgrowth, low muscle mass, and such low fat stores that they appeared emaciated?

Now operating his own research laboratory, Dietz set out to refine the understanding of the disease. Some of his early research focused on studying lung dysfunction in mouse models of the disease. He assumed that he would observe mouse lung tissue worsen over time because of structural weakness. Instead, Dietz found structural abnormalities *at birth* with no signs of inflammation or destruction. It appeared that some molecule had failed to guide normal lung tissue development. In experiments conducted over several years, the team led by Dietz showed convincingly that in a mouse model of Marfan syndrome a deficiency of fibrillin leads to an *increase* of another molecule called TGF-β (transforming growth factor-beta) that might be the *real* culprit in causing aortic dilatation. The research suggested that normal fibrillin-1 attaches to TGF-β and leads it to connective tissue that needs it. If the body lacks normal fibrillin-1, then free-floating TGF-β causes harm to a wide variety of cells. In a way, this was great news. Why? Because a safe, widely used blood pressure drug called losartan (which is known as an angiotensin II receptor blocker or ARB) was known to *lower* TGF-β activity. Perhaps, reasoned Dietz, this drug could provide a medical therapy for Marfan syndrome that, if started early enough, could avert severe dilatation and eliminate the need for surgery.

In 2004 Dietz enrolled 17 children with Marfan syndrome (in whom other medical therapy was not slowing the advance of already detectable aortic root dilation) in a clinical trial. He placed the children on losartan and used imaging technologies to measure the rate of change of their aortic root dilation over time. In 2008 the research team reviewed the data. In the years just before enrolling the children in the study, the aortic dilation had increased on average 3.54 millimeters per year. After starting losartan the annual rate of increase slowed remarkably to an average of 0.046 millimeters per year, about a 90% reduction. Given the immensely complicated biology of the cells that make up the medial layer of the aortic root, Dietz and others were cautious in interpreting the results. They did not want patients, parents, or their physicians to jump to conclusions.

Nevertheless, the report stimulated researchers in a number of laboratories around the world to attempt to show that other, similar drugs also slowed aortic root dilatation. A couple of groups seemed to offer encouraging news. For example, in 2009 a group in Asia reported that by treating affected mice with both losartan *and* a widely used antibiotic called doxycycline (which was chosen because it was known to inhibit certain members of an enzyme family called matrix metalloproteases) the aortic roots of the animals essentially stopped dilating.

Based on the encouraging results of the first trial, Dietz and his colleagues launched a much larger, prospective study to assess the impact of losartan on the aorta over the course of several years in about 600 patients with Marfan syndrome. They divided the group, aged 6 months to 25 years, into two groups: one to receive the standard therapy of an antihypertensive drug called atenolol (which was a standard therapy but that had not been rigorously studied in a randomized, prospective clinical trial) and the other to receive losartan. It would have been more helpful to compare the effects of losartan against those of a placebo, but parents adamantly refused to permit their children to take a 50% chance of not receiving any therapy for such a long period as several years. The results of the 3-year trial, reported in *The New England Journal of Medicine* in November of 2014, were deflating. Losartan was no better than atenolol in slowing the dilatation of the aorta. Experts still believe that both drugs provide some benefit, but no study has been undertaken that proves that is true. Dr. Dietz quickly acknowledged his disappointment. But his efforts to help patients with Marfan syndrome continue unabated.

Over the last decade, the group led by Dietz and other research groups around the world have made major advances in understanding the pathophysiology of several additional rare disorders that affect the integrity of the aorta and other major vessels. For example, Dietz' work led to a much deeper understanding of the role of TGF-β in building and maintaining vessel walls. In 2005, another physician at Johns Hopkins, Dr. Robert Loeys, and Dietz were lead and senior authors, respectively, on a paper that described a new genetic disorder (Loeys–Dietz syndrome) that is similar to Marfan syndrome, but that arises owing to mutations in the gene that codes for TGF-β.

The work led by Hal Dietz offers an important reminder about the value of grasping the molecular basis for rare disorders. Sometimes, such research will generate insights about the value of using *existing* drugs for *new* therapeutic purposes. This is often by far the *most* rapid road to new therapies.

Friedreich's Ataxia

In 1863, Dr. Nikolaus Friedreich, a 38-year-old neurologist on the faculty of the medical school at the prestigious University of Heidelberg, began publishing clinical observations of several of his patients, each of whom had what appeared to be the same curious early-onset neurodegenerative disorder. Eventually, he published five papers (the last in 1876) that delineated the course of this rare disorder, now known as Friedreich's ataxia, or FA. The five papers were so comprehensive and clinically astute that, except for new discoveries that flowed from advances in radiology, little more was added to our understanding of this orphan genetic disorder until 1996 when the gene responsible for the disease was cloned.

Friedreich's ataxia is an autosomal recessive disorder arising because of a triplet repeat mutation in the FXN gene that results in sharply reduced levels of a protein called frataxin. This causes the cells in the dorsal root ganglia of the spinal cord and in the heart to gradually degenerate. Frataxin seems to do its most important work in the mitochondria—the organelles that act as power plants in cells. In the mitochondria, frataxin plays a key role in the formation of iron-sulfur complexes. FA patients have low levels of frataxin, which over time debilitates their mitochondria. This in turn leads to an excess of molecules that badly damage cells, eventually killing them. Autopsy studies have confirmed that in patients with FA certain columns in the spinal cord, the dorsal root ganglia, and an area of the brain called the dentate nucleus all degenerate over time.

Even though FA is widely regarded as a neurological disease, most affected persons die in their late 30s or 40s of heart failure. It is likely that the cardiac disease, which usually progresses over a couple of decades, is also a direct result of the abnormalities in the mitochondria owing to the faulty iron metabolism that is caused by an inadequate amount of frataxin. Many relatively young patients with FA also develop diabetes, presumably because of damage caused by a lack of frataxin in the islet cells of the pancreas.

In 2012, the year I began to do research for this book, I had the good fortune to meet Nick Johnson. Nick, who was diagnosed with FA during his teenage years, has led a remarkable life despite the burden the disease has placed on him. He is a mechanical engineer who is an expert in designing buildings to have the most efficient climate control systems. He is also an inspirational leader among patient groups that struggle to advance the national research agenda for FA. His life history, which Nick shared with me, captures FA better than any summary I might prepare from the medical literature.

Born in 1963, Nick had a great childhood in which he excelled at school and sports, especially baseball at which by age 13 he had already earned much notice. When I had my first long talk with Nick at his home near Boston, I pressed him to recall in as much detail as possible the early days of his struggle with FA. Not surprisingly, he set the discovery in the context of sports. The first memory that he called up was of a baseball game in which he failed to get a single on a slow grounder that he was sure he could beat out. Later that season, playing right field, Nick did not get to a fly ball that he was sure he could gather in. By the next spring, he was no longer on the squad's first string, which he remembers as reflecting a stunning decline in his skills, even though he was still well enough to play.

His first visits to doctors (in the late 1970s) did not lead to a diagnosis, but one local physician sent him to a prominent neurologist at a prestigious Boston hospital. During this part of the story Nick, an upbeat guy, spoke with obvious bitterness. "The doctor examined me carefully," he recalled, then stood up, and told Nick he would be back in a few minutes with a diagnosis. When he returned, said Nick, "He looked me in the eye and said, "You have a disease called Friedreich's ataxia. There is no cure for it. You will be in a wheelchair within 5 to 10 years." Then the doctor told him that there was no need for a follow-up visit as there was nothing that he could offer. "That," said Nick, who was then 15, "was the worst day of my life."

Somehow Nick rallied from the depression that came with the diagnosis, and he became determined not to let FA limit him. He could no longer play sports, but he could still walk. Nick doubled down on his studies and got admitted to the prestigious Worcester Polytechnic Institute, where he chose to major in mechanical engineering. During his college years his motor skills declined steadily. Every few months, he lost yet another degree of freedom. He walked first with a cane and then with a walker. Nick recalls how hard it was to make the trip from one class to another on the large campus. Although FA is slow in its progression, certain moments mark a new level of decline. Especially tough was the morning that Nick awoke to discover he could no longer hold a pen to write.

"That really hurt," he recalled. But again he persevered. Although it took a major effort at a time when college campuses were not especially friendly to disabled students, Nick graduated, and landed a good job. In general, over the years his several employers were supportive. But one day, a boss with whom he was quite friendly called him into his office and asked him to give up his walker and give in to a wheelchair. The owners of the company were worried, he said, about a lawsuit should he fall. Another tough moment, but Nick recalls that he could see his employer's point of view. Life in a wheelchair was inevitable, anyway, and he had put it off almost as long as humanly possible.

One of the common consequences of having FA is to develop a severe scoliosis (curvature of the spine) that can compromise many aspects of life, including how effectively one uses a wheelchair and how one breathes. While still in his 20s, Nick underwent orthopedic surgery in which long metal rods were affixed to his spine to counter the scoliosis (which arises in part when muscle groups on one side of the spine are working better than their counterparts on the other side). The first surgery did not go well, but eventually the intervention stopped the progression of the scoliosis. But if one looks at Nick's X rays, one sees nearly as much metal on his spine as there is bone in it.

In his 30s, wheelchair bound and single, Nick refused to quit. He helped to start one of the major patient groups to further the goals of people with FA, and he was a ready source to anyone who wanted to learn or report about the disease. Despite his physical constraints, Nick loved work, and he took pride in emerging as a local leader in the environmentally sensitive design of new buildings. Then, one day, he got lucky. At a visit to his dentist, he was assigned to a dental hygienist named Sue with whom he enjoyed talking. He really liked how she reacted so matter-of-factly to his disability. He asked her for a date and, to his surprise, she said yes. They have now been married for 15 years! In the year in which they married, Nick had already lived beyond the then median life expectancy of people with FA. When I hear such stories, my faith in human resiliency is reinvigorated.

Now more than 50 years old, each day Nick meets challenges that I find unimaginably daunting. He cannot get into or out of bed on his own. His wife must dress him and transfer him to the motorized wheel chair. It is difficult for him to be alone because of his inability to respond to any problem that might arise (such as a fire alarm in his apartment building). He now works from home, but it is getting harder and harder for him to produce, a situation that he finds greatly frustrating. His voice is failing (the nerves that control those muscles are also affected by the disease), but his mind is more than sharp. He has mild diabetes, but his heart functions well. This is unusual; ~70% of FA patients die of heart failure, usually before age 50. In March of 2013 Nick e-mailed me to tell me that his younger brother, who also had FA, had died. There are only about 10,000 people living with FA in the United States today, and about 15,000 persons with the disease live in Europe, but when you get to know someone like Nick, FA does not seem like a rare orphan disease any longer.[1]

Today, we know a fair amount about the cause of FA. At the DNA level, it is one of about 20 unusual single-gene conditions that are called "triplet repeat disorders." In these conditions, molecular processing errors during meiosis (the formation of the egg or sperm cells) cause a stretch of DNA within the gene that is composed of a repeating trio of nucleotides of a certain length to greatly *expand* in number. In normal people the gene that codes for frataxin has about 30 (or fewer) repeats of a trinucleotide called GAA. Affected persons are born with many more (typically 500–600, but sometimes more than 1200) trinucleotides in both copies of the gene that they inherited from their parents. The parents, who each have one expanded gene and one normal gene, are unaffected.

The molecular pathology of FA is different from that of other triplet repeat disorders (a group that also includes Huntington's disease and Fragile X syndrome). In FA, the expansion occurs in a region of the gene called intron 1, a stretch of DNA that does *not* code for the protein itself (the expanded RNA is edited out during the production of messenger RNA). In fact there are usually *no* mutations in the portions of the gene that code for the needed protein. But the expanded repeat weakens the cell's translational machinery to make enough frataxin—presumably because

[1] Nick died on February 20, 2015. As he was an inspiration to so many people with Friedreich's ataxia, I know he would want his story to remain in this book.

the extra DNA interferes with the efficiency of the various enzymes that transcribe DNA into RNA, but what is made is *normal.* Most patients with FA have about 10%–15% of the normal protein.

This fact has prompted scientists to wonder if they might use existing drugs or newly generated small molecules to coax the cells of patients with FA to operate more efficiently and partially overcome the frataxin deficiency. Because we know that persons with just one abnormal frataxin gene (the parents of affected children) do *not* develop any signs of disease, it is reasonable to hope that if we could double the amount of frataxin made by the cells in affected persons, we might significantly ameliorate the disorder. Because persons with FA usually are diagnosed around age 12 and typically live a life of long, slow decline (more than a decade may pass before they become wheelchair dependent), there would seem to be ample time to intervene with new drugs.

Over the last decade clinical researchers have tried two other, less dramatic, but somewhat easier approaches: (1) they have repeatedly tried to use a small molecule called idebenone that has antioxidant properties (very much like the popular nutritional supplement, coenzyme Q) to slow the rate of damage to the cells caused by the excess of oxidants (a neuroprotective strategy) and (2) they have administered iron-chelating drugs in an attempt to pull excess iron from the mitochondria. The efforts to use drugs with a high affinity for iron that easily remove excess amounts from *blood* have shown no benefit to patients with FA. Existing iron-chelating agents cannot reach intracellular compartments. There are several early research efforts trying to develop other ways to do this.

There have been about 10 clinical trials to investigate whether idebenone could delay the downward spiral of patients who have FA. Because there are no reliable biomarkers or radiological measures to use in studying the impact of this intervention, neurologists who lead these clinical trials follow patients clinically and periodically assess them with one of several clinical rating scales. Unfortunately, the idebenone studies have involved a relatively small number of patients who were followed for a relatively short period of time (often a year or less) in a nonrandomized and uncontrolled way. Because the time course of decline in patients is slow and varies among patients, even in a large, multiyear study it could be difficult to discern a statistically significant clinical benefit unless it is quite robust. The only well-conducted randomized trial found that a year's treatment with idebenone offered no benefit over a placebo drug as measured by scores on a widely used rating scale. A subscore analysis suggested one positive benefit—that regular use of idebenone decreased the mass of the heart's left ventricle, a potentially important finding. Because idebenone is safe to consume, inexpensive, and available without prescription, many patients will continue to consume it, even though the evidence that it confers a benefit is weak. One website offering the product makes the highly dubious assertion that idebenone "supports higher brain function," and helps to ameliorate memory deficits and "brain toxicity issues." Confronting claims like this one is a common problem faced by families dealing with rare genetic disorders. Too often, weak data are used to sell questionable interventions to people in desperate situations.

Another major effort to increase intracellular levels of frataxin has centered on erythropoietin, a drug that has long been used to stimulate the production of red blood cells in cases of severe anemia (often arising secondary to cancer). Experiments with cells from Friedreich's ataxia patients showed that epoetin-alfa *does* raise frataxin levels. A small open-label study in which the drug was given to patients showed increased levels of frataxin 8 weeks later. However, half of the patients also developed unsafely high levels of red blood cells. As yet no dispositive, controlled clinical trial has been conducted, so it is uncertain as to whether epoetin-alfa could be of any clinical benefit. However, the fact that scientists have shown that one or more drugs can measurably elevate cellular levels of frataxin is encouraging.

There are several ongoing areas of research. One involves the use of drugs called HDAC inhibitors with the hope of reducing transcriptional repression and thus, indirectly up-regulating the production of frataxin. For a few years, a company called Repligen focused on that approach, but by 2013 had insufficient resources to continue the program. Isis Pharmaceuticals, the pioneer in the commercialization of antisense technology, is investigating the possibility of using small stretches of modified RNA to up-regulate frataxin. California-based Edison Pharmaceuticals is undertaking a large program aimed at improving mitochondrial function in FA.

In 2013, our growing interest in Friedreich's ataxia as a possible target area for a new company led me and my colleagues to interview many of the world's leading experts on the disorder (sadly, one unifying feature of efforts to develop new therapies for orphan disorders is that this is usually not too difficult to do as the number of experts is small). Among the extraordinary clinician scientists I met, Arnulf Koeppen was especially helpful. Now in his 70s, Dr. Koeppen, a diminutive man who moved to the United

States shortly after finishing medical school in Germany, began focusing his clinical and research activities on helping people with FA soon after he evaluated his first patient with the disorder in 1969. He spent most of his career working as a neurologist at the VA hospital in Albany, New York. Old school and courtly, opinionated yet a good listener, Koeppen recalls the iconic early 20th century European professor of medicine.

Almost invariably, when our team gathers in a room to interview a scientific expert, the visitor boots up a computer and presents his opinions through 60 to 70 PowerPoint slides. Not so, Dr. Koeppen. At our first meeting, once we had exchanged pleasantries, he pulled out a sheaf of reprints of his publications on FA, leaned back, and asked, "What do you want to talk about?" Three hours later, I had acquired a much deeper grasp of the natural history, cellular biology, and neuropathology of Friedreich's ataxia than I had garnered from days of reading. Perhaps most surprising to me was that Koeppen, who had spent decades studying FA the old-fashioned way—by treating patients and studying their tissues after they died—was so enthusiastic about using gene therapy to attempt to ameliorate the disease. Because (especially during the first decade after symptoms develop) much of the disease's relentless course involves declining function in the nerves emanating from the dorsal roots of the spinal cord, he found the idea of delivering (through injection into the spinal fluid as in a spinal tap) viral vectors carrying a normal version of the frataxin gene to those cells in patients who were young and still mobile, to be a highly plausible therapy.

Fragile X Syndrome

In 1943, two London physicians, J. Purdon Martin and Julia Bell, published a small paper in the *Journal of Neurological Psychiatry,* describing an extended family in which there were 11 boys and men with moderate to severe mental retardation scattered across three generations, but in which all the women were normal except for two with mild cognitive disabilities. They cautiously suggested that this was the *first* firm evidence of an X-linked form of mental retardation. By today's standards the report is remarkably simple. They did not, for example, indicate that they had noticed most of the particular physical abnormalities that would eventually be tightly associated with the disorder. However, they did observe that some of the affected boys and men were socially withdrawn and typically averted their gaze to avoid eye contact. They also grasped that the modest disabilities in two of the women suggested that some more complex factor was at work. Their observations marked the beginning of an ongoing effort to understand what was for two decades called Martin–Bell syndrome, but what we now call Fragile X syndrome, by far the *most* common monogenic form of inherited intellectual disability.

The case report stimulated physicians to ask if they have evaluated patients who might fit into the new diagnostic entity, and whether they might refine it. Over the next two decades, other clinicians elucidated some physical features that were clearly associated with Martin–Bell syndrome, especially an elongated face, a large head, and (after adolescence) unusually large testicles (perhaps due to excess levels of FSH, a brain hormone that stimulates the Sertoli cells). Subsequent studies of extended families with this disorder also usually found a few mildly affected women, thus confirming that the inheritance pattern of the defective gene did not strictly follow the rules for an X-linked disorder—that is, that women should not be affected (because of the presence of a normal copy of the gene in question on the other X chromosome, which usually is enough to counter the risk).

The next chapter in the history of our understanding of Martin–Bell syndrome opened when a clinical geneticist named Herb Lubs reported in the *American Journal of Human Genetics* in 1969 that when white cells from certain men with mental retardation were grown in a culture medium that had not been supplemented with folate (which was then the usual practice), that under the microscope the end of the long arm of the X appeared broken apart from the rest of its chromosome. In those days before DNA analysis, the ability to detect consistent variations in human chromosomes under the light of a microscope was extremely limited, so this was a most unusual finding. Lubs realized that the variant he had discovered in the course of evaluating a young boy from a family with several mentally retarded male relatives could be a marker for the disease and, perhaps, explain the cause. Before publishing his paper, he looked for the abnormality among the white blood cells of 4000 consecutive newborns, but did not find it again. He may have just missed finding a presymptomatic infant. Today, we know that the condition that he was seeking to understand, now named Fragile X syndrome (after another geneticist, Fred Hecht, coined the term "fragile site" to describe the apparent breakage seen under the microscope), affects just about one in 4000 boys.

During the period from 1970 to 2000 clinicians and basic scientists made steady progress in understanding the complex clinical nature of this relatively common disorder. By the mid-1970s it became possible to perform chromosome analysis on men and boys who (based on family history and physical examination) doctors suspected might have the condition. But the cell culture techniques were difficult and, in retrospect, they probably failed to induce the appearance of a fragile site about half the time. Still, the ability to diagnose the condition led some researchers to explore to what degree it explained the long recognized excess of men then living in the many state-operated long-term care facilities for mentally retarded persons that were so common until a generation ago. Across all societies there is about a 5%–6% excess of mental retardation among men over women. To what extent might this skewing be due to single-gene disorders on the X chromosome? Could just one monogenic condition explain much of it? About that time, several population surveys of Fragile X syndrome suggested that it alone might account for about 20% of this disparity.

I became familiar with this condition in the late 1980s when I was medical director of the Fernald State School in Waltham, Massachusetts. In those days even the most astute physicians were able to diagnose the cause of the developmental disabilities in at most one-half of patients. Many older patients (who had been admitted as children during the 1920s) were labeled with inaccurate diagnoses made decades earlier.

One day when I was making rounds, I saw a man in his 50s with a large head and large, cupped ears. As I walked past him, he struck out at me, grazing my shoulder. When I later read his medical record, I noted that he had been admitted to Fernald when he was 4 years old, and that the admitting physician attributed his severe developmental delay to a head injury from a fall from a crib when he was a year old. Based on the man's appearance, I ordered a chromosomal test for Fragile X syndrome, which turned out to be positive! In addition to sharply altering our clinical perspective on this man (for example, the diagnosis required rethinking how to manage his behavioral disorder), the finding carried immense importance for his extended family, some of whom could be affected with or carriers of the Fragile X mutation. The diagnosis led to an extended outreach effort and, as expected, the discovery of several more family members with developmental disability who currently carried no diagnosis, but who almost certainly also had Fragile X syndrome as well as women of childbearing age who did not then know of the high risk they had of bearing a child with the disability.

A major scientific breakthrough occurred in 1991 when scientists in Rotterdam used newly developed tools for DNA analysis to track down and clone the causative gene. In so doing, they discovered the remarkable fact that the Fragile X syndrome is caused by an abnormality in DNA copying during meiosis (the formation of the germ cells) that greatly expands a normal part of the gene from less than 45 copies of a 3 base pair (called CAG) repeat to more than 200 copies. This expansion occurs in a part of the gene called the promoter, and resulted in silencing of the gene. Lacking a functional promoter, the abnormally expanded DNA was no longer able to make what became known as the FMR1 protein. Subsequent studies over the next decade showed that this protein played a role in binding RNA and was especially prominent in normal brain tissue, a finding consistent with the phenotype, but there was still much to learn about its function. This discovery soon led to the development of a DNA test that was easier and more accurate than had been the chromosomal analysis.

A better understanding of the phenotype of Fragile X in affected young boys, the growing understanding that the mutated gene was a relatively common cause of intellectual disability, the realization that some boys labeled as autistic actually had Fragile X syndrome, and the new availability of a DNA test to confirm or rule out the diagnosis should have resulted in a prompt reduction of the age at which boys were diagnosed. But, at that time, relatively few pediatricians were well trained in the evaluation of children with cognitive delay. As the yield (likelihood of a positive result) of the test was—because it was being used to screen boys who could have any one of hundreds of different conditions—certain to be low, it took about a decade for the new test to be as widely used as it should be. I remember occasionally giving seminars to groups of pediatricians in which I advocated for the Fragile X test as a screening tool and of being met with skepticism. This is understandable, as it was then an expensive test and it yielded few positive results. In medicine, change is not as rapid as the news media often suggests.

The discovery that Fragile X syndrome arose because of the expansion of a naturally occurring triplet repeat that silenced a gene helped to open up a new field in molecular genetics. Today, scientists have identified about 20 orphan genetic disorders that arise via an expanded triplet repeat error, and that all cause some form of neurological disability. In all of them, the cellular pathology seems to arise for one of three reasons: the failure to make enough normal protein, damage caused by an improperly organized protein, or the toxic effects of the abnormal messenger RNA (the intermediate between the DNA and the protein) on the cell. In Fragile X syndrome there is no normal protein. Despite the similarity in molecular errors, the diseases manifest at quite different times and in quite different ways. Fragile X syndrome is a developmental disorder that manifests in early childhood. Huntington's disease usually does not become apparent until adulthood.

The discovery of the defect responsible for fragile X syndrome was a watershed event. There are more than 25,000 men and boys in the United States who suffer from severe intellectual disability because of this syndrome, and many thousands of girls and women are burdened with milder impairments. The isolation of the causative gene offered the first glimmer of hope that scientists might someday be able to develop a treatment for the disorder. Everyone realized that it would be a long, hard road, but now that the gene had been identified one could begin a deeper, molecularly based study.

A critical next step was to develop animal models that mimic Fragile X syndrome in humans. In 2000, a group at the University of Utah developed a model of the disorder in *Drosophila melanogaster* (the common fruit fly that for a century has been used to understand gene location and function). Although one must travel back in time hundreds of millions of years to find the last common ancestor of flies and humans, the structure and function of their neurons (brain nerve cells) are remarkably similar. Since 2000 a growing number of research groups have used the fruit fly model (which also does not have a functioning gene) to understand the pathophysiology of the disease as well as to search for small molecules that might be used to ameliorate or reverse the disorder.

Researchers have also developed a disease model in zebra fish (attractive in part because their transparency eases the study of organ development) and at least five mouse models of the condition. A vast amount of work has shown that the *FMR1* gene makes a protein that plays an important role in forming the synapses (connections) between brain cells. Specifically, it acts in a regulatory manner, binding certain kinds of RNA that help to guide the growth of the spiny extensions of some neurons. Fruit fly models of Fragile X syndrome actually have a phenotype that recalls the human condition, and in some of the mouse models the affected animals have abnormalities that are quite similar to those seen in people (including, for example, having large testicles).

By 2003 scientists had determined a potentially crucial fact about the role of FMRP (as the protein was called) in brain development. As is the case with many developmental pathways, there was a second protein called mGluR5, a cell surface molecule known as a G-protein-coupled receptor, that worked to *counter* and modulate FMRP. In 2004 this led Mark Bear, a neuroscientist at Massachusetts Institute of Technology (MIT), to develop a new therapeutic hypothesis: Because animals and people who lack FMRP develop the neurological disease, down-regulating the activity of mGluR5 might restore some balance to the brain and thus reduce the severity of the disease. In 2005 he and others, notably a physician named Randall Carpenter, formed a company called Seaside Therapeutics to attempt to develop small molecules that act as mGluR5 *antagonists*. They were fortunate to have the support of a major philanthropist who ensured the company could move quickly.

About the same time ongoing studies of the brains in mouse models of Fragile X syndrome led to a second closely related therapeutic hypothesis. Evidence emerged that many of the behavioral abnormalities could be traced to abnormalities in the area of the brain called the amygdala. In mice, scientists found a deficiency of a neurotransmitter called γ-aminobutyric acid type B (GABA-B). This led them to hypothesize that patients might be helped by a drug that acted as a GABA-B agonist. Because GABA-B had been shown to antagonize or compete with mGluR5, it made sense that up-regulating the production of GABA-B might indirectly dampen the action of mGluR5. Seaside Therapeutics undertook a drug development program that focused on GABA-B agonists. Helped by solid funding from the private source, eager patient organizations, and committed clinical investigators, by 2007 Seaside had advanced a program involving each target into clinical development.

In July of 2010 at an annual research meeting on Fragile X, the principal investigators reported the results of a Phase II trial of an existing drug called *arbaclofen*, a GABA-B agonist, that the team hoped would reduce the symptoms of "social withdrawal" that were such a major problem for patients and their families. The study, which involved three groups of patients divided by age, was conducted at 17 different clinical sites in the United States (an approach that facilitated enrolling the needed number of patients as quickly as possible). Placebo controlled and blinded, it was designed to assess the impact of the arbaclofen on various scales that rated behavior and irritability. Although the drug did not show a significant benefit on its primary end point—the reduction of irritability—overall the data seemed to suggest a trend toward improvement, and offered hope that a larger, longer trial might give positive results. The early results from the arbaclofen trial in Fragile X syndrome stimulated much interest among the families of affected children. In 2012, Autism Speaks, a foundation that provides major financial support to research projects with merit, awarded a $2 million grant to Seaside Therapeutics to pursue studies in the closely related field of autism spectrum disorders.

Unfortunately, in mid-2013, Seaside announced that the large trial that it was conducting to assess the efficacy of arbaclofen in patients with Fragile X syndrome had failed to meet its end points. In addition, it disclosed that a similar Phase II trial had failed to meet similar end points in patients with autism spectrum disorder. In June of 2013, the Swiss drug giant, Roche, decided not to exercise its option to in-license the drug. Out of money, Seaside Therapeutics ceased operations. This was especially disappointing as the Fragile X trial had come quite close to meeting some of its predefined clinical end points. When I talked to Mark Bear in December of 2014, he was still optimistic that arbaclofen would one day be proven to be of significant benefit to children with Fragile X syndrome. The Simons Foundation Autism Research Initiative seems to agree. In the spring of 2015 it purchased the rights to arbaclofen with which it may conduct a larger clinical trial. Hopefully, Seaside's research effort will contribute to the eventual development of a safe, orally delivered drug that will free children and adults with Fragile X syndrome from psychiatric medicines currently being used that offer modest clinical benefit and carry some risk.

Seaside had worked to develop drugs to mitigate *symptoms* of the disorder, *not* to correct the *underlying molecular* cause. Unfortunately, this indirect approach characterizes most of the drug research under way for the autism spectrum disorders (ASDs), a somewhat misleading term that probably encompasses scores of discrete genetic entities, each of which may be successfully treated only with the development of a drug targeted to the particular cause.

We are still a long way from seeing the approval of a new treatment for Fragile X syndrome. In October of 2014, I reviewed 44 entries listed on ClinicalTrials.gov that came up from a search for "therapeutic trials in Fragile X syndrome." Of that number, researchers had completed 20 (several involved arbaclofen), of which none had shown a convincing therapeutic effect. Of the remaining 24, four had been started and terminated early, one had been suspended, the status of several was listed as unknown, and only nine were recruiting patients. Of these, all involved well-known small molecules that were being investigated to determine if they could have modestly beneficial effects on behavioral abnormalities. This is not too encouraging. But, as we will see in later chapters, some powerful new approaches to treating orphan single-gene disorders are emerging that may be applicable to Fragile X syndrome. Perhaps the most hopeful approach would be to develop a small molecule that reactivates the expression of the *fmr1* gene.

Spinal Muscular Atrophy

After obtaining his medical training, Guido Werdnig (1844–1919), an Austrian physician, served for nearly a decade in the armies of the Austro-Hungarian Empire as it struggled to put down rebellions in the region we know today as Bosnia. His early publications addressed the clinical challenges of moving wounded soldiers across mountainous terrain. In 1888, Dr. Werdnig returned to civilian life in Graz, began practicing neurology, and won an appointment at the respected Institute of Clinical Pathology. It was there, in 1891, that he examined an infant boy named Wilhelm Bauer who had been born with poor muscular tone and who was having much trouble breathing. Dr. Werdnig, who grasped that the respiratory problems must be a consequence of muscular weakness in the diaphragm, which in turn arose owing to an as yet undescribed neurological disease, followed the boy's clinical course until his death at age 5. The case report he wrote about the child (that included an autopsy study) was among the first ever published about what we today call spinal muscular atrophy, or SMA. After performing an autopsy and studying tissue samples with a microscope, Werdnig noted that most of the *anterior horn cells* in the boy's spinal cord had degenerated. (The term describes tracts of motor neurons that descend from the brain through a portion of the spinal cord that on cross-section bulges outward like the horn of Africa.) In 1896, still building his career, Werdnig moved to Vienna. Ironically, in 1907 he too became afflicted with a rare neurological disorder—spastic paraplegia—from which he died in 1919.

Werdnig's observations were replicated and confirmed by Johann Hoffmann (1857–1919), a young German neurologist, in 1892. Born in Hahnheim, and educated in Worms, Hoffman (as was common at the time) studied at several medical schools, graduating with a grade of "superior" from the University of Heidelberg where he wrote his student thesis on tetany (probably a case of Guillain–Barré syndrome). Hoffman had the good fortune to spend his career at Heidelberg where he worked for decades under the tutelage of William Erb, one of Europe's foremost neurologists. In 1892, Hoffman published a paper similar to Werdnig's under the (translated) title *Familial Progressive Muscular Atrophy.* He maintained a lifelong interest in this new disorder, closely following seven patients from four families with 20 affected members, and extending Werdnig's neuropathological investigations. Over the next decade he published five papers about the natural history of the disease. Although he lived in Erb's shadow, Hoffmann seemed not to mind. He rose through the ranks, receiving full tenure in

1914 (not unusually late in that era). During World War I he was entrusted to oversee the entire clinical enterprise in Heidelberg. After Hoffmann's death in 1919, possibly due to sepsis from a smoldering jaw infection, Erb wrote an obituary that lionized his colleague's contributions to neurology.

An understanding of what became known as Werdnig–Hoffmann disease grew slowly. It was not until 1961 that researchers unanimously agreed that Werdnig–Hoffmann disease is an autosomal recessive disorder caused by severe degeneration of the anterior horn cells of the spine. At that time they had no idea what caused the relentless deterioration. Today, medical education is moving away from the use of eponyms, and there is no longer any space in the curriculum to pause and romantically recall the contributions of the clinical giants of the 19th century. Werdnig–Hoffmann disease is now almost universally known as spinal muscular atrophy type 1 (SMA1).

For an orphan disorder, it is relatively common, affecting about one in 10,000 children, and it should immediately come to the mind of every pediatrician who examines an infant with unusually poor muscle tone. Because SMA1 is an autosomal recessive disorder, the incidence of live births means that about in one in 50 persons is born carrying a mutation in *one* of their two SMN (survival motor neuron) genes. This frequency is inferred as follows: An affected child must be born to two healthy carriers, each of whom has one normal and one mutated form of the gene. For that couple, each conceived embryo carries a one in four risk of having inherited both of the chromosomes on which the mutations sit; $1/50 \times 1/50 \times 1/4 = 1/10{,}000$. The disease is called SMA1 because, although the full function of the defective protein is unknown, it is clearly essential to the survival of the anterior horn cells in the spinal cord. The fact that about one in 50 of the persons reading this page carries a mutation in the *SMN* gene may seem surprisingly high. Given that affected children do not reproduce, why is the mutation this common in the human population? No one knows. It is possible that having one mutation is in some subtle way beneficial, but if so, the benefit is a mystery.

The fate of infants born with deleterious mutations in each of their survival motor neuron (*SMN1*) genes is ultimately determined by another quirk in the human genome. Fortunately, evolution provided humans with two or more copies of an extremely closely related gene (located on a different spot on the same chromosome [number 15]) called *SMN2*. *SMN2* makes a short form of the normal protein that is partially active. For reasons that we do not yet understand, many persons are born with three, four or even five copies of the *SMN2* gene. The more copies of *SMN2* that a baby has, the less severe the consequence of also having the mutations in both copies of *SMN1*. Babies born with mutations in their *SMN1* genes, but who have *three* pairs of normal *SMN2* genes, are less severely afflicted than those with just two pairs, often not needing a ventilator. Affected children who were born with *four* pairs of the *SMN2* gene usually have milder disease. The few children with *five* copies of the *SMN2* gene are able to lead a relatively normal life. Unfortunately, about 65% of children born with SMA have the most severe form of the disease.

Most SMA1 infants present with obvious muscle abnormalities and they are usually diagnosed within the first few months of life. Diagnosis may take this long because there are many rare diseases of infancy associated with poor muscle tone, and among SMA1 babies the disease varies in the speed with which it advances. DNA testing is now routinely used to confirm clinical suspicion. Sadly, many SMA1 babies become dependent on ventilators before their first birthday, and many die by the age of 3 or 4. In the United States, at any given time, more than 1000 children are living with SMA1, with about 400 more being born each year. Despite the fact that the disorder is relatively common, it is not subject to newborn screening. This is in part because of the lack of an inexpensive test, in part because the disease is usually diagnosed relatively quickly, and in part because the interventions available to help the affected children are limited to generalized nutritional and respiratory support.

Given that SMA1 is relatively common and very severe, it is no surprise that over the last 15 years or so there has been an explosive growth in drug discovery research for this disorder. Many strategies have been tried, but the most compelling fall into one of three categories: efforts to incapacitate the abnormal protein (which may have toxic effects on the motor neurons), efforts to greatly increase (up-regulate) the production of the (helpful) SMN2 protein, and, recently, gene therapy (using vectors to directly deliver a normal copy of the *SMN1* gene directly to the motor neurons). In 2014, the most advanced clinical research findings for children with SMA1 were being driven by the use of "antisense" technology—the injection of molecules designed to shut down production of SMN1.

The fact that higher levels of the SMN2 protein correlate with having milder disease has led some researchers to search for ways to up-regulate its production. One approach relies on high-throughput screening—the massively parallel review of literally hundreds of thousands of small molecules to ascertain if one or more of them have a beneficial effect on the survival in culture of cells that were obtained from affected patients. Large-scale screening usually finds more than a few molecules that give a positive signal, but which often fail to generate positive results on retesting. This part of drug discovery is known as trying to proceed from a "hit" to a "lead" molecule. Many early compounds that show promise in cell lines fail when tested in mouse models of the disease.

A common exercise in early efforts to develop new therapeutic approaches for rare disorders is to use the set of small molecules that have already been approved by the Food and Drug Administration (FDA) as a "screening library." Among these 4000 or so molecules, a few might up-regulate the production of the SMN2 protein or extend its half-life. In research on SMA, scientists have studied hydroxyurea (a drug that is widely used to up-regulate production of hemoglobin in patients with sickle cell disease), sodium phenylbutyrate, valproic acid (an antiepilepsy drug), and several other compounds with the hope of making cells produce more SMN2. Thus far, despite some tantalizing early results, no clinical trial has shown a clear benefit for this approach. This is in part because infants with SMA1 are chronically ill and medically fragile, so it is difficult to enroll them in a study in a timely fashion, especially if it means that the child and his or her parents must travel to a distant medical center. Despite the eagerness of parents to help their children, the logistical challenges not infrequently delay the enrollment of an adequate number of subjects in the trial. Sometimes trials are terminated for that reason.

About two decades ago, Dr. Stanley Crooke, a former professor of pharmacology, helped to start a biotech company called Isis on the hunch that he and his colleagues could develop a new kind of drug that could be tailor-made to target a specific DNA sequence and inactivate it—thus the name "antisense." The plan at Isis was to become a platform company; that is, to make antisense drugs in partnership with much larger pharmaceutical companies that have the immense financial resources needed to shepherd the most promising drugs through clinical trials. Although it has been a long and sometimes bumpy road, after 20 years of effort, Isis has made impressive progress. Two of its drugs have gained regulatory approval, it has 25 antisense drug programs under development, and investors have rewarded it with a market value of more than $6 billion.

Isis has been working to develop its antisense therapy for patients with all three types of SMA for about a decade. By 2010 it had produced convincing evidence that its lead experimental drug could modulate the formation of RNA molecules in a manner that *up-regulated* the production of the SMN2 protein, a process that might ameliorate the disease. By 2011 its work showed enough promise that *Biogen Idec*, one of the nation's top-tier biotech firms, acquired worldwide rights to the company's SMA program under terms that could (should the program meet the negotiated milestones) result in payments of $300 million and double-digit royalties on product sales to Isis. In March of 2013, Dr. Claudia Chiriboga, a clinical investigator at Columbia University, reported data from an open-label, escalating dose, Phase Ib study (essentially, the study of the safety of three different doses given each to a few children with SMA2 or SMA3) of the Isis drug at the meeting of the American Academy of Neurology. Although it was not planned primarily to do so, the study showed that children had a positive clinical response to the drug, and that it might need to be given (by injection into the spinal fluid) only once or twice a year!

On May 9, 2013, Isis announced that it had earned a milestone payment ($3.5 million) from Biogen Idec on dosing the first infants with SMA1 in a Phase I safety trial. More good news followed. In September of 2013, Isis reported further data on a follow-up of the 24 children that had been reported on in March. Those results indicated that the children receiving the two highest doses of the drug showed improvements in muscle function up to 14 months after receiving a single injection! This is exciting news but, given the novelty of antisense technology and its uncertain long-term safety risks, the FDA voiced concerns about further treatment of children with the *milder* forms of SMA. In 2014, Isis set to work to develop a structure for future clinical trials that the FDA will accept.

In October of 2014 at the World Muscle Congress in Berlin, investigators who are leading clinical trials of the Isis drug in young children with SMA1 reported the hopeful news that in about half there was, when compared with the trajectory of the disease in children who were not treated, a substantial delay in the time to either needing ventilatory support or to death. Currently, Isis is moving forward with a large study that is enrolling 110 patients with SMA1. In this randomized, double-blind trial in which the drug will be injected via lumbar puncture, the clinical researchers will follow the children into 2017 to monitor their respiratory status and overall muscle function.

As our understanding of human biology grows, other creative approaches to ameliorating orphan genetic disorders will emerge. This chapter briefly highlighted several: repurposing an existing drug to counter the effects of a mutation (Marfan syndrome), searching for or creating small molecules that will increase the production of a normal protein when the disease arises because there is too little of it (Friedreich's ataxia), developing molecules that rebalance a defective biochemical circuit (Fragile X syndrome), and using a novel chemistry to alter the production of a messenger RNA that tricks the cell into making more of a desired protein (increasing levels of SMN2 to ameliorate SMA1). It is scientifically plausible to believe that these and other interventions will lead to therapies that meaningfully reduce the burden of a particular disorder. But they are unlikely to lead to complete cures. Most genetic disorders ultimately manifest owing to a gene's failure to command the production of the right amount of a protein in the proper cells at the right time in the life of the individual. The best way to treat SMA1 may well turn out to be gene therapy. Currently several small gene therapy trials are underway. Such studies could demonstrate clinical success or failure in less than a year, so we may have at least some sense for the likelihood of success by 2017.

In the next chapter I describe yet another novel possible therapy—an efort to make and deliver a *structural protein* (typically these are large molecules that help to maintain the architecture of a tissue, but which do not drive metabolic pathways) to ameliorate a truly terrible orphan disorder. It is an effort in which I had the good fortune to be directly involved.

9

Butterfly Children

Rebuilding the Skin

Dystrophic Epidermolysis Bullosa

AMONG THE MANY HUNDREDS OF SEVERE single-gene disorders, few are more horrible than the autosomal *recessive* form of a disease called dystrophic epidermolysis bullosa (rDEB). This life-threatening skin disorder is caused by a failure to make a large protein that plays a critical role in anchoring the epidermis (outer layer of the skin) to the dermis (deeper layer). The consequences for patients, as I shall illustrate by telling the story of one young man and his family, are so severe that the resiliency that they display challenges understanding—but first some history.

Our understanding of rDEB begins with careful clinical descriptions compiled by physicians in the late 19th century. A French dermatologist, François Henri Hallopeau (1842–1919), who is today remembered for original contributions in studying a wide range of disorders including leprosy and trichillomania (a severe behavioral disorder in which chronic hair pulling causes profound hair loss) was among the very first physicians to describe the many complications of rDEB and to recognize that it was probably a heritable disorder. One of the most revealing elements of Hallopeau's life is that after a brilliant career that included election to the French Academy of Medicine, he chose to spend his final years providing primary medical care to the poor at a charity hospital in Nanterre.

The other physician whose name is most closely related to rDEB is Hermann Werner Siemens (1891–1969), a German dermatologist who after studying in Berlin and Munich eventually became chair of the dermatology department at Leiden in the Netherlands. Siemens was one of the first physicians to distinguish identical twins from fraternal twins and to use twin studies to dissect the relative contribution of genetic and environmental factors, an interest he first brought to bear in trying to understand the frequency and distribution of skin moles. It was at Leiden that he had occasion to write up cases of patients with rDEB. Siemens is less celebrated today than he might otherwise be because of his open support of Nazi eugenics. In the 1930s he wrote a book that called for the voluntary sterilization of persons with serious disease, asserted that genetic predisposition affected many disorders, and suggested that patients had a duty to the state to avoid passing on the risks for genetic disease of which they were aware. Known for several decades as Hallopeau–Siemens syndrome, during the 1970s the eponym was displaced by the clinical term we use today.

I made a trip from Boston to Nobleboro, Maine on a sunny morning in February of 2012. For the last hour I drove the coast road north, admiring the occasional ocean panoramas, remembering the old three-masted ship that over the years I had seen slowly rotting away in Wiscasset harbor, and taking great pleasure as a bald eagle flew low across the road. The Hibbard house is, I suppose, easy enough to find . . . if you have been born and raised in Nobleboro! After making wrong turns at a couple of small intersections, I swallowed my pride, placed a call, and memorized the directions that Tim's dad, Glenn, provided. A few minutes later, as I pulled onto a dirt drive between a small barn and a modest farmhouse, a remarkably friendly black cat greeted me.

Glenn answered the door. A tall, articulate man, who makes his living as a navigator on a large tugboat, Glenn motioned me to a seat at the kitchen table and called his family. Tim walked in a moment later. By habit, developed over the decades as a physician interested in people with rare disorders, I made an instant appraisal. Dressed in a T-shirt and pajama bottoms, Tim, who struck me as about 18, looked as thin as a concentration camp survivor. He had flaxen hair, pale blue eyes, and a wry smile. He started talking about his life with dystrophic epidermolysis bullosa before I had taken off my coat, as if he had been looking forward to doing so for a long time. I immediately noticed his language. He used complex sentences with an impressive vocabulary. But he spoke with an unusual intensity, in a pressured way, making me think that he was barely containing a deep anger. Living life with a horrible disease had not yet defeated him, but it seemed to have taken its toll.

Over the next 3 or 4 minutes, Tim's mom, Ellen, a warm, direct woman with a Long Island accent, and Tim's cousin, Cody, a quiet 20-year old who had recently moved to Maine to live with the Hibbards, joined us. We filled the chairs at the kitchen table. The best way to learn about a disease is to spend time with a person who lives with it. Tim was not the first patient with rDEB who I had ever met, but over the next 2 hours, he and his family taught me much more about it than I had gleaned from all my reading and other, briefer contacts with families. Listening to his story was one of the most moving experiences of my life.

Tim, who was 23 years old when I met him (the fact that I guessed wrongly on his age is in part due to the chronic malnutrition with which he lives), was the first child born to the Hibbards. Ellen's pregnancy had been uneventful, and as they were admitted for labor and delivery at a well-regarded New York hospital, she and her husband were eagerly anticipating the routine birth of a healthy baby. Her labor and delivery were not overly difficult, but the joyous moment ended when the nurse gently slapped the newborn baby's bottom and a large swath of skin came off in her hand. The look of terror on her husband's face told Ellen something was very wrong. Instead of handing the baby to the mother, the nurse rushed out of the delivery room where she wrapped all but the baby's face in soft bandages. Fortunately for the Hibbards, despite the rarity of rDEB (only about one in every 300,000 births!), doctors diagnosed their son's condition within a few hours. The obstetrician called the chief of pediatric surgery, who took one look at Tim's skin and made the diagnosis. What he did not tell them was that the disease would redefine their family, and that with each passing year they would face ever-greater challenges.

rDEB is caused by a genetic mutation that renders the skin cells unable to produce the normal form of a protein called collagen VII. Skin cells called keratinocytes make and secrete this protein, which takes up residence in a matrix that keeps the outer skin layer attached to the inner skin layer. Without the anchoring (trident-shaped) fibrils that are formed by collagen VII, the slightest trauma can cause the outer skin to fall off. In clinic when pediatric dermatologists are assessing milder forms of the disease (there are many different genetic "bullosa" disorders, a word that indicates easy blistering), they sometimes take a pencil eraser and gently rub the skin. If the patient is affected, even this mild trauma will pull the layers apart, causing a blister like the one a runner can get from much more traumatic and repetitive slipping of a toe against a sneaker.

If that is the case, why did Tim at age 23 not look more disfigured? When I met him there were no blisters on his face, nor any obvious ones on his hands. Fortunately, the disease often spares much of the face from bad lesions. This is probably because it is one area of the body that is not routinely subjected to many abrasions or mechanical tension. Although any part of the skin can be affected by rDEB, it is the areas most subject to pressure—the hands, the feet, the buttocks, and the back (during the simple act of sleeping as the body moves about, the skin is often damaged) that are particularly prone to developing chronic skin lesions that do not easily heal.

On his arrival into the world, Tim changed the rhythm of family life far more than does the arrival of a healthy child. Ellen's life became one of constant vigilance. She spent *hours* each day wrapping her son in special, very expensive protective bandages (which can cost more than $50,000 a year). She had to teach everyone who came in contact with her family the strict rules for touching her child. As Tim grew, she even had to create an entirely new approach to cooking!

rDEB does not just devastate the skin; it ravages certain parts of the child's digestive tract—especially the upper part of the esophagus and the rectum. This causes almost unrelenting pain for the child, and sorrow for the parents, as it leads to many unpleasant medical procedures, especially esophageal dilatations. It is not uncommon for patients to enter a hospital each year so that doctors can use special tubes to widen an upper esophagus that is scarred and narrowed by the body's response to injuries due to the trauma of merely swallowing food! Ellen had to train herself to prepare only the softest foods, yet make sure her son received adequate nutrition.

As the 2 hours allotted for my visit unfolded and the natural barriers of privacy gradually fell, Tim became steadily more open with me, and his conversation became more emotionally charged. Not knowing how he might answer, I asked Tim what bothered him most about living with rDEB. At that moment I was recalling a similar meeting with a 16-year-old girl in Cincinnati who had severe damage to her feet—her toes were fused together (a condition called pseudosyndactyly) so she was wheelchair bound. Despite many other problems, she had answered right away, "If only I could walk, so I could be with my friends." Tim answered differently, speaking almost violently. "Taking a shit!," he said. Tim has severe skin lesions on his upper inner thighs, around his anus, and *in* his rectum. Even a soft bowel movement causes agonizing pain. Tim takes heavy pain medications on a rigid schedule around the clock, often taking an extra one when he feels the need to have a bowel movement. The pain makes him not want to eat, the drugs make him constipated, and the constipation causes more pain. It is a vicious, unending cycle. That is why he is so thin. Yet, difficult as it is to believe, Tim is lucky. Some young patients cannot eat at all. They have to have special tubes (gastrostomy tubes) surgically placed through their abdomens and sewn into the walls of their stomach so that they can receive feedings by that route for the rest of their lives. As Tim spoke those raw words, he started to cry, and lowered his face into his hands. Tears welled in Ellen's eyes and a silence settled over the table. I waited a moment and then continued my interview.

I asked Glenn and Ellen to tell me about Tim's school experiences. During the last 20 years the Hibbard's lived in New York, North Carolina, and Maine with relocations largely shaped by Glenn's job changes. Not surprisingly, Tim's education from kindergarten through high school was harmed by the extraordinary number of days he was *unable* to attend school. His school absences became steadily more common as he grew older; 1 year he missed 59 of the 180 days. During grammar school, Ellen, through tireless efforts to educate mostly sympathetic teachers, was able to move Tim, a good student, along the conveyor belt of American public education. In middle school it got a lot tougher. One year Tim missed so much school that the principal (a sympathetic soul) avoided state regulations that would have required holding Tim back by placing him in a special education classroom. Tim remembered that experience bitterly, saying he felt that it caused him to be labeled as being intellectually disabled. But he graduated from high school on time, and he was even able to maintain some male friendships.

With the end of high school Tim fell into the abyss of social isolation. Life in rural Maine, living with a family with an income (although modest) too large to qualify for financial support from the state, unable to live on his own without substantial support, and intellectually paralyzed by pain medications that left him in a "constant fog," Tim knew college was a long shot. His one effort in a small local community college was just too physically demanding. His few high school friends drifted away. Tim fell into a cycle of spending almost all his days at home—reading, watching "too much" TV, and dreaming of a cure for his disease.

As the minutes ticked past, and Ellen and Glenn and Tim let down more of their natural guard, I felt I could ask Tim if he would show me some of the lesions that had trapped him in his painful prison. When I asked, Tim told me he was quite willing to strip naked if I wanted him to. Glenn broke in and began to reminisce about the many misadventures they had had with physicians who did not know the disease, who recoiled in horror from the lesions, and who sent them away with a feeling of hopelessness. And no wonder! In addition to the unending pain, problems eating, limited mobility, multiple surgeries, isolation, and lack of opportunity that mark the lives of patients with severe EB, in midlife many develop an aggressive form of skin cancer that is intractable and often fatal. Tim told me that when he was younger and attended a special summer camp for rDEB teenagers, one of the first questions they asked each other was, "Have you been diagnosed with cancer yet?" The horrible fact is that as yet there is no therapy for rDEB. Without one, many patients will die of an aggressive cancer during their 40s.

Tim stood up. With Ellen's help he peeled off his black T-shirt to reveal a large white bandage that encircled his torso from his belt line to just below his nipples. Without his shirt, he was ghostly thin. He turned around, and Ellen gently undid a Velcro strap. From just below the lower tip of his scapula to his belt line, Tim's entire back was red and raw. It looked as though he had been flayed open with a medieval instrument of torture. As Ellen pulled the bandage away, I could see that it was sodden with green-yellow pus and some granulation tissue. Although I was a few feet away, I could smell the odor. I was at a loss for words.

Tim broke the silence, "If you think this is bad you should see my butt and my rectum." Ellen told me that Tim's skin had looked like that—even worse—for many years. The upper middle of his back was one big healed scar much like what you see on the skin of a burn victim. We had been at the table for nearly 2 hours. Tim did not sit back down. He burst into tears, abruptly left the room, and did not return. Ellen explained that to sit that long on a kitchen chair was extremely painful to Tim; he had only been able to do it because of the oxycodone he had taken before I arrived. He now needed to sleep. Our visit had come to its end.

As I drove back to Boston, I thought about the challenges of building a company to treat such a rare disorder. Because rDEB is so rare and so difficult to treat, it has until recently not gotten much attention from the medical research community. As is the case with many orphan disorders, there are just a handful of basic scientists and research clinicians in the U.S., European, and Japanese medical schools who are trying to take our understanding of the causes of rDEB to a deeper level. In the United States in a good year, the National Institutes of Health (NIH) might award grants worth a few million dollars to study this disease, of which most is spent on understanding the molecular aspects.

I first thought about whether it might be possible to develop a new treatment for rDEB in 2008 when a bright young doctor suggested that I consider it as a focus for a new company. But after reading the relevant medical literature, I concluded that the disease was so rare that even if a new company developed an effective drug, it might never recapture its investment. However, over the ensuing 3 years, I changed my mind.

First, I realized that there was a good chance that the incidence (births per population each year) and prevalence (total patients) of rDEB were, although very low, perhaps several-fold higher than what was stated in the medical literature. If instead of being a disorder that affected only one in 900,000 persons at birth (as some of the papers asserted), it affected one in 300,000 persons, there might be enough patients to support a commercial venture. As I talked to the few clinical experts in the disease, especially in European countries, where care is often centralized and where patient registries are often better maintained than in the United States, I found that they often were following more patients than I anticipated, suggesting that the disease was more prevalent than had been reported. Second, I was repeatedly told that it was typical of drug development for rare diseases that as soon as a new effective therapy entered clinical trials the patient census doubled. When there are no treatments and people have no hope, they tend not to go to doctors (or at least not to doctors at academic centers) and are therefore not counted. Research on new therapies brings patients with rare, chronic disorders back into the health care system.

Third, and most important, I met and began to work closely with Mark de Souza, a biochemist who had recently left a position at a biotech company with the dream of building his own. Mark and his partner, a seasoned former venture capitalist named Jim Fordyce (who for many years had been the Chairman of the Board of the Mary Lasker Foundation, the United States' equivalent of the Nobel Foundation), had in their search for attractive projects, become highly interested in rDEB, largely because of the work of two scientists at the University of Southern California named David Woodley and Mei Chen. They asked if I would help them assess and build out their plan for a new company with the hope that it might in time attract funding from Third Rock Ventures. After closely reading the scientific papers by Woodley and Chen (a research dermatologist and a cell biologist, respectively), who working as a team, have spent most of their professional lives trying to help patients with severe orphan skin disorders, especially rDEB, I agreed.

On average, an adult has about 1.7 square meters of skin. In addition to many other functions, this amazing organ provides the main line of defense against the many invisible dangers lurking in the environment. When a protein that is essential to maintaining the structural integrity of the skin is absent or not working, it poses a tremendous biological problem. Think of the pain one feels when a fall causes one "to skin" one's knee. The abrasion that shears the epidermis from the underlying dermis sends a wave of intense warning signals to the brain that the body's defensive wall has been breached. Over millions of years skin evolved into a complex structure with several layers of cells organized in a manner that has led dermatologists to conceptualize groups of them as reactive units, a variety of embedded structures (such as hair follicles and sweat glands), a complicated network of tiny blood vessels, a dense array of sensory nerves and a complex extracellular matrix that—if normal—holds all the parts together. Even though we do not usually think so of it, the skin is a complex, dynamic organ that is immensely busy every second of an individual's life.

As recently as the early 1990s, scientific understanding and treatment of rDEB had—with the exception of improved protective bandaging and pain medications—barely advanced from the days of the original case reports. The first big scientific breakthrough came in 1991 when a team in Dr. Jouni Uitto's laboratory at Thomas Jefferson University School of Medicine in Philadelphia cloned the gene for collagen VII and confirmed that defects in it were associated with rDEB, a step that opened the door for a much deeper understanding of the cause of the disease. One member of the team was David Woodley, then a young professor of research dermatology who was spending a year on sabbatical with Uitto. In the two decades that followed, Woodley and his colleague, Mei Chen, on their own or with colleagues, published a series of elegant papers that explored different approaches to treating the disorder. For more than a decade Mei Chen was possibly the only scientist in the world who was able to use cell culture techniques to produce (in small quantities) collagen VII.

The biggest problem Woodley and Chen faced was that, unlike most drugs (small molecules that can be synthesized by medicinal chemists), collagen VII is a behemoth protein. When fully assembled it has a molecular mass that is much larger than the largest "drug" (an antibody) yet approved by the Food and Drug Administration (FDA). In its natural form, collagen VII is organized as a trimer—a protein structure composed of three separate units. When a molecule is that big and complicated, protein chemists have to harness the cell's production capabilities to make it. Scientists have developed techniques for doing this—the most widely used is to insert a normal copy of the human gene into cell lines derived from Chinese hamster ovary (CHO) cells and coax them into manufacturing the human protein along with their own proteins. Even when one successfully makes the protein in small batches, there are many hurdles. Two of the major ones are making sure that these very big proteins organize into the correct three-dimensional pattern, and to figure out how to scale up the manufacturing of the proteins to have enough to treat patients. The emergence of Chinese hamster cells as a tool in drug development merits a brief aside.

Since early in the 20th century, the mammal that geneticists have most intensively studied is the mouse (*Mus musculus*). It is small in size and easy to care for, so researchers can maintain colonies of animals with special traits of interest; it breeds easily and has a short (23-day) gestation period; and its physiological systems are similar to those of humans. At leading mouse centers, especially the Jackson Laboratory in Bar Harbor, Maine, scientists have been assiduously studying mice with diseases caused by genetic mutations for 75 years. Many of these mutations arise spontaneously. Astute technicians notice something different about the mouse, which if confirmed through breeding experiments to be caused by a unique gene defect, result in a new addition to the catalog of murine genetic disorders. About 25 years ago scientists developed techniques to "knock out" individual mouse genes, thus allowing them to investigate the affect of mutations on a mouse's health on a gene-by-gene basis.

One of my most interesting experiences each summer is serving on the faculty in the Short Course of Mammalian Genomics and Human Genetics that the Department of Genetics at Johns Hopkins School of Medicine and the Jackson Laboratory jointly run for 2 weeks in July in Bar Harbor. One afternoon each year the laboratory sets up a giant tent on the lawn behind the research building. Inside, dozens of technicians station themselves behind scores of cages each of which contains a mouse with an orphan genetic disorder that arose because of a mutation in the same (cognate) gene that causes a similar disease in humans. The scene is reminiscent of a bizarre county fair. Crowds stand three deep to admire unusual specimens. One can stroll among the long tables and ask the technicians about the animals. There are mice with neurological disorders that make them tremble, morbidly obese mice, mice with Fragile X syndrome, deaf mice, and mice with brittle bones, to name just a few.

The Chinese hamster (*Cricetulus griseus*), which originated in the highlands of Mongolia, is another rodent that has also played a critically important role in improving human health. Scientists first used it in medical research in 1919 when experts in infectious disease studied its immunological response to the streptococcal bacteria that cause many human pneumonias (a process called "typing") that helped in the understanding of bacterial virulence. About four decades later after they learned that it has only 11 pairs of chromosomes (compared with 23 in humans), biologists became interested in developing the Chinese hamster into a well-characterized research animal.

In the mid-1950s, Theodore Puck, a scientist who spent much of his career at the University of Colorado and who is one of the fathers of human cytogenetics and cell biology, developed a cell line derived from Chinese hamster mouse ovaries, part of the work for which he was awarded the Albert Lasker Basic Medical Research Award in 1958. It was also in that year that he and another scientist, J.H. Tjio, published the paper that definitively established the correct number of human chromosomes. In announcing the award, the Lasker Foundation predicted that new techniques permitting the creation of pure cultures derived from single cells would revolutionize work in genetics and cancer, which they did. I had the pleasure of meeting Puck several times in his later years. At our first meeting, he struck me as the "ideal" Hollywood version of the dedicated scientist. He was a fiercely intellectual person who was passionate in his search for discovering fundamental truths about biological systems. Naturally skeptical, Puck would not hesitate to challenge another person's assertions. He also was a passionate advocate for the proposition that federal funding of basic science was the key to improving human health (and was not hesitant to argue that it was chronically underfunded).

With the dawn of genetic engineering in the 1970s, it became possible to use bacteria (notably *Escherichia coli*) to mass-produce important molecules like human insulin and growth hormone. Scientists did this by splicing a normal copy of the desired human gene into *E. coli*, fermenting it in giant chambers, and purifying the protein of interest mainly by running it over special columns. These became the first genetically engineered drugs. This approach permitted drug makers to greatly reduce the risk of immunological problems associated with using proteins derived from animals (such as insulin from pigs) for delivery to humans. But bacteria lack the molecular machinery that is necessary to make more structurally complicated proteins. To be clinically effective, proteins must be made in a biologically active form, which requires that they undergo the proper folding pattern and the complex "posttranslational" modifications, often involving the addition of scores of sugar molecules at precise addresses.

Cell lines derived from Chinese hamster ovaries have the molecular machinery to manufacture big human proteins (from human genes that have been spliced into the host cells) in forms that are very similar to the normal three-dimensional structure. In addition, CHO cells are remarkably resilient to manipulation, and can be grown in very high density while in suspension, a critically valuable trait if one is trying to mass-produce a protein at a reasonable cost.

During the period from 1960 to 1990, thousands of scientists worked with Chinese hamster cell lines, and those lines became the method of choice in which to explore the manufacture and purification of human proteins. The field became so familiar with them that CHO cells eventually earned the important distinction from the FDA of being generally regarded as safe (GRAS), a recognition that eases the path of manufacturing and approval. In 1987 the FDA approved the first drug for human use derived from production in CHO cells, a molecule called tissue plasminogen activator (TPA), which is used to treat people with strokes. The drug, known as Activase, weighs 59,042 Da, which is impressively large, but far smaller than the fully assembled collagen VII.

Since 1987, scores of drugs have been developed after screening genetically engineered CHO cells to select clones with the best yields for protein production. In 2011, pharmaceutical companies that sold proteins made from CHO production systems generated $30 billion in worldwide sales. In addition to eight drugs for orphan genetic disorders, the list includes many drugs developed to treat cancer, stroke, and autoimmune disorders. Currently, >70% of all proteins produced for use in humans are made starting with CHO cultures. The decades-long experience we have with CHO has generated much evidence that it does not harbor dangerous viruses, an assertion that one cannot make about other cellular production systems.

Perhaps even tougher than using CHO systems to *make* large, stable proteins is to figure out how to ensure that the purified protein will reach the destination(s) in the body where it is most needed. Collagen VII exists outside the skin cells in a region called the extracellular matrix (ECM). In healthy persons, it protects the skin from shear forces. If one looks at a electron micrograph of a rDEB patient's skin lesion, one finds that the matrix of intercalated proteins that hold the epidermis to the dermis lacks most of the trident-shaped molecule that physically unites the two layers. What is the best route by which to administer collagen VII? Will it stay intact long enough to make its journey? Will it cause a serious allergic reaction in the patient? These and other tricky questions confronted my colleagues and I as we contemplated building a company to manufacture and deliver collagen VII for rDEB.

Conducting research that stretched back about 20 years, Woodley and Chen had greatly advanced our knowledge of collagen VII. In 1993, Woodley had worked on the team that cloned the gene (known as *ColVIIA*), a critical first step. Several years later—years devoted to painstaking research—Chen had characterized the structure of collagen VII and showed that it must self-assemble as three connected chains to achieve its functional form. Several years after that she had genetically engineered a *human* cell line called

293 (derived from embryonic kidney cells and widely used in research) to make small amounts of human collagen VII.

Once they were able to make small amounts of the normal form of the protein, Woodley and Chen could take a crucially important step—run experiments to see if administering the protein could help heal lesions in mice with the murine form of the disease. One of the most important advances in biomedical research in the last 30 years has been the development of ways to selectively "knock out" genes in mice and then produce litters of animals with genetic forms of human diseases. Woodley and Chen were among the first to use these genetic engineering techniques to "make" an rDEB mouse.

In 2008, Woodley and Chen were able to show that if they injected recombinant collagen VII directly into the skin lesions of mice with DEB, the treated lesions healed much faster than did the untreated ones. This was a great step forward. But man occupies a limb on the evolutionary tree that is quite far from the spot where the mouse resides. Although collagen VII plays the same structural role in mice as it does in humans, nothing was known about how much of the protein would be needed to ameliorate the disease in humans. Positive experiments with a few mice were a hopeful signal, but certainly did not promise a cure.

Fortunately, nature provided a lucky break. About that time a veterinarian who was studying for a PhD at a veterinary college in Lyon, France discovered a *dog* with a canine version of rDEB—a puppy that was born with signs so similar to those seen in human babies that he strongly suspected that the pup had mutations in the dog version of that gene. Further investigation proved him right. This was bad news for the dog, but wonderful news for humans with rDEB. It would now be possible to breed and study a few dogs with canine DEB (which is less severe than rDEB in humans). Because of their relatively large size and relatively close evolutionary proximity to people, dogs can be immensely important in helping researchers to learn if they are on the right track as they move from mouse toward man. In the case of rDEB, the big problem was that Mei Chen was the only scientist in the world who had been able to make the protein, and she had made it in only small amounts.

Lotus Tissue Repair

In the winter of 2010, I began to work closely with de Souza and Fordyce to see if we could develop a compelling proposal to fund a biotech company to develop a protein replacement therapy to treat children with dystrophic epidermolysis bullosa.

In addition to becoming intimately familiar with the "product engine" that would constitute the company's breakthrough technology (in this case protein replacement therapy), we had to become experts in the natural history and clinical course of rDEB, evaluate all other potential therapies that were in early stages of development, assess the feasibility of working effectively with the two key scientists who had developed the recombinant protein, negotiate with their medical school over issues of intellectual property and the in-licensing thereof, develop a deep understanding of the true number of patients who actually have the disease, establish amicable relations with the leading patient groups, painstakingly determine the true birth incidence and life expectancy of children born with rDEB, interview many of the key clinical experts in the United States and Europe to assess their views of our plan, assess the current costs of caring for persons with rDEB, develop a detailed drug development plan, assess the regulatory hurdles, determine the price range in which a successful drug might be sold, and model how the new company might acquire value in the marketplace.

In dozens of meetings over the next 9 months, Mark (who led the team), Jim, and I took on the various tasks I have listed. As the months passed, several key issues emerged. rDEB is an ultraorphan disease. When we started, we were not sure if there were even 200 treatable patients in the United States. The sparse scientific literature suggested that about one in 1,000,000 people were born with this disorder. If that figure was correct, even if the chances of developing a highly new therapy were high, the market would be so small that it would be exceedingly difficult to build a commercial case for investing in a new treatment for the disorder.

We began to compile a thorough patient census. Mark de Souza did this by traveling to Europe to conduct meetings with many of the physicians who directed the major clinics in each country. He focused first on Europe because of the way the various national health care systems are organized, typically centralizing care for very rare disorders in only one or a few centers. A regular feature of our monthly meetings was to review a world map on which the rDEB patient count for each nation was updated. In less than a year, Mark established that there were hundreds more rDEB patients than the incidence and prevalence figures quoted in the medical literature suggested. We became confident that in Europe rDEB was diagnosed in about one in 250,000 births—still quite rare, but four times more common than the literature taught. Perhaps because of its fractured and inefficient health care delivery system, in the United States our survey effort moved the numbers only a bit. Yet, as most Americans were of European background, there was no basis to think the incidence of rDEB would be lower in the United States. We could not prove it, but we were fairly certain that we would eventually find more treatable patients in the United States. In that regard, it was interesting that the Hibbard family did not appear on any clinical registry of rDEB patients. They had "dropped out" of the system because it had so little to offer.

But even if there are enough patients to justify developing collagen VII as a drug, the proposed company would succeed only if it could accomplish two goals: (1) successfully manufacture (under stringent conditions) a gigantic protein, one that would be considerably larger than any drug ever approved, and (2) develop a method to deliver the protein to its destination in a safe and efficacious manner.

In thinking about how to treat patients with the severe skin lesions caused by rDEB, there are three possible modes of delivery: a topical cream applied directly to the wounds, intradermal injections (to make sure the molecule reaches the area between the epidermis and dermis where it could do its work), and intravenous therapy. For obvious reasons, intravenous therapy was the Holy Grail. If one could slowly infuse the protein into a peripheral vein from which it would circulate throughout the body with a sufficient amount of it reaching the wounds, it might be that just a few treatments each year would cure existing lesions and prevent the development of new ones! The reason for this hopeful speculation is that, in general, collagen molecules are highly stable and in their natural state they have a slow turnover rate.

Few of the scientific and clinical experts that we interviewed believed that the intravenous delivery strategy would work. They had several objections: (1) the therapeutic protein might be inactivated or destroyed by other molecules that it would encounter in the circulation, (2) if it did remain intact it was so large that it was unlikely to be able to move easily through capillary walls to reach the key extracellular space where it was needed, and (3) there was no basis for thinking that it would home to the wounds, so even if it got out of the capillaries, it was by no means certain that it would diffuse far enough to reach the wounds where it was needed. Among the eight scientific experts with whom I spoke, the message was almost uniform. They believed that there was a fair chance that if the man-made protein reached its target it would greatly help patients, but they also believed that the odds of successfully delivering it by intravenous injection were low. Nobody in this group, many of whom had spent their careers trying to understand how to treat rDEB, gave intravenous delivery more than a 20% chance of succeeding.

On the other hand, to a person, the experts felt that the development of rCVII as a drug to be delivered directly to the lesions by either topical cream or intradermal injection with tiny, almost painless, needles would be a transformative event in treating the disorder. Such a drug would hold out the possibility of protecting the mouth, the esophagus, and the anus from the destructive lesions that incapacitated Tim. Equally important, periodic injections into the webbing of the fingers and toes might prevent the fusion of digits that rendered many children wheelchair bound and unable to use their hands. Despite the great appeal of developing a drug to prevent such devastating wounds, we worried that a topical treatment would not generate sufficient economic value to justify taking the risk of investing millions of dollars into a research effort.

Given the rarity of this orphan disease, we were surprised that several academic groups and at least two companies were pursuing competing therapeutic approaches. In May of 2010 a leading clinical research group at the University of Minnesota published a paper in *The New England Journal of Medicine* reporting on the use of stem cell transplants in an attempt to cure rDEB. If allogeneic stem cell therapy were shown to work, despite the risks associated with this procedure, it might well constitute sufficient reason to weaken interest in our project. But, sadly for the patients, the report from Minnesota suggested that stem cell transplant was not the therapeutic answer for rDEB. Two of the seven human subjects in the study died from events related to the bone marrow transplants and two more had experienced life-threatening side effects. Furthermore, the end point used to measure the beneficial impact—reduction in use of bandages over a year—was not impressive.

In 2015 several research groups are conducting or planning trials to determine if various drugs (polyphenon E and thymosin B) will accelerate wound healing. Another is exploring the use of phototherapy for the same purpose. Much effort is being devoted to cell-based therapy. Researchers at Stanford University are recruiting patients with rDEB into a clinical trial in which they will remove unwounded portions of the patient's skin, use a gene vector carrying a normal version of the ColVII gene to transduce those cells, grow them into sheets in culture, and then transplant them to the patient's wounds. They call these sheets LEAES, an acronym that stands for "LZRSE-Col7A1 Engineered Autologous Epidermal Sheets" (with the first term referring to the vector). The hope is that the sheets of cells will have a sufficient number of anchoring fibrils to restore the barrier function of the skin. The most recently proposed trial (at King's College in London) will be a Phase I/II study that will infuse cells, called mesenchymal stem cells (MSCs), obtained from third-party donors into patients. MSCs are nonhematopoietic multipotent cells that reside in the bone marrow. The hypothesis is that after infusion some MSCs will differentiate into skin stem cells, which will provide new cells to heal chronic wounds.

After reviewing the allogeneic stem cell therapy, early projects to develop gene therapy, and efforts to genetically engineer skin cells, grow them, and then use sheets of them as a poultice to overlay the wounds of rDEB patients, we remained convinced that our approach had the best chance of clinical success.

After careful due diligence, TRV funded Lotus in late March of 2011 as a "virtual company." Armed with sufficient funds, Mark de Souza would work full time and Jim Fordyce about half time to outsource the key early work: learning how to manufacture rCVII in Chinese hamster cell lines, studying the impact of rCVII in healing wounds in the three different mouse models that academic laboratories had developed to study the disease, arranging similar experiments with the few affected dogs, preparing the "preclinical toxicology package" (the FDA requires that a potential drug be administered to two species of animals followed by extensive study of their tissues to look for signs of potential danger to humans), developing a patient-oriented website, and expanding our count of patients to other continents. TRV partner Neil Exter and I would have seats on the board and meet regularly with Mark and Jim to discuss ongoing challenges.

In December 2011, just 9 months after company launch, we received gratifying news from the laboratory of Dr. Leena Bruckner-Tuderman. She had created a hypomorphic mouse (one with a less severe mutation) that was burdened with a somewhat milder form of the disease. Basically, the cells in her mouse made about 5%–10% of the usual amount of rCVII, leaving it with lesions similar to humans. Working meticulously, a postdoc in her laboratory had succeeded in injecting rCVII into the animals' tail veins (an extremely challenging task given the tiny size of mouse pups). After the animals were killed, he was able to show that the molecule stayed intact and traveled to the basement membrane of the skin. To learn that the rCVII could be systemically administered was great news. But the next part of the report was really exciting. By using a technique called immunofluorescence, he was able to show that rCVII traveled preferentially to the areas of the skin where there were lesions. When one looked at skin biopsies under the microscope a glowing green line marked the presence of the rCVII right at the basement membrane! It was traveling to where it was most needed!

Each year at TRV we recognize International Rare Disease Day (the last day of February) by inviting a patient with an orphan disease to visit and share his or her story about living with one of these terrible conditions. There is no more powerful reminder of the importance of creating new companies to take on rare genetic disorders. In 2012, Tim Hibbard and his family joined us. After the usual chatter had quieted and people had taken their seats, I began my interview of the family by asking Ellen to recall how she first learned of her son's diagnosis. From the earlier visit, I knew that Ellen, like most mothers of children with severe disorders, had acquired a toughness that comes from rising each day to face challenges about which the rest of us never dream, and that she was willing to answer painful, intimate questions to educate strangers. "I knew something was wrong right away," she said. "As soon as I delivered, the nurse picked him up and walked out of the room. I took one look at my husband and his face said it all. Timmy had a big problem." In coming through the birth canal, most of the skin on Tim's leg had sloughed off.

Over the next hour, to an audience that was leaning forward to catch every word, Ellen and her husband, Glenn, and Tim recounted the last 23 years of their family's life. An especially poignant moment was when Tim started to talk about his younger (unaffected) brother, and started to cry. His brother, Kris, had been emotionally scarred by living each day of his childhood thinking that his older brother might die (a warning the family had heard all too often from doctors), largely because of the ongoing risk of serious infection. Tim said that he knew Kris loved him, but that as a young adult his brother now kept his distance, as he was involved with his girl friend and his schooling. Kris saw that the disease was crushing his brother and realized that he could not help.

Ellen fought back tears as she told how each day she struggled to find a ray of hope for her son (who is unable to attend college or hold a job, has been largely abandoned by friends, and has little hope for a girlfriend). When I asked Tim to describe his life, he said he drifted through each day on painkilling drugs, terrified of eating because of the pain he experiences when he does so. If it were not for the fact that his cousin, Cody, had come to live with them, Tim said, he would be so depressed he did not know what he would do. Glenn spoke angrily about the medical establishment and poor-quality state disability programs. Tim qualifies for little nonmedical financial support and many of his needs—such as for special bandages, which cost thousands of dollars a month—are not well covered. Glenn told us of traveling to a major medical school for advice. He recalled that for most of their time at the hospital his son was paraded before medical students, that the experts spent only a few minutes alone with the family, and that a month later he received an exorbitant bill. He laughed bitterly as he recounted how he returned it unpaid with his own invoice for the educational seminar his family had offered to the pediatric residents.

The hardest moment of the visit was when I asked Tim (who had agreed to this in advance), who at six foot one inch weighs only a bit more than 100 pounds, to stand and bare his back to the small audience. I then asked Ellen to remove the large bandage that ran from his scapula to his hips. An audible gasp flowed through the room. The lower two-thirds of Tim's back was one large red ulcerating lesion, as though he had just survived a fall in a fire pit.

When the hour was over people just did not want to leave. Tim seemed almost eager to keep fielding the questions, mostly now about what he liked to do and what he dreamed of doing if there were only a drug to control his disease. Though he really should not do so (because of the pressure sores he gets from sitting), Tim loves to drive, especially dirt bikes (which he did far more when he was a teenager than he does now). In answering one question, he said that he spent a lot of time just reading about adventuresome activities that he would never be able to undertake. Ellen closed the session with a plea: "I have investigated every experimental approach ever tried to help people like Tim who have DEB. When I read about your new company, I said at last, someone is trying something that to a mother makes sense. I speak for all the mothers out there when I say we are rooting for you."

When groups invest in biotech start-up companies, they do so with an eye on the calendar. The investors who place large sums in their hands do so in the hope that those companies will create significant clinical benefits for patients, benefits which will later translate into commercial value, either through sale to a larger company or by taking the company public (thus creating liquidity for investors). By their very nature, venture firms and their limited partners engage in high-risk, high-reward activities. Ever mindful of the calendar, from day one newly created biotech firms are running a race to create value.

Lotus faced three immediate challenges: develop cell lines in which to manufacture the recombinant collagen VII protein, identify as many patients as possible in the world who suffered with rDEB, and conduct "preclinical experiments" in animal models that might provide signals of efficacy. Under Mark's leadership, the company enjoyed success on all three. By the fall of 2011 Lotus was ready to have a round of "informational" meetings with potential suitors—meetings at which it could present solid evidence that it had cell cultures in which to manufacture the protein, that it is possible to administer the protein intravenously to animals with DEB in a way that produced wound healing, and that it had located significantly more patients (mostly in Europe) than the medical literature would predict existed. Of course, the company was still at least 3 years from being able to test the protein in humans, by far the most challenging hurdle. Thus, a high degree of uncertainty would cloud any calculation of the potential value of rColVII as a drug. But it was almost certain that rColVII would have to be administered regularly (perhaps about six to eight times a year) for life. If it had a transformative effect on the disease, life expectancy might soar, and each year, the number of treatable patients would grow, making the possible commercial market even more attractive.

After about 15 months of operation, Lotus began to hold exploratory discussions with a few companies that might become interested in our project. One company, Shire, through its subsidiary, Shire Human Genetic Therapies, became really interested in the possibility of acquiring Lotus. Throughout the fall of 2012, a team from Shire spent much time and effort performing a "diligence" review of Lotus. By mid-December, Shire had decided to attempt to acquire the company. After negotiations led on the Lotus side by Neil Exter of Third Rock, the parties agreed that Shire would acquire Lotus for an "up-front" fee (that recognized work thus far), and other payments that would be made over time, if and when certain developmental milestones were met. The largest payment would occur if and when Shire initiated a "pivotal" clinical trial, one intended to support an application for new drug approval with the FDA.

On January 8, 2013, during the large, annual JP Morgan biotech conference in San Francisco, Shire announced that it had acquired Lotus. If Shire ultimately succeeds in gaining FDA approval for rColVII, its acquisition of Lotus may be remembered in the commercial world for the speed with which Lotus reached milestones that triggered the purchase. Hopefully, on the scientific side it will be remembered as the transaction that led to the first approval by the FDA of a structural protein as a therapeutic agent, an event that may open the door for developing dozens of other structural proteins as treatments for orphan disorders.

Merosin-Deficient Congenital Muscular Dystrophy

In the spring of 2012. I had the good luck to start working with James McLaughlin, a talented young graduate of Harvard Business School. James and I began meeting every day to brainstorm about novel therapeutic platforms, and to review recently published research that might constitute the product engine for a new company. About 6 weeks into our effort, we became interested in a life-threatening, ultraorphan muscle disease historically known as "merosin-deficient muscular dystrophy" (MDC1A, for short).

Most people are familiar with Duchenne muscular dystrophy, a relatively common X-linked, ultimately fatal disorder that affects about one in 4000 newborn boys. Typically, young boys proceed from having difficulty walking to becoming wheelchair bound by adolescence and, typically, die of heart disease in their 20s. During my childhood, the comedian and actor Jerry Lewis began hosting televised annual fundraisers that brought immense attention to Duchenne muscular dystrophy (DMD). Similar fund-raising events in France each year are also hugely successful. In the United States, the Muscular Dystrophy Association (MDA) is the largest not-for-profit organization devoted to driving research on genetic muscle disorders forward. But the MDA encompasses many different rare disorders, so it is not surprising that devoted parents sometimes start new organizations to focus exclusively on the disease affecting their children. This has been the case for MDC1A, now the focus of a group called CureCMD.

As genetic disorders go, MDCA1 is not very well known. Until the mid-1990s, physicians who were called to evaluate babies with unusually poor muscle tone could only use the physical examination and a few simple tests to study infants with poor motor function to decide if they were most likely affected by a congenital disorder that damaged a key element of the dystrophin complex—an assembly of proteins that work as a unit to power muscles—or if they fell into a large cluster of similar but undefined muscle disorders. In 1994, a French scientific group led by M. Fardeau and F.M. Tomé which had already made significant contributions to better understanding genetic muscle diseases, reported that they had discovered that a few children lacked a key protein known then as merosin (but now known as laminin M or laminin-111). Infants born with this disorder rarely walk (even with assistance), often cannot stand, and often die quite young of respiratory failure due to the weakness of their diaphragms and intercostal muscles. The report stimulated several groups interested in congenital muscle diseases to refine the molecular understanding of this new form of muscular dystrophy. In 1995 another group reported that affected children also have abnormalities of the brain's white matter, adding a second means to distinguish children with merosin deficiency from other children with muscle diseases.

We became interested in MDC1A when a man named Richard Cloud, who lives in Minnesota, contacted us. Richard is the father of a little girl who is afflicted with MDC1A. Frustrated by the minuscule amount of research being undertaken to study the disease, he had joined forces with a scientific colleague, Brad Hodges, who had founded a company called Prothelia. They were trying to jump-start a basic research program that could generate the data needed to provide the basis for clinical trials aimed at ameliorating the course of MDC1A. Cloud was supporting research by Hodges, an expert in muscle physiology who was working in rented laboratory space just a few miles west of Boston. Collaborating with scientists at the University of Nevada in Reno, he had produced impressive data showing that by delivering the protein one could substantially extend and improve the lives of mice that had been genetically engineered to lack the gene that causes the disease in humans. These data were encouraging. But it would cost tens of millions of dollars to develop a therapy. If that happened, how many patients would there be to treat?

Intrigued by our initial meeting with the two men, James and I plunged into the limited literature on merosin-deficient muscular dystrophy, and I arranged to interview Brad at this laboratory. As we read the relevant scientific papers, we became more and more intrigued with the possibility that if one could deliver enough laminin-111 to enough muscle tissue early enough in the course of the disease, that one might be able to avert the respiratory failure that is the usual cause of death in the disorder. But there was a paucity of evidence concerning the number of living, treatable patients. Over the next few weeks, we challenged Richard and Brad to define the prevalence (total number of living affected persons) in the United States of the patient population. At one point, Richard asserted that there were 1200 kids in the United States with the disorder. We responded that if there were even *half* that many, we would engage in the extensive technical diligence that would be required to assess the feasibility of manufacturing and delivering the protein.

I assumed that because the disease is so severe and a relevant diagnostic test had been available to pediatric neurologists for a decade that the vast majority of affected children would be correctly diagnosed. Assuming that most affected children only lived 10 years (a reasonable guess) then to maintain a point prevalence of 1000 persons, each year there should be about 100 new patients born in the United States. If that is the case, it would mean that the disease affected about one in 40,000 newborns, a frequency that seemed much too high.

If there are 1000 persons in the United States with the disease, then there are probably about 1500 in Europe. These numbers are sufficiently large for one to imagine that the vast millions of dollars needed to win FDA approval for a new drug would be worth investing. The harsh reality is that to make clinical advances in the world of orphan genetic diseases, one must weigh the terrible burdens on a few lives against the millions of dollars that could be deployed to pursue many different useful projects. No one really knows the economic limits on developing drugs for rare disorders. Assuming the drugs are highly efficacious, will society tolerate payments of $300,000 a year, $600,000, $1,000,000? More? I return to this question in the final chapter.

It is no easy matter to determine how many treatable patients exist. For all but a handful of orphan disorders, the disease incidence (live births per year per population) and point prevalence (living patients at a given moment) numbers either have *never* been studied or are extrapolated from one or two small regional estimates or are merely educated guesses by expert clinicians. One is not awarded grant funds to count patients with rare disorders, young professors do not get promoted for doing so, and journal editors have little interest in any resulting publications. When one pressure-tests incidence and prevalence numbers, far more often than not there are few data to support them.

In the case of MDC1A only a handful of papers discussing the incidence and/or prevalence have ever been published, and none of these have been based on studies in the United States. The best study is from northern England where clinical experts ascertained every patient in a catchment area of 3,000,000 persons. They diagnosed 20 patients, giving a point prevalence of about one per 150,000 persons. But an epidemiological study in white rural northern England could well generate different findings than one in Spain, let alone in the ethnic melting pot of North

America. There is only one way to find out the true number of patients—count them one at a time. For ultraorphan diseases there are, in Europe, usually just a few major tertiary care clinics or research teams focusing on the disease. If one is persistent and obtains directly from the leading physician an actual patient count at each clinic, one can reasonably approximate the national or international point prevalence. Our challenge to Brad and Richard led them to send scores of e-mails to clinical experts in the United States and Europe asking how many MDC1A patients they followed in their clinics.

The key finding that should alert pediatricians to ask if a baby has a congenital muscular dystrophy is if he or she from the first days of life shows poor muscle tone. In severe disorders, this weakness can be life threatening because it involves the diaphragm and chest muscles and often compromises the ability to breathe. In the merosin-deficient form of the disease a complicated protein named laminin-211 is not able to perform its structural role, making the muscle membrane weak and liable to rupture. With such a severe phenotype, you would think that it would be easy to count all the patients, but it is not.

Although some CMDs are exceedingly rare, with the molecular description of each new one, questions about the incidence of the prevalence of the others arise. In the sparse literature on the epidemiology of CMDs, the percentage of the total said to be owing to merosin deficiency is quoted from as high as 75% to as low as 8%. The most agreed on number—40%—may or may not be correct in Caucasian populations.

Even more difficult than establishing the true patient population is the challenge of guessing how many people with the disorder comprise the *treatable* population. People with orphan single-gene disorders often fall loosely into two groups: those who seem to produce no detectable amounts of the protein and those whose cells make some misshapen or foreshortened forms that often provide some function. Those with no functional protein, called nulls, are usually severely affected. The severity of disease in patients with dysfunctional proteins can vary greatly; some may be only mildly affected. Generally speaking, about one-third of patients are nulls and two-thirds are not. If one-quarter of the latter have mild disease and may not need a really expensive therapy, the treatable patient count may be 15% less than the patient count implies. When one is dealing with small numbers to start with, a reduction in the estimate of treatable patients will weaken the argument for taking the substantial risk that accompanies every drug development program.

To address this uncertainty those who work to develop new therapies for orphan disease almost always embark on a natural history study of the disorder that frequently involves more patients studied in more ways than has ever before been undertaken in academe. This is not a criticism. The NIH and other funding agencies rarely pay for natural history studies. Performed properly, the natural history study will provide a much clearer picture of the range of severity of any single disorder over time. For example, in the case of MDC1A, the study might find that 30% of infants born with the disease are at high risk for respiratory failure and early death, that another 30% would be able to breathe on their own but would be wheelchair bound by age 10, that 20% would suffer less severe muscular problems, such as lifelong weakness that impaired many functions, and that 10% have such mild impairment that they might not need therapy.

Richard and Brad could only provide solid evidence that there were about 350 patients with merosin-deficient CMD living in the United States. We reluctantly concluded that the treatable patient population was too small.

Brad Hodges and Richard Cloud and their academic colleagues made substantial progress in the years that followed. In just 3 years Brad and the scientists in Reno succeeded in making human laminin-111 in a cell line. Then, they showed that if they give laminin-111 to mice with the murine form of MDC1A, that they *double* the animals' life expectancy, and increased their strength. Scientists have devised ingenious methods to test the strength of mice. One test makes the little guys do the equivalent of chin-ups with the disincentive that if they let go of the bar they fall into a dunk tank. Another measures how long they can stay on a rotarod, a bit like timing humans as they stand on a log in a lake. By studying the muscle of treated mice and comparing it to tissue taken from untreated mice, the scientists also showed that laminin-111 reduces both the inflammation in muscles and the decline in muscle fiber number and size. Some of their research even suggests that the therapy reverses the damage done by the disease.

These are, to be sure, impressive findings. But it is always right to be skeptical that a drug that is safe and works well in mice will be safe and effective in humans. It would be helpful to see the mouse results confirmed in large animals, but no one has yet figured out how to knock out genes in dogs to create canine versions of human orphan diseases. No veterinarian has yet found MDC1A in dogs and, although it has been described in cats, no one is studying the disorder in that species.

Early in 2014, I called Brad to inquire about his progress with merosin-deficient CMD. He happily told me that a few weeks earlier Prothelia had entered into strategic agreements with the University of Nevada in Reno and a large biotech company named Alexion. The terms were not publicly disclosed, but Alexion has agreed to fund research on laminin-111 at the university and has an option to acquire Prothelia if certain research milestones are met. Once again, the passion and determination of the parents of a child with an orphan disease had successfully accelerated the research and development efforts for an orphan disorder.

10

Ligands

Turning Genes On

X-Linked Hypohidrotic Ectodermal Dysplasia

FOR MARY KAYE RICHTER, A MIDWESTERN FARM WIFE, everything seemed fine at first. Her new baby arrived into the world with a lusty cry. He was cute, of normal size, and doing the things that newborn babies do. Neither the pediatrician nor the nurses noted any problems. The relatives were smitten. For months life with her new son was what life with a new baby was supposed to be.

It was around Charley's first birthday that Mary started getting concerned. This is about the time that mothers begin to compare the development of their infants with others of the same age. Mary found it curious that her son had very little hair, which had not been the case with other babies in her family or among the children of her friends. She became more concerned when at the age of 15 months not a single tooth had yet erupted. She took Charley to the dentist. He quickly suspected that Charley had a genetic disorder, and he advised her to take her son to the university dental school to have an expert in pediatric dentistry examine him. Mary promptly made the appointment.

After examining her son, the academic dentist asked Mary a few questions about his medical history and about his hair. He asked if anyone else on her side of the family had been born with missing teeth. Then he told her that his physical examination found little evidence of tooth buds and that the little boy had an unusually small alveolar ridge (the part of the jaw in which teeth form). He said it was difficult to be certain because there was no test to perform, but he thought that Charley had been born with a genetic disorder called hypohidrotic ectodermal dysplasia.

Charles Darwin, the father of evolutionary theory, was among the first scientists to learn about this disorder. In the mid-1840s he received a letter from a medical officer in the British Indian army, describing families in Punjab in which grandfathers and their grandsons lacked teeth and hair and could not sweat, a dangerous condition for peasants who toiled in the hot sun. In his work, *The Variation of Animals and Plants* (1875), Darwin wrote "I may give an analogous case, communicated to me by Mr. W. Wedderburn, of a Hindoo family in Scinde, in which ten men, in the course of four generations, were furnished, in both jaws, taken together, with only four small and weak incisor teeth and with eight posterior molars. The men thus affected have very little hair on the body, and become bald early in life. They also suffer much during hot weather from excessive dryness of the skin. It is remarkable that no instance has occurred of a daughter being affected . . . though the daughters in the above family are never affected, they transmit the tendency to their sons; and no case has occurred of a son transmitting it to his sons. The affection thus appears only in alternate generations, or after long intervals." Surprisingly, given his acute powers of observation and inference, Darwin missed the chance to discover X-linked inheritance, which was not fully described until about 1910.

A British physician, John Thurman, deserves credit for publishing perhaps the first clinical case report (of two affected first cousins) of XLHED in 1848. Over the next century, a few physicians published additional case reports. These early academic efforts resulted in a tripartite eponym—Christ–Siemens–Touraine syndrome—that physicians used until the 1980s. Josef Christ (1871–1948), a German dentist who practiced in Wiesbaden, was among the first to describe the dental aspects of the disease. Hermann Werner Siemens (1891–1969), who was Chairman of Dermatology at the University of Leiden and who is associated eponymically with describing *five* rare skin disorders, was among the first to undertake research on XLHED. As he was also a pioneer in using twin studies to parse the genetic and environmental contribution to disease, his name should be much more readily recognized than it is. But during the late 1930s and early 1940s, Siemens embraced Nazi eugenic thinking. He wrote a book called *Foundations of Genetics, Racial Hygiene and Population Policy* that called for voluntary sterilization of "pathological persons." A later edition praised Hitler's racial policies. Siemens must have had a change of heart because in 1942 the Nazis removed him from his teaching post, and for a time he was incarcerated for resisting German occupation policies. But although he obtained an academic post after the war, he did not regain the high regard that he once held in the academic community. During the middle third of the 20th century, Albert Touraine (1883–1961) was perhaps the most erudite and accomplished of all French dermatologists. He earned his doctorate of medicine at the University of Paris in 1912, writing a thesis on syphilis. After serving in the military during World War I and in several rural hospitals thereafter, he was named to a senior position at the Hôpital Saint-Louis in Paris. During his career, he published 120 original articles and edited more than 1000 contributions to a leading French dermatology journal.

Perhaps the most interesting early cultural account of XLHED appeared in 1938 when a writer named Charles Graves who was working for the federal Works Progress Administration (WPA) published an extensive description of the "Whitaker Negroes," a large Mississippi family in which many members had features strongly suggestive of XLHED. In describing the group, he wrote, "Frequently the Negroes take buckets of water to the field with them, turning the water over their head to soak their clothing . . . their hair is fine and silky but thin and short . . . their peculiarities seem to be inherited only by the male children, the females being normal."

In the mid-1970s, a few academic physicians, notably Newton Freire-Maia and Marta Pinheiro in Brazil, began a sustained effort to delineate the many rare genetically caused ectodermal dysplasias (conditions in which one or more tissues derived from skin progenitor cells fail to develop properly), attempting to sort them by the differing constellations of signs and symptoms involving hair, teeth, nails, and glandular structures. It was an immensely complicated and frustrating process that resulted in the creation of at least 11 clusters of putative single-gene disorders. Among other findings, their efforts confirmed that XLHED was by far the most common among the 117 distinct entities that they had delineated by 1984. Today the known number of distinct forms of ectodermal dysplasia exceeds 170, each due to a derangement in a particular gene that affects the formation of skin and its appendages.

In the mid-1980s, Dr. Angus Clarke, a clinical geneticist at the University of Wales College of Medicine, undertook a comprehensive study of the natural history of XLHED in 56 families with affected members, including detailed histories for as many generations back as possible, home visits with physical examinations, and blood drawing for DNA analysis. The most surprising (and disturbing) discovery was that among these families, about one out of four boys born with XLHED died in childhood, an extraordinarily high mortality rate. These deaths were usually caused either by respiratory illnesses in the years before the advent of antibiotics or by severe *hyperthermia*. Although there is no basis to dispute these early findings, clinical experts currently agree that the mortality risks, although higher than in unaffected children, are *much lower* than what Clarke found four decades ago.

Naturally, on hearing the term "X-linked hypohidrotic ectodermal dysplasia," Mary Kaye's first words were more or less, "Could you please say that in English?". After making clear that he was not an expert on this rare condition, the dentist explained the signs that suggested that someone was affected: very few teeth, sparse hair, and few sweat glands. Many months earlier, Mary Kaye had noticed that her son's skin always seemed dry, but as with some of her other observations, she had not thought it to be cause for alarm. She thought about her family, but could recall nobody with the same findings.

It look a while for Mary and her husband to find a doctor who could explain the genetics of the disorder. After some two or three frustrating trips to see various experts, they finally learned that Charley was burdened with an X-linked disorder. This means girls and women (who have two X chromosomes—one with the disease-causing mutation and one with the normal gene)—are usually only mildly affected, but boys and men are either unaffected (because they inherited the X carrying the normal copy of the gene from their mothers) or affected (because they have inherited the X with the mutated gene).

Mary Kaye asked the two crucial questions: What causes this? What can be done for my son? The answers to both questions were painfully short. Although scientists knew the causative gene resided on the X chromosome, they had no idea about its function. It would not be until the late 1980s that Jon Zonana, a clinical geneticist at the University of Oregon, and others began a sustained effort to map and clone the causative gene (called *EDA1*), a goal they reached in 1996. But in 1981, almost nothing was known about the underlying pathology. The doctor could say little more than that it was due to a rare genetic mutation.

As for helping the little boys who presented with the condition, the answer was only slightly better. The biggest medical concern for parents and caregivers was to be wary of overheating. Because boys with XLHED have very few, if any, sweat glands, they can easily develop hyperthermia. There had been a few instances when boys had died suddenly on hot summer days while on the playgrounds when their temperatures soared and they could not cool themselves by sweating. Mary Kaye learned that she, her family, and the schoolteachers would have to be vigilant about the need to keep her son cool. As for the sparse hair and few teeth, there was little to do but wait and see. In most affected children a few teeth would form; in others there would be none. From early childhood Charley would probably need dentures, a very difficult issue for a child. As a consequence of the failure of the teeth to develop, Charley's face would look slightly unusual due to flattened cheekbones. Depending on its severity, the cosmetic issues could be particularly hard to deal with in adolescence. Mary Kaye is a tough woman. She met the news head-on, and began to think about what she could do to make a difference in the strange new world into which her family had entered.

The prominent anthropologist Margaret Mead famously said, "A small group of thoughtful people can change the world. Indeed, it's the only thing that ever has." As with many mothers who learn that one of their children has a medical problem for which there is no treatment, Mary Kaye set out to change the world. First, she searched for other families with affected children. After she had found a dozen or so, in 1981—literally working at her kitchen table—she (along with two cofounders, a dentist named John Gilster and a minister named Charles Sheffield) started the National Foundation for Ectodermal Dysplasia (NFED) with the slogan, "We Can't Smile Without You." I first met Mary Kaye in the spring of 1983 when she attended a conference on birth defects in Baltimore at which I spoke. In May of that year we corresponded about whether the refusal of insurers to pay for the expensive dental care that children with XLHED need might be grounds for filing a class action lawsuit asserting insurance discrimination. Sadly, 30 years later, lack of coverage for expensive dental care remains a serious problem for boys with this disorder. Mary Kaye served as Executive Director of the NFED for nearly 30 years until 2010. In the early (pre-Internet) years she wrote a monthly newsletter that was a warm and engaging method to build the community she served.

My interest in XLHED crystallized in August of 2008 when I read an article that had been published in November of 2007 in the *American Journal of Human Genetics* about efforts to treat puppies with the canine form of XLHED. Given evolution's conservatism, it was no surprise to learn that the protein (known as ectodysplasin A or EDA) performs the same function in the dog as it does in humans—acting as a signal to tell various embryonic cell lines to differentiate into hair, sweat glands, and teeth. In the article, scientists at the University of Pennsylvania School of Veterinary Medicine and the University of Lausanne reported that by injecting the EDA protein soon after the birth of affected pups, they had been able to significantly ameliorate birth defects in 4 out of 5 dogs that were afflicted with XLHED. I became intrigued because large mammals such as dogs that are born affected with a cognate version of a human genetic disorder often provide a much better model than do mice in which to assess a potential therapy that is ultimately intended for humans. The positive results constituted a kind of proof of principle, suggesting that it might be possible to use the same approach to avert or ameliorate similar congenital malformations in children.

A critical early discovery was serendipitous for research on this rare disorder. In 1953 a German scientist named Falconer, who was an expert on mapping X-linked genes in the laboratory mouse, reported that mice with the phenotype known as "tabby," had a mutation in a gene that caused a developmental disorder strikingly similar to the XLHED in people. More than three decades later, molecular biologists (by then armed with tools that permitted them to do so) began to create and maintain colonies of animals to use for molecular studies relevant to the human disorder. The easy availability of a mouse model for XLHED was an important advance, but unfortunately, the animals differ significantly from humans in how they manifest the disorder. The two most important were that the tabby mice did not have serious problems in temperature regulation, and that all mice are born with *one* set of teeth to serve them throughout life, whereas humans are born with two. Still, the mouse model opened up a new pathway to understand the disease. In 1990, for example, a team showed that by administering a protein called epidermal growth factor it could induce the development of sweat glands in the mice.

The paper that initially intrigued me grew out of a clinic visit made several years earlier by the concerned owner of a sick dog. In 1994, vets at the University of Pennsylvania were asked to examine an 8-week-old male German shepherd that had been born with patches of hairlessness and missing and misshapen teeth. By examining biopsied tissue under the microscope, the doctors determined that the dog also lacked both hair follicles and sweat glands (which in dogs are on the foot pads). In conducting a detailed analysis of the reproductive history of the affected pup's parents, the scientists found that among the 33 pups that the father had sired, there was only one affected individual. The doctors concluded that the affected puppy's problems were caused by a new mutation. In the fall of 1996 they published the discovery and partial characterization of a canine form of XLHED. The discovery was well timed, for it coincided with the fine molecular work by Jon Zonana and his colleagues that resulted in cloning (isolating and identifying) the gene in which the causative mutation arose in humans.

When establishing a dog colony for research purposes, it is much easier to work with smaller breeds. At Penn, a veterinarian scientist named Dr. Magi

Casal, who had earned both a DVM and a PhD and who had, serendipitously, worked in a laboratory that studied the relevant protein, led the team that pursued the study of XLHED in dogs. They moved promptly to breed a strain of smaller dogs (more easily adapted to cage life) in which to study the disorder. They bred the affected German shepherd with female keeshonds and basset/beagle crosses that were part of the breeding colony maintained by the school. Such colonies are operated pursuant to strictly enforced federal regulations that demand that a high standard of care be provided to the dogs.

According to the Mendelian laws of inheritance, each of the daughters born to the *affected* male German shepherd must carry a copy of the mutation, despite the fact that they are not affected (even though they inherited the mutated X from their father, they also inherited a normal X from their mother). Among the offspring of these female carriers, on average one-half of the male pups will be born with XLHED (each pup has a one in two chance of inheriting the mutation depending on which X chromosome from the mom is in the egg). In 1997, Dr. Casal published the first comprehensive description of XLHED in dogs in the *Journal of Heredity.*

While Casal and her team continued studying the dog model of XLHED (the colony at Penn was the only one in the world), at the University of Lausanne a talented biochemist named Pascal Schneider, who was an expert on a group of proteins called TNF ligands (a ligand is a molecule that binds with a receptor on a cell surface and triggers a cascade of orders to various biochemical pathways within those cells), was studying how the mutation in the *EDA1* gene caused the birth defects, a project that he hoped someday would provide the foundation for developing a treatment for boys with the disease. By about 2000, Dr. Schneider had become intrigued with the idea that if he could figure out a way to deliver the "effector" portion of the protein that was mutated in XLHED in a timely fashion to the cells that needed it, he could avert the birth defects that arise if the cells do not receive the signal when they should.

By 2002 he and his team had learned to make a "fusion protein" (a combination of the key EDA receptor-binding domain to the Fc portion of the immunoglobulin *G1* gene). Basically, this biochemical wizardry resulted in a protein that could be injected into pregnant tabby mice and would be able to cross the placenta to reach mouse fetuses afflicted with XLHED. In a paper published in 2003 in *Nature Medicine,* the scientists in Schneider's laboratory, especially Olivier Gaide who was then at the University of Geneva Hospital, showed that if they delivered the protein early enough that they could ameliorate virtually every aspect of the disease. Remarkably, it appeared that *just a few* injections of the protein would be enough to permanently correct the cellular deficits!

Collaborating with the group at the University of Pennsylvania, Schneider turned his attention to Casal's newly developed dog model. Treating dogs would almost certainly offer immensely important insights in regard to efforts to treat humans, especially in regard to what constituted an appropriate dose of the protein. But the dog posed new challenges as well. One crucial difference was that unlike the situation in the tabby mouse, the newly created fusion protein (called EDA1 for ectodysplasin 1) could *not* be administered to pregnant dogs because the cells in their placentas do not permit its uptake. This meant that the experiment would have to be performed on *newborn* pups. But given that the cell lines that lead to the development of teeth, hair, and glands normally develop *during the middle of gestation*, treating right after birth might be too late to see a significant beneficial effect.

In November of 2007 the teams from Penn and Lausanne reported on their efforts to treat infant dogs with XLHED with recombinant (meaning a genetically engineered protein that had been produced by cells in culture and harvested therefrom) EDA1. The results were striking. Most of the nine treated animals showed a remarkably positive clinical response to the protein when compared with the features of untreated affected animals. In addition to normalization of tooth number and structure, the treated animals developed normal lacrimal (tear) glands, and improved sweating ability (in dogs this is measured on the surface of the paws). They also showed that the treated animals had normal resistance to respiratory and eye infections, whereas the untreated ones were burdened with recurrent infections.

The results were even better than had been the results of treatments of affected mice! Perhaps most significant was the fact that four of five dogs that were treated with the highest dose on four occasions over the first 2 weeks of life had the best outcomes in dental development. This provided hope that there might be a sufficiently large window of opportunity for treating newborn boys, perhaps as much as several months. The treated dogs also showed some evidence of developing sweat glands. Because hair development in dogs is complete about 10 days before birth, it was not surprising that the experimental therapy did not appear to have much impact on that feature.

In the history of medicine there has been only one approach to birth defects—surgery. Surgeons repair cleft lips and cleft palates, they close the spinal cord lesions of spina bifida, they insert shunts to reverse hydrocephalus, and they patch holes in babies' hearts. Now, the success with the dogs offered a *new paradigm* for treatment of birth defects: Administer a man-made version of the crucial missing protein in time for it to prevent or reverse the development of abnormalities in *physical* structures. This work suggested that there might come a day in which fetuses destined to have birth defects could have physical problems ameliorated *in utero* by the timely administration of the normal version of a defective ligand.

Reading the 2007 paper, I noticed that three of the coauthors listed their affiliations as with a Swiss biotech company called Apoxis. This meant that before exploring the possibility of building a scientific and commercial case for developing a therapy for XLHED, I had to learn if there was intellectual property to which a new venture would need access. I called the company, Apoxis S.A., in Lausanne, only to learn that just a few months earlier a Danish company named TopoTarget had acquired it. When I called TopoTarget, I had a second surprise. It had acquired Apoxis mainly to gain access to some chemical compounds it hoped to develop for treating cancer. TopoTarget had no interest in developing a therapy for XLHED, and just *3 days* before my call, it had licensed rights to the EDA1 protein to a newly formed biotech in Lausanne, Switzerland called Edimer Biotech S.A. When I finally reached folks at Edimer Biotech S.A., I realized that it was essentially a shell company that owned the license for clinical use of the molecule, but had not yet raised sufficient funds. The three owners were a biochemist who was a coauthor on the paper reporting a success in treating the dogs, a young Swiss businessman, and a small Swiss venture firm that had helped them to acquire the license to the protein.

It soon appeared that the owners were interested in having another party develop the potential therapy. Now the challenge was to assess the scientific and commercial effort that would be needed. For the next few months I had the good fortune to work with Nick Leschly, at that time a partner at Third Rock Ventures, and a consultant named Neil Kirby, who is a highly experienced drug developer. We set out to assess the scientific, clinical, and commercial feasibility of building a company to treat XLHED.

To build a company with the goal of developing a novel, transformative therapy for an orphan genetic disorder one must understand the disorder and its impact on families at a level that can never be attained simply by reading medical journals. It is essential that those who wish to create such a company work diligently to earn the trust of the small, often beleaguered, patient community, so when Mary Kaye Richter (who I had talked with by telephone) invited me to attend the NFED's eighth annual Halloween fund raiser in Manhattan, I eagerly accepted. By 2008 the NFED had grown from an idea in Mary Kaye's head to a well-established patient advocacy and support group that served about 5000 families burdened with any one of scores of genetic forms of ectodermal dysplasia, of which persons with XLHED accounted for ~70%. Ruth and Keith Geismar, New Yorkers with a child with a different genetic form of ectodermal dysplasia, had invigorated fund-raising and were hosting the event.

As I purchased my ticket and entered the Gotham ballroom, the NFED party struck me as much like every other fund-raising event I have attended. The crowded room was lavishly decorated (in a Halloween theme), a jazz combo played lively music, well-dressed men scooped expensive hors d'oeuvres from mountains of food at stations scattered about the room, the women shimmered in eye-catching cocktail dresses, and the hall throbbed with conversation. But I quickly noticed that there were quite a few children scattered among the adult crowd. They ranged in age from infants through late adolescence, and virtually all of them were burdened with various birth defects. A quick look revealed that some of them had been born with one of the many other rare (and usually severe) forms of ectodermal dysplasia. Most had wispy, sparse hair, some had disturbingly misshapen faces, a few had horribly misformed hands, many seemed emaciated, and some were wheelchair bound.

The young people with XLHED were pretty easy to recognize and looked remarkably alike. In some cases, I later learned, they looked more like each other than like their unaffected brothers. By mid-childhood, XLHED boys, who are typically as bright and energetic as unaffected boys, have a characteristic face. The major features are thin, sparse hair, a beaked nose, missing teeth, a small jaw, dark circles around the eyes, and dry skin. Perhaps because it is so common now for young men to shave their heads, the young adult men with XLHED in the room did not look too unusual. In addition to spending time with Mary Kaye that evening, I also talked for a while with Charley, her now adult son, who spoke frankly with me about living with the disorder.

How does one quantify the burden of living with such conditions? In the case of XLHED for many patients the major issue is the impact of living life with very few teeth. The absence of teeth affects how the face develops, often leading to an unusually small lower jaw. Affected children begin wearing dentures before they reach school age. During childhood, dental bills sometimes surpass $100,000. Kids tease other kids; having little or no hair makes one an easy target. As the children approach adolescence, some have serious problems with self-esteem. The problems that arise due to a lack of sweat and other glands pose health risks that are (if carefully monitored) manageable. He described his life with XLHED as relatively normal. A stoic, he took the optimistic view that everyone has problems, and he had just happened to draw the XLHED card from fate's deck. However, many other severely affected patients do not share such an optimistic view.

That evening I also talked at length with a woman in her late 20s who was a teacher. She was born with a different form of hypohidrotic ectodermal dysplasia that also caused her to have almost no sweat glands and very few teeth. She was so open and articulate about how the disorder had affected her life that I decided to invite her to Boston to talk to my colleagues about her experiences. When she visited a few months later, she left us with no doubt that being born with few or no teeth *does* constitute a major problem. She recalled adolescence as very difficult. For example, during high school, she was so embarrassed by having dentures that she shunned sleepovers with her girlfriends and was terrified of dating. In college she dated one young man for some time before she told him that she was born without teeth, a disclosure that did end the relationship. She sobbed as she recalled how her parents had spent their retirement savings to pay for the full set of dental implants (not covered by insurance) that she had recently undergone, an intervention that she described as life changing. Thinking about our project, I asked her how society should value a new treatment that if given as a few injections in infancy could restore the development of teeth, hair, and sweat glands. At once she replied that she would have paid "any price" to have had such a treatment.

Just a couple of weeks after the NFED Halloween bash, I attended the annual meeting of the American Society of Human Genetics in Philadelphia, mainly so I could talk with Magi Casal. I thought that her research work on the dogs with the canine version of the disease could be of great value in assessing whether or not to build a biotech company to drive development of a drug for this disorder. I met Magi at her station in a large hall divided into corridors lined with hundreds of posters about recent scientific work. Because she at first seemed reticent, I wondered if she was naturally shy or was she being cautious in responding to an alien being from the world of venture capital. Happily, as I told her of my hope to help build a new company to develop a drug to treat humans with XLHED, she became more animated. Toward the end of our talk, the conversation turned to Pascal Schneider, the Swiss biochemist who knew more than anyone else about the molecular cause of the disease. Magi had trained with him. "Pascal Schneider," she said with conviction, "is a brilliant biochemist."

In the winter of 2009, Nick, Neil Kirby, and I flew to Switzerland. One of our key goals was to meet with Pascal Schneider, the world's leading expert on the ectodysplasin A, at the University of Lausanne. Except for a brief stint as a postdoctoral fellow elsewhere, Schneider, who was then in his late 40s, has spent his scientific career at the University in Lausanne in a small beautiful city nestled along a lake that separates Switzerland from France.

On first meeting, Pascal struck me more like a romantic poet than a brilliant scientist. Tall, lean, and so soft-spoken that I had to focus intensely to understand his heavily accented English, Pascal has a beatific smile. Only later, when he and his team met with us for several hours in a small seminar room, did I witness the exuberance that I so often associate with great scientists when they discuss their research. When Pascal started talking about EDA and its potential importance to patients with XLHED, he raised his voice and his body language shifted from quiet reserve to passionate engagement.

On the second evening of our visit, our hosts took us to a traditional Swiss restaurant on a mountainside high above the city with extraordinary views of Lake Lausanne. I took a seat next to Pascal, intending to learn more about him. When I asked about his life outside the laboratory, he told me that he enjoyed collecting certain old Swiss coins. I found this in keeping with my sense of his reserved nature. He did surprise me when he next said that during summer vacations he spent his time spelunking—a dangerous and physically demanding sport!

Pascal's laboratory developed methods to make enough EDA (from cell cultures) with which to experiment on tabby mice. The key question they sought to answer was: If we give the mouse enough of the manufactured EDA at an early enough time, can we make up for its natural absence, and convince the appropriate mouse cells to differentiate into tooth buds, sweat glands, and hair follicles? This was no easy task. First, they had to make sufficient amounts of the protein, and then they had to administer it to very tiny newborn mice (weighing just a few grams). Further, they had no idea of the correct dose to give, how many doses should be given, or as to the time window in which the drug would work. They would have to find their way by trial and error.

The EDA gene makes a protein that self-organizes in groups of three. A single strand cannot produce the signal that the waiting cells need. Ectodsyplasin A is large and complex with three important functional domains, which poses substantial challenges when one attempts to manufacture it. In normal individuals, embryonic skin cells secrete a soluble form of EDA, which interacts with a receptor molecule (called EDAR) on other skin cells. That interaction activates a protein called NF-κB, which then instructs still other molecules to deliver another important activation message. They interact with cells that are genetically programmed to form teeth, sweat glands, and hair follicles. If all goes well, the signal started by the EDA protein conveys the message that the time is now right to let development proceed. If the signal is incompletely given, the structures either do not form or form improperly. In essence, XLHED is a set of birth defects that arise because cells do not receive the molecular command to do the work for which they were programmed.

Edimer Pharmaceuticals

In mid 2009, after successful negotiations with the small group in Switzerland that owned the intellectual property covering the use of EDA as a therapy for XLHED, TRV launched Edimer Pharmaceuticals Inc., which would be built to focus on moving the protein that Schneider's laboratory had worked on for so long through the complex, expensive, and highly risky process of drug development. If it succeeded, Edimer could be the first biotech company in the world to win approval for a drug to ameliorate birth defects through treatment with a ligand.

There could be several other impressive firsts as well. EDI200 (as the molecule was dubbed) might also become the first drug to treat a genetic disorder by giving *one or just a few* injections of a protein, perhaps over a span of just a few weeks. And, if it were eventually approved for human therapy, EDI200 would be the largest molecule ever marketed. Most drugs are small molecules, typically weighing a few hundred daltons (the word, coined to honor the British chemist Joseph Dalton, is about the weight of one hydrogen atom). EDI200 has a molecular weight of about 250,000 Da, a size so large that some medicinal chemists might be concerned about whether it could be delivered intravenously, and whether it could stay intact long enough to reach target receptors on the cell surface. Further, because its structure would vary slightly from the natural form of the ligand (laboratory production methods led unavoidably to variations in the small side chains on the protein), despite success with animals, one could not predict its efficacy in humans.

Along with the challenges of figuring out how to make the purified protein at a reasonable price, moving it through the standard preclinical toxicity studies in two animal species, and trying to determine what might be the proper dose to someday administer to infant boys, the small team at Edimer (which was planned to operate as a "virtual" company, meaning it would outsource much of the work that needed to be done) faced a daunting commercial challenge.

No one had yet conducted a thorough study of the incidence (live births of affected boys each year) or prevalence (number of living affected persons) of persons affected with XLHED. Existing incidence estimates ranged from a low of one in 100,000 births to a high of one in 10,000 births. If the higher projection was correct, the potential treatable population would be sufficient to justify investing in the development effort. In contrast, if the lower number was more accurate, the road to commercialization would be much more challenging. Further, no one really knew what percentage of girls who carried the mutation also developed physical abnormalities. It was known that some carrier girls were quite severely affected, but it seemed that most carriers were only mildly so, almost certainly not enough to justify administering an extremely expensive drug to avert the physical effects of the mutation. On the other hand, if the fraction of girls with significant health problems turned out to be fairly high, it might substantially increase the treatable population.

The most crucial question was whether the company could identify newborn boys with XLHED early enough so that it could enroll them in the key clinical studies within the first 2 weeks of life. About two-thirds of the time women who are at risk for bearing affected boys will have a positive family history. Young women in XLHED families have a one in two chance of being a carrier, depending on which X chromosome they inherit from a carrier mother. Many women with the mutation tend to have physical features (such as narrow noses, dry skin, and thin hair) from which one can infer that they carry the mutation.

Neil Kirby, a pragmatic, no-nonsense Englishman who had spent 20 years in the drug development business and who had worked as a consultant on the XLHED project, became CEO of newly formed Edimer Pharmaceuticals in the spring of 2009. It was a smart choice. Over the years, Neil had worked for both big pharmaceutical and small biotech companies, including on programs that addressed orphan genetic disorders, and he was well schooled in the complexity of drug development. He also knew many scientists who were expert at solving particular steps in this complex process. Kirby hired several key employees, but virtually all the preclinical work—such as developing cell lines in which to make the protein, devising the methods to purify it, scaling up the producing cell lines to yield the protein in sufficient quantities to enable toxicity studies in animals, and later, human trials—would be undertaken by companies that specialize in doing such work. Two other members crucial to the small Edimer team were Dr. Kenneth Huttner, a neonatologist who had earned an MD and a PhD in genetics, and Ramsay Johnson, who was a veteran in the area of clinical and regulatory affairs. Together, this trio would take on the daunting task of learning as much as possible about the disorder, an effort that would in time greatly inform the design of the human clinical trials.

As with all new companies dedicated to developing drugs for rare disorders, Edimer could not hope to generate revenues for many years. After it consumed the millions provided by the major Series A investor, it would need to raise more money to pay for further clinical trials. To do so, it would have to convince investors that it had made solid, timely progress *and* that the marketplace would eventually award it with reasonably good revenues.

Four years into its existence, Edimer had made excellent progress both in developing the protein for use in humans and in advancing understanding of the nuances of the disorder in children. Ken Huttner, a soft-spoken, thoughtful, and meticulous clinical researcher, had become a leading expert on the natural history of XLHED, advancing knowledge on heretofore under-appreciated aspects of the clinical course in adulthood. In a relatively short time, he and Ramsay Johnson had generated new data that led to about 10 abstracts (formal presentations at scientific meetings), 10 poster presentations, and three articles in peer-reviewed journals.

But although knowledge about the disease grew, uncertainty about the number of treatable patients remained high. Few scientists conduct incidence studies of rare disorders, top journals do not publish the results of those that do, and public health agencies understandably focus on common disorders like asthma and obesity. Yet, for the companies created to develop new treatments for rare disorders, these data are absolutely crucial. The companies usually need several infusions of capital over the years to complete the drug development marathon. To raise that money, the start-up has to convince potential investors that a market exists for the product. To satisfy investors, one must have a convincing answer to the question: If the drug is approved how many patients will receive the therapy each year?

In 2010, the NFED launched an International Ectodermal Dysplasias Patient Registry (open to anyone affected with any one of its many forms) into which it hoped many patients would enroll. In its first 4 years, enrollment in the registry grew steadily, but it did not lead to the identification of many young women who were carriers of XLHED and who, if they became pregnant, might want to enroll in Edimer's clinical trial. Realizing that resolving that question would be crucial to its future, Edimer commissioned a group of epidemiologists, population geneticists, and clinical experts to guide it in developing the most accurate possible assessment of birth incidence. Drawing on large public health databases, including the comprehensive medical registry in Denmark, the consultants eventually generated data indicating that the best guess (actual numbers can vary by region) was that one in 30,000 males is born with this disorder, a figure that constituted reassuring news for the commercialization effort.

Edimer faced the unique challenge of developing a method to ensure that it would be able to identify affected newborn boys in time to treat them. Studies in animals suggested that human infants would likely respond only if the protein was injected during the first few weeks of life. But although newborn boys with XLHED do have physical signs of XLHED at birth, few pediatricians routinely recognized the condition, and the average age at which the diagnosis is made is typically about 12–18 months, when it becomes clear that the teeth have not properly developed. Today, most young women who are obligate carriers (that is, they have an affected father) or likely (1/2 risk) carriers (because they have a mother who is a carrier) can discuss XLHED with their physicians. However, about one-third of children with XLHED are born to women with no history of the disorder, but in which a new mutation has arisen in a germ cell.

The solution to the problem that Edimer faced was novel: build a database of women known to carry a mutation in the EDA gene who are between 18 and 40 and who might at some time consider having a child. Edimer could then keep them informed of its march toward a clinical trial, and—if a carrier woman became pregnant and found that she had a male fetus—invite her to participate in the investigational study. But the means by which to accomplish this was not so clear. From its inception, Edimer had strived to develop close relationships with the NFED and comparable groups in Europe. Among other things, early on the company made an arm's length grant of significant funds to NFED so that it could computerize and upgrade a database of affected families. In addition, Edimer created on its own website a voluntary registry. It also developed a program to support DNA testing of women who were at risk for being carriers.

Such activities were crucial to the company's most important objective—conducting clinical trials in humans. The regulatory algorithm requires that the first such effort be a Phase I safety trial, one that is usually conducted with healthy adults who volunteer. Regulators at the Food and Drug Administration (FDA) and its counterparts around the world are understandably wary of permitting a new drug to be given to babies unless its safety has been assessed in adults. After several interactions with the FDA, in 2012–2013 Edimer successfully recruited six adults (four adult men affected with XLHED and two adult female carriers) and treated them in groups of two with increasingly higher levels of the newly made recombinant EDI200. The standard detailed medical and laboratory analysis that was conducted on each subject showed no evidence of serious side effects, so after the data safety monitoring committee gave its permission, Edimer was able (under the Investigational New Drug application it filed with the FDA) to initiate a Phase I/II trial in newborn boys in whom DNA testing had confirmed that they were affected with XLHED.

In the argot of drug development, the trial is officially known as a "Phase 2, Open-Label, Dose Escalation Study to Evaluate the Safety, Pharmacokinetics, and Pharmacodynamics/Efficacy of EDI200, an EDA-A1 Replacement Protein Administered to Infant Males with XLHED." Over a period of 12–18 months, Edimer planned to enroll six to 10 newborn boys in two groups, the first to receive the protein at a low dose of 3 mg/kg of weight and the second to receive 10 mg/kg (which might be a therapeutic dose). It sought to evaluate efficacy by looking at sweat gland development, salivary gland function, and the quality of lubrication on the surface of the eyes—each one known to be abnormal in affected children. Compounding the immense challenges posed by the rarity of the disorder was the fact that the study intended that the five-dose treatment regimen be initiated *between days 2 and 14 of life*, a narrow window. To help manage the rarity of the disease, Edimer opened clinical trial sites in Germany and England, as well as on both coasts of the United States.

The study would be logistically challenging and, on a per-subject basis, quite expensive. It necessitated identifying women who were proven by DNA testing to be carriers of an XLHED mutation, ascertaining (by ultrasound) that they were carrying male fetuses, obtaining their consent to participate in the study, managing the transportation of the expectant mother and her family to a trial site, obtaining a DNA test at birth to confirm whether or not the infant had inherited the mutation (this could be done prenatally via amniocentesis, but most women are reluctant to do so), and keeping the affected boy in hospital for 3 weeks (for safety monitoring purposes). If a woman did agree to travel, it was only natural that she would want her spouse and her other children to accompany her. The company would have to pay all the expenses associated with this disruption in their lives.

There were days when the team at Edimer despaired that they were connecting with so few women who were carriers of XLHED, who were planning to become pregnant, *and* who would agree to enter the clinical trial. This worry was complicated by the fact that only one in four pregnancies to a carrier woman results in the birth of an affected boy. This means that to successfully enroll just six affected infant boys, that the physicians who were conducting the trial sponsored by Edimer should be tracking about 24 pregnancies. Fortunately, as word of the successful adult Phase I safety trial spread through the patient and caregiver communities, Edimer began to hear about an increasing number of "candidate" pregnancies.

In the winter of 2013, physicians in Germany treated the first infant baby boy with the experimental protein, which binds to the EDA receptor on human cells in just the same manner as does the normal protein. Although it would take more than a year to ascertain the efficacy (if any) of this new drug, the early signs were cause for hope. The first two boys seemed to be growing better than was typical of XLHED infants, but this is a highly subjective assessment that was of no real predictive value.

Although it remained a challenge to find and enroll pregnant carriers, by late 2013 Edimer was tracking about 10 pregnancies and was learning of more at a much greater rate than it had in earlier years. A second affected boy who was born in California underwent treatment in January of 2014, and a third entered the study in March, thus completing the three-person low-dose cohort. After assessing the clinical data on the three boys, the Data Safety Monitoring Board permitted the trial to advance to the next phase—treating the next cohort of infant boys with a higher dose.

In early 2014 Edimer was following several pregnant XLHED carriers with male fetuses in nations as far-flung as Australia who had expressed interest in enrolling an affected son in the clinical trial. Working with the family in Australia shows the complexity of the study. If the woman were to give birth to an affected son and decide to enroll him in the study, Edimer would have to fly the woman and her family to Los Angeles, the location of the nearest clinical trial site. There they would have to live while their infant son received five doses of EDI200 and was observed for adverse effects—a stay of 42 days.

Based on toxicity studies in which the recombinant protein was given to dogs and monkeys without significant adverse effects, it seemed unlikely that the EDI200 protein would harm the infants who are enrolled in the Phase II study. But it is also not possible to predict whether the protein will help them. Although animal studies suggest that it might be possible to successfully treat human infants just after birth, there is a considerable risk that the therapy will be too late. In normal babies, EDA-A1 tells cells to begin to develop during the second trimester. If it could be done safely, it makes good biological sense to treat affected *fetuses* with EDI200 as soon as they are identified, rather than waiting until they are born, a delay that may greatly narrow the treatment window. It could be that if EDI200 was given prenatally (perhaps after 18 weeks of gestation) that it would have a much more beneficial approach. Indeed, successful treatment could help open up a new era in fetal medicine.

The Phase I/II trial illustrates another major problem in developing drugs for ultrarare disorders. A study that seeks measures of efficacy in just a handful of children cannot generate meaningful statistical support for its findings. If, for example, the drug shows no benefit when given at the lower dose to three children, but shows moderate efficacy (as reflected, say, in an increased sweat gland count) when given at the higher dose in four of six infants, is that a persuasive sign of efficacy? Because a small biotech company has limited resources, and its future depends on the time it takes to show efficacy, it may never be able to conduct a study large enough to "prove" (statistically) that the drug is highly efficacious in newborns. Scores of companies will eventually face this problem in their efforts to develop drugs for other rare genetic disorders. It seems inevitable that the FDA and its sister agencies, when assessing efforts to develop drugs for rare disorders, will have to amend traditional methods of judging new therapies for such conditions.

As 2014 came to a close, Edimer was nearing the completion of enrollment in its Phase IIb clinical trial, and was gearing up to conduct an even larger trial using an even higher dose. The good news was that the company had made substantial progress in identifying women from XLHED families who were pregnant. The likelihood of finding and timely enrolling up to 30 affected infant boys looked feasible. The clinical readout of the children who were treated in the Phase IIb study will probably become available in 2016.

Assuming the Phase IIb trial provides sufficient evidence of efficacy to justify a "pivotal" Phase III trial, it will require millions of dollars to bring EDI200 across the ultimate goal line of FDA approval. Ease of access to high-risk capital is a will-o'-the-wisp. But, even if the capital window is open, investors will invest only if they believe they will get an acceptable return in a reasonable time. In that regard, despite its impressive success in producing EDI200 and in advancing clinical studies, Edimer must continue to gather evidence that the number of potentially treatable patients is enough to bring the saga to a happy ending.

Few would have imagined a decade ago that one might be able to avert a complex set of birth defects with the timely delivery of a recombinant protein with just five intravenous injections. Of course, the current clinical trial may fail to show sufficient signs of efficacy. But despite the tough hurdles it faces, Edimer's attempt to ameliorate XLHED may someday be remembered as the first successful effort to avert physical abnormalities with molecular, as opposed to surgical, therapies. This is the kind of result that parents, physicians, scientists, and biotech entrepreneurs dream about. If the strategy to treat neonates is moderately successful, it may be plausible to treat affected males in utero. If that strategy shows even greater therapeutic benefit, research that began decades ago in a mouse, advanced through research of a brilliant Swiss biochemist, and benefited from the observations by a clinical veterinarian about a dog might open the door to a new chapter in fetal medicine.

11

Mending Broken Proteins

THE HUNDREDS OF DISTINCT SINGLE-GENE DISORDERS manifest in myriad ways and require many different therapeutic approaches, ranging from dietary manipulation as in phenylketonuria (PKU), purifying and supplying a missing blood factor as in hemophilia, and creating complex biofactories to produce large enzymes in a form that can be delivered to the cells of patients with lysosomal storage diseases, as Genzyme has done so well. But there are many single-gene disorders that currently cannot be attacked by any of the approaches that I have discussed thus far. Fortunately, unlike the case in many cancers and in most forms of heart disease, single-gene disorders usually permit drug developers to identify the targets at which they should aim. The pathological consequences associated with severe single-gene disorders usually arise because of the absence or, more often, malfunction of a single protein. To develop breakthrough therapies researchers must become ever more creative in developing innovative ways to overcome the consequences of having the damaged protein.

It is not easy to convey the immensity of the challenge of overcoming damaged proteins. To succeed, one must develop a novel, safe chemical compound that can be delivered to billions of cells often in many different tissues at a time when the disease is not so advanced as to be untreatable. Many efforts to develop new drugs fail early because in preclinical animal studies the selected chemical compound turns out to be dangerously toxic. Those (relatively few) compounds that meet the safety requirements often fail because they cannot be delivered efficaciously to the proper cells. This is especially true in regard to the brain because of the "blood–brain barrier," a special cordon of cells that make it very difficult for comparatively large molecules to cross from the circulation into the brain. For example, no enzyme replacement product has yet been able to treat the central nervous system (CNS) aspects of the various lysosomal storage disorders. The molecules are too big to cross over.

Over the last decade or so, molecular biologists, medicinal chemists, and pharmacologists have taken a new approach: rather than try to provide the normal version of the needed protein via protein replacement therapy, they have tried to use small molecules to coax some function out of the patient's mutated protein. In this chapter, I will briefly discuss three promising efforts to overcome mutations that lead to dysfunctional proteins: chaperones, exon skipping, and the use of small molecules to alter the shape of a large molecule so as to restore some of its function (these drugs are sometimes called "correctors" or "potentiators"), each in the context of a particular orphan disease.

Lysosomal Storage Disorders: Chaperone Molecules

During the late 19th century the world's top medical schools were in the great European capitals, especially Berlin and Paris. Unlike medical education today, in that era to earn their medical degrees students were required to write a thesis. In 1882, Philippe Charles Ernest Gaucher (1854–1918), who had been born and raised in Nièvre, France, satisfied this requirement by writing up his research on a 32-year-old woman who had come to clinic in Paris with a mass in her abdomen that turned out to be an enlarged spleen. In those days a massively enlarged spleen was almost always seen in the setting of advanced leukemia, but this woman did not have a blood cancer. Gaucher's thesis constitutes perhaps the earliest written record of an effort to understand this disorder (which he mistakenly considered to be a nonmalignant tumor). He was probably the first to note that under the microscope the *cells* taken from the woman's big spleen after surgery were much enlarged. He did not, however, infer that she had a storage disorder. Although Philippe Gaucher enjoyed a highly successful medical career, including a professorship of dermatology at the still famous Necker Hospital in Paris, after medical school he did not publish studies of any other patients with the disease that now bears his name. Thirteen years elapsed before the second written report of a patient with a closely similar clinical picture, in that case a 6-year-old boy who also had a massive spleen and anemia.

By the early years of the 20th century, physicians had published a sufficient number of case reports about this odd disorder that it stimulated a few academic experts, including the great William Osler at The Johns Hopkins Hospital in Baltimore, to try to make sense of "splenic anemia" (all the patients had low red cell counts as well as a big spleen) as it was then called. In 1904 the New York Pathological Society published the summary of a conference on the topic, mainly centered on findings by Dr. N.E. Brill who in 1901 suggested that in these patients the nonmalignant splenomegaly arose because of a hereditary disorder, and who showed (from autopsy studies) that the disease involved grossly enlarged cells in the liver, lymph nodes, and bone marrow. His work showed irrefutably that this rare disorder was not an atypical cancer.

Many began calling it "splenomegalie primitive," the second word implying that it was of early onset, perhaps congenital. Because studies of several large families found that more than one child was affected, the experts agreed that it could well be a hereditary disorder, but the concept of recessive inheritance was not yet well understood, so the question remained speculative. They agreed that among all cases of "splenic anemia" there was a "Gaucher type" that constituted a distinct subset. During the first half of the 20th century occasional case reports clarified that there were three distinct clinical forms of Gaucher disease. In 1920 a paper described a severe early-onset (infantile) form of the disease that is today called type 2, and in 1955 a group of physicians described an exceedingly rare form that included serious neurological aspects (today known as type 3 or the "neuronopathic" form of the disorder).

As I discussed earlier, our deep understanding of Gaucher disease (as well as other lysosomal storage disorders), can be traced directly to the great work of Roscoe Brady. Beginning in 1955, Brady was first among many in deconstructing the disorder and in developing an effective therapy for it. In 1964 he and others elucidated the enzyme (glucocerebrosidase) deficiencies that cause the disease, and crystallized his idea that if one could deliver a purified form of the normal enzyme to patients that it could reduce the amount of harmful stored material in cells and ameliorate the disease. Over a 6-year period extending into the early 1970s, he and others developed a painstaking method to purify the normal form of the enzyme from human placentas, an exercise that required industrial-scale methods. In 1973, under an approved experimental protocol, he treated the first patient—a young boy—with enzyme replacement. Success as measured by reduction in storage material was modest; most of the precious enzyme was not entering the cells that needed it.

It was Brady's colleague, John Barranger, a physician–scientist then also at the National Institutes of Health (NIH) (later moving to the University of Pittsburgh School of Medicine) who correctly reasoned that if one could remove some of the sugar side chains from the purified enzyme to expose a molecule called mannose, that receptors on the cell surface would recognize and grab it and pull the protein into cells. The development of a method to move the exogenous enzyme into cells helped to convince Henri Termeer, the CEO of the newly formed Genzyme Corporation, to commit many millions of dollars to the then arduous process of purifying the enzyme from thousands of placentas. The development of enzyme replacement therapy (ERT) for Gaucher disease remains one of the great successes in genetic medicine. Around the world several thousand patients with type 1 Gaucher disease lead far better lives because of enzyme replacement. But the therapy is extraordinarily expensive, often costing $300,000 a year (the cost is a function of the amount of the enzyme that each patient needs, which is in turn a function of particular mutation, as some mutations cause more severe deficiencies than do others, and of the patient's weight). Given its extraordinary expense (the drug must be taken throughout life) and the fact that <1% of the dose actually enters the cells, researchers have long sought to develop a less expensive and more efficacious treatment.

Proteins are three-dimensional structures the efficient function of which can decline dramatically if even a slight conformational change (due to a gene mutation) alters the binding site (the key spot where the basic action that the enzyme has evolved to perform occurs). If the enzyme is "*misfolded*" (which is the cause of many diseases), it will often fail to emerge from a part of the cell called the endoplasmic reticulum (ER) that acts as a protein quality control site. This puts the body in great danger, as the material that the defective protein was meant to metabolize now accumulates in toxic quantities.

In many genetic disorders, the misshapen protein is caused by a very small genetic error. In a gene that codes for a protein that is composed of hundreds of amino acids, a mutation that alters a single base pair of DNA might be enough to seriously disable the protein. But the focal nature of the deficit also provides a target for researchers. It has encouraged efforts to develop novel small molecules that can bind to a key spot in a defective enzyme in a way that stabilizes it or partially corrects the misfolding, thus allowing it to escape from the ER so that it could do at least some portion of the work it was programmed to do. Such small molecules are called *chaperones*, a word meant to convey that they stay close to the enzyme, helping it to avoid being recognized as defective and being degraded by quality control systems as it travels through the several cellular compartments. A key aspect to being a good chaperone molecule is that it should bind *reversibly*, that is, in a way that once outside the cell, it falls away so that the enzyme can do its job by acting on the proper molecules.

The idea of designing small molecules to help misfolded proteins escape from the ER and exit the cell so that they can perform at least part of the function they evolved to do is not new. It arose in part from the recognition over the last 30 years that many healthy human proteins naturally depend on chaperones to help them acquire the correct three-dimensional structure that allows them to exit the cell and travel to where they are needed. In addition, during that era scientists discovered that many common human disorders arise at least in part because of the inability of naturally occurring chaperones to do their assigned work. The list of diseases involving misfolded proteins, topped by Alzheimer's disease, is so long that it may include one-half of human disability. Chaperonopathies (as they are sometimes called) are associated with the accumulation of unwanted material in the cell, as is the case in lysosomal storage disorders such as Gaucher disease.

More than a decade ago, scientists began to study several lysosomal storage diseases—in particular, Fabry disease, Pompe disease, and Gaucher disease—to see if they could develop chaperones that would either (1) increase the efficacy of ERT by getting a greater percentage of the enzyme into the cells or (2) offset the defects of naturally produced but misfolded enzymes by helping them exit the cell and do at least some of their work, albeit with reduced efficiency.

About 2007, a group at the Hospital for Sick Children in Toronto made the insightful decision to screen more than 1000 Food and Drug Administration (FDA) approved drugs to determine if any of them improved the function of defective forms of glucocerebrosidase (the defective enzyme in Gaucher disease) in cell cultures obtained from patients. They found one, a small molecule marketed under the name Ambroxol (it is used to break up mucus and is often found in cough syrups) that enhanced enzyme function in many patient-derived cell lines. Interestingly, some patients (having heard about the scientific research with cell lines) were already using it. Because Ambroxol is available as an over-the-counter purchase, no one knew how many were doing so. A few experts have reported anecdotally that it provides modest clinical benefits to some patients, but no formal clinical trials have yet been conducted. Studies of patients' cells offer no hint concerning the proper dose of a drug to give to people so the Ambroxol tale remains unresolved.

Although still under development, chaperone therapy to treat patients with Fabry disease looks hopeful. Originally called Anderson–Fabry disease, this was the second lysosomal storage disease that doctors recognized. In 1897 two young physicians, Johannes Fabry and William Anderson, independently described patients with similar constellations of signs and symptoms. At the time, Fabry, who would become one of the leading dermatologists in Germany until his death in 1930, was working at a hospital in Berlin. Dr. Anderson, a British dermatologist, trained at St. Thomas' Hospital in London. Not surprisingly, in their initial case reports, both men focused on the skin findings now known to be characteristic of the disease. The first patient who Fabry described was a 13-year-old boy who started developing unusual skin lesions at the age of 9. These so-called *angiokeratomas* are small purple-red flat or raised lesions that we now know arise because of the accumulation of a waxy chemical in cells named globotriaosylceramide (or GL-3). This is the most obvious, and among the earliest, signs of Fabry disease.

Fabry disease, which arises because of mutations in the gene that codes for an enzyme called α-galactosidase A, is unlike most storage disorders in that it is an X-linked disorder, so it primarily affects men. The disorder is comparatively slow to develop, and it is often not diagnosed until adolescence or early adulthood. Early signs that may be missed by pediatricians include pain in the hands and feet, fatigue, and exercise intolerance. The disease progresses relentlessly. Over the years, the GL-3 accumulates in the heart, kidney, and capillary cells and can cause heart failure, renal failure, and stroke. Before the advent of kidney dialysis, patients often died in their early 40s. Even with advances in methods to treat kidney and heart failure, as recently as 2005 life expectancy among affected men was only about 50 years. Since the approval of Genzyme's enzyme replacement drug (Fabrazyme) in 2003, median life expectancy has increased; it is now estimated to be nearly 60, a big improvement, but still more than 15 years short of that for the average American.

Several biotech companies, notably Amicus Therapeutics, have focused on the development of chaperones to treat rare disorders. Amicus took one of its compounds, isofagomine tartrate, through Phase II clinical trials, but it failed to meet the hoped for clinical end points. Amicus is conducting several large clinical trials attempting to assess the efficacy of two other chaperone molecules to treat Fabry disease and Pompe disease. Of particular interest, in 2013 scientific work by Amicus on a small molecule that can be taken orally, called migalastat, showed promising results. Migalastat works by binding to the α-Gal A enzyme, thus stabilizing it and helping it to be secreted by cells. When given to a genetically engineered Fabry mouse, the drug sharply reduced the levels of GL-3 in kidney, heart, and skin. When given to six adult men with Fabry disease, migalastat moderately lowered levels of GL-3 in three, but not in three others. Similar effects were found in adult women carriers. This is not surprising; given the many mutations that can be found in the gene, any chaperone is likely to be of more help in some patients than in others. Amicus has developed an assay to predict which patients are more likely to benefit from taking its chaperone. Further clinical results could arrive by late 2015. If regulatory agencies ultimately approve migalastat for human use, the pharmaceutical giant, GlaxoSmithKline, which owns 19.9% of the company (certain tax rules make it prefer to stay under 20%), will have control of its manufacture and sales. Relationships like this are relatively common in efforts to develop orphan drugs.

Because rare monogenic disorders may arise owing to mutations at many different spots in a gene, a major challenge in developing chaperones is to find ones that ameliorate misfolding caused by several different mutations. In most cases, it is not economically possible to develop mutation-specific drugs. Some orphan diseases constitute exceptions to this problem, either because one particular mutation accounts for a large percentage of patients (as in cystic fibrosis) or many mutations have similar effects (which is the case in many disorders).

Duchenne Muscular Dystrophy: Exon Skipping

Although several other physicians recognized and wrote about the disease before he did, it was a comprehensive clinical account written by Guillaume Duchenne de Boulogne, an unconventional French physician, near the end of his life in 1868, that tied his name forever to this relatively common form of inherited muscle disease. Duchenne's distinguished career in medical research might never have occurred had it not been for a bizarre personal event. After completing his medical training in Paris, he returned to his home city to open a practice. But the death of his wife in childbirth 2 years later, an event at which he was the sole attending caregiver, stimulated hateful rumors about him that drove him back to Paris.

At that time leading academic physicians were becoming fascinated with the role of electrical currents in cell function. Duchenne was intrigued, and he soon began using electric currents to study muscle action. Among the earliest of the many such studies he pursued was investigating electrical currents in facial muscles to understand the physiology of facial expression. One of his most interesting discoveries was that a truly spontaneous smile involved contraction of muscles about the eye as well as the mouth (to this day known in esoteric medical circles as a Duchenne smile)! Over the years Duchenne, harnessing the new technology of photography with that of electricity, generated hundreds of photos of facial expressions, some that one can occasionally find for sale on the Internet. Duchenne and others used electrophysiology to show that central and peripheral nerves functioned normally in muscular dystrophies, thus proving it was a disease of intrinsic muscle fiber degeneration. Today, experts in neuromuscular disease have described about two dozen inherited muscular dystrophies.

Debates over who really first recognized a disease for the first time are common among historians of medicine. In regard to Duchenne muscular dystrophy (DMD), a strong case may be made that decades before Duchenne's report, a British physician named Edward Meryon described the disorder and correctly deduced that it was a muscle disease, not a neuromuscular disorder. His conclusions about the disease pathology were more accurate than those of Duchenne, who thought there were abnormalities in the spinal cord, and who did not grasp that deficits in a part of the muscle called the sarcolemma were a critical feature of the disease. Nevertheless, Meryon's name will never be more than a footnote to that of Duchenne.

Because the disease is caused by a defect in a gene on the X chromosome, it affects only boys (the women who carry a mutation on one X chromosome are protected by the normal copy on the other X). With each pregnancy, carrier mothers have a one in four chance of having an affected son (this risk is calculated by multiplying the odds of having a son which is one in two times the odds of transmitting the mutated X in the egg cell, which is one in two). Throughout the world, about one in 4000 boys is born with DMD, making it a quite common orphan disease.

Affected boys appear normal for the first two years of life, but around then their parents begin to get concerned that their sons are not developing motor skills as well as they had when they were younger. By age three, most affected boys have a markedly abnormal gait about which much has been written. One of the best known clinical observations about young boys with DMD was made by a British physician named William Richard Gowers who pointed out that to rise from a lying or sitting position, the boys put their hands on their lower thighs and push themselves up to a standing position (still called "Gowers' sign").

The course of DMD is one of slow, steady decline in motor function, leading inexorably over 5–10 years to becoming wheelchair bound. The disease has other manifestations that are difficult to correlate with the muscle disease. About 10% of the boys are intellectually disabled, a rate that is several-fold higher than the background risk. During adolescence many patients develop a dilated heart. As the years pass, the larger limb muscles continue to degenerate (facial muscles are spared), and the patients become ever more dependent on others. Today, patients often die in their late 20s from respiratory failure and pneumonia or heart failure.

Lou Kunkel, a leading molecular biologist at Children's Hospital in Boston, cloned the "*dystrophin*" gene in 1987, generating much hope that its discovery and the characterization of the protein would open up important new avenues for therapy. In the ensuing 30 years, we have learned a great deal about the disorder, including the structure and function of the protein, but therapeutic gains have been small. Despite the expenditure of hundreds of millions of research dollars and the efforts of thousands of scientists, no medicine has yet been developed that much alters the course of the disease. Perhaps the most important *clinical* advance took place about 1990 when it was shown that the use of ventilators to assist breathing *at night* extended life expectancy. In 2002 a large study in Newcastle in the United Kingdom found that this relatively simple approach had increased the chance of living to age 25 from ~10% to ~50%.

An exciting new era in DMD research began about 1997 when an Australian research group noticed that, in a mouse called *mdx* that is genetically engineered to lack the full dystrophin protein, some functional dystrophin protein was unexpectedly present in some of the muscle fibers. This surprise finding was best explained by occasional *reverse* mutations that permitted partial transcription of the messenger RNA. In turn, the RNA was able to direct the assembly of a short, but partially functional, protein. The researchers and their colleagues around the world quickly grasped that if one could develop a drug to cause such reversions, that it might be possible to produce enough partially functional protein to significantly improve the fate of boys with DMD—in effect by converting it to a milder form of the disease. The main support for this idea came from studies of patients who had long been thought to have a different disease called Becker muscular dystrophy (BMD), but it turned out that this disorder is caused by *milder* mutations in the *same* dystrophin gene. In boys with BMD, the disease is milder because the mutations are less devastating to the process of producing proteins, allowing cells to make forms that were partially functional.

To better understand this, consider the structure of the dystrophin gene. Dystrophin is one of the largest of our 20,000 genes, requiring about 2.5

million bases of information and having its coding region distributed across some 65 functional units called *exons* that are interspersed among 67 non-coding regions called *introns*. In the process of transcription (the creation of the messenger RNA, which journeys from the cell nucleus to the cytoplasm), the enzymatic machinery cuts out the introns. About 15% of boys with DMD have large deletions in the gene that result in no functional protein, but the other boys have mutations that essentially cause the transcriptional enzymes to read the code out of frame. This suggests that if one could skip over the exon in which the mutation occurs in a manner that would make the cell's transcriptional tools return to reading the message "in frame" that it might be able to make a partially functioning protein.

As with many biotech companies, Sarepta Therapeutics, a Cambridge, Massachusetts based biotech company that is leading the effort to develop exon skipping as a therapy for patients with DMD, grew out of work by academic scientists who were trying to develop methods to control the cellular machinery. About 1985, Dr. James Summerton developed small RNA-like molecules (called morpholinos) that can cover a specific, short length of the DNA molecule to prevent its transcription. Over the years scientists working with him and others have developed many different morpholinos (basically modifications of the four arms of the structure), many of which are now being studied in animal systems. Because in DMD many mutations involve the disruption of only one or two exons within this huge gene, restoration of reading frame could lead to the production of functional (albeit truncated) protein. Unfortunately, DMD arises because of scores of different mutations, and different drugs must be developed to treat various clusters of them. Still, if clinical trials showed that a particular morpholino successfully ameliorated DMD in children with a particular type of mutation, then it should be possible to develop other morpholinos to correct frame reading problems caused by other mutations by permitting production of partially functional proteins.

For the last few years Sarepta focused on developing a morpholino that could help boys with deletions that account for ~13% of all patients. Analysis of the DNA letters in the gene showed that a drug that forced the cell's enzymes to block exon 51 could restore the normal reading frame. Working closely with clinical researchers at Nationwide Children's Hospital in Columbus, Ohio, scientists at Sarepta who had developed such a morpholino that they called eteplirsen, sponsored several clinical trials in an attempt to show that intravenous injection of the compound was both safe and effective. Early small studies suggested that eteplirsen was safe.

In 2012, Sarepta conducted a 24-week randomized, double-blind, placebo-controlled trial Phase IIb (dose ranging) study. The clinical team enrolled 12 boys with DMD who were between the ages of 7 and 10 who had mutations that could be treated by a drug that forced the skipping of exon 51. Divided into three groups of four, the boys either received no drug (placebo) or one of two different doses of drug by intravenous injection once a week. After 24 weeks, all 12 boys were entered into a preplanned extension study in which each received one of two drug doses. The researchers hoped that by studying tissue taken by muscle biopsy that they would be able to show that the boys who received the drug would, after 6 months of therapy, have more *dystrophin-positive fibers* in their muscle cells than they did at the start of the trial. In addition, they investigated whether the treated boys did better on the widely used 6-minute walk test than did the boys in the placebo group. At the conclusion of the extension study (48 weeks after the initial trial began), the data analysis showed that both trials met their primary end points. Both the eight boys in the first study who were treated and the four boys who had started with placebo and then switched over at 24 weeks to drug showed a significant increase in dystrophin fibers.

In June of 2013 Sarepta reported clinical results through 84 weeks of treatment. The eight patients who had been treated in both the studies and who could perform the 6-minute walk test were able on average to walk about 46 meters further than could the four boys in the initial placebo-controlled group. In both groups the decline in walking ability was less than would be typical of untreated patients. The report drew international attention, and Sarepta's stock price soared.

Although the studies involved only a handful of patients, the encouraging results led the company to conclude that the FDA might consider an increase in the quantity of dystrophin protein in muscle after treatment as "an acceptable surrogate end point," a decision that would open the gate to a more rapid drug approval process. However, the FDA, considering a novel therapy, responded cautiously, refusing to commit itself to so doing, but suggesting it would consider the matter. On that news, the price of Sarepta's high-flying stock fell sharply.

The Sarepta trials, although understandably inducing much excitement, had yet to show convincingly that eteplirsen is an efficacious drug. Why? The trial only involved 12 subjects, and two of the four patients treated with the lower drug dose in the first study had rapid disease progression and could not participate in the extension study, it is difficult to accurately measure levels of dystrophin in muscle tissue, and it is hard to know what small differences in the 6-minute walk test might actually indicate for the disease trajectory. Still, the exon skipping technology that Sarepta is developing could become the first approach to demonstrably slow disease progression.

Just weeks after Sarepta announced that its exon-skipping technology had generated positive findings in a small trial of patients with DMD, a rival, a Dutch company named Prosensa, disclosed that its large Phase III trial of another exon-skipping drug called Drisapersen (a synthetic oliognucelotide) in patients with DMD had *failed* to meet its primary end point—the amount of distance a subject could walk in 6 minutes. The double-blind, placebo-controlled study of 186 boys (two-thirds treated and one-third untreated) followed the boys for 48 weeks. When the data were analyzed, there was *no difference* between the two groups across several measures of mobility. The news was a crushing blow to the 13% of DMD families in which affected children have a mutation possibly treatable by drisapersen. With the announcement made jointly with its partner, the pharmaceutical giant, GlaxoSmithKline, Prosensa's stock plummeted, losing ~70% of its value on the opening bell.

But the team at Prosensa continued to negotiate with regulatory agencies about end points, and to conduct clinical trials with drisapersen. In 2013, the FDA awarded an accelerated approval pathway to drisapersen. Over time, the company gathered data on results in more and more patients. Although no one study was overwhelmingly positive, in aggregate the data supported claims of efficacy. Encouraged by several new findings from its clinical trials, BioMarin, a much larger biotech company, began the complex dance of determining whether or not to acquire the smaller company. On November 24, 2014, Prosensa agreed to be acquired by BioMarin for $680 million up front and the possibility of two additional payments of $80 million each if Prosensa meets certain timelines concerning its clinical trials. The offer was nearly double the stock's trading price. BioMarin has made a big bet, but it is not out of line with the awards the market has given for other new drugs for orphan disorders.

The exon-skipping story nicely illustrates the roller-coaster ride of drug discovery. Even though both regulatory authorities and a big pharmaceutical partner reacted skeptically to positive data generated on a handful of patients, overall the approach taken by Sarepta and Prosensa is scientifically elegant. Because BMD is so much less severe than is DMD, a drug that modestly improved muscle function for an extended period of years would constitute a major advance in therapy. Perhaps, exon-skipping therapy will eventually do just that. In the meantime, parents of affected children are lobbying the FDA for approval of exon-skipping drugs.

Cystic Fibrosis: Modifying the Chloride Channel

Just before I started Yale Medical School more than three decades ago, I spent a year at Yale Law School on a fellowship to study public policy issues at the interface of law and medicine. There I became friends with a young woman who was working as the secretary for the program. One day, as we talked over lunch in a cafeteria, she asked me about a paper I was writing on genetic testing (this was long before the era of personal computers and one of her chores was to type the manuscripts in final form). When she persisted with her questions, I asked why she was so interested. Her response floored me. During her childhood three of her siblings had died of cystic fibrosis.

Cystic fibrosis (CF), a disease that severely harms the lungs and the pancreas, is one of the most common orphan disorders, affecting about one in 2500 Caucasian children. Although it is an orphan disease, CF is quite common among persons of northern European descent. In a few locales as many as one in 15 people are carriers, an allele frequency that predicts that one in 900 children will be born with the disease. It is less common among African-Americans. In the United States today there are more than 30,000 people living with CF, a prevalence that reflects the progress that has been made in treating this disorder over the last 50 years.

The struggle to understand CF has ancient roots and the quest to cure it has been exceedingly difficult. The impressive gains over the last quarter century in life expectancy for patients with CF have largely been due to aggressive efforts to prevent infection, the development of powerful antibiotics for treating recurring lung infections, better nutritional support, and (for patients with very advanced disease) lung transplants, but none of these interventions address the basic cause of the disease. In 2012 the FDA approved the first drug developed to improve the function of the broken protein that is the underlying cause of CF. Approval capped far more than a decade of intense scientific and clinical research. The clinical benefits provided by the new drug, Kalydeco, to a subset of CF patients are the result of a strikingly novel "work-around" of a defective protein.

Physicians did not give a name to the constellation of findings associated with CF until 1938 when Dr. Dorothy H. Andersen, a pathologist at Columbia University in New York, called it "cystic fibrosis of the pancreas." It is too bad that the name was shortened to CF because the acronym misleads. For roughly 350 years, doctors thought of CF almost exclusively as a disease that ravaged the *digestive tract,* not as a lung disease. Before about 1950 most children with CF died in childhood of severe malnutrition (due to pancreatic disease), and those physicians who recognized that the kids also had respiratory problems thought that they were secondary to the severe nutritional problems. The term "cystic fibrosis" refers to the blebs and scarring seen in the pancreatic ducts at autopsies of children. The reason that people now think of CF as a fatal respiratory disease is that in the developed world children with CF no longer die in early childhood from malnutrition. As CF has become a chronic disease now often taking four decades to kill its victims, the respiratory problems are the focus of research and treatment.

The earliest discussion in the medical literature of a person with CF may be the report of an autopsy performed in Leiden in 1595 on an 11-year-old girl. The meticulous physician noted that the child was severely malnourished and had a severely scarred pancreas, and damage in the lungs. How painful it must have been for the grieving parents to allow an autopsy on their little daughter be performed in their home! By 1606 physicians were teaching medical students that a newborn child with unusually salty sweat was not likely to live to the age of 5. There are some folktales of ancient origin that make the same observation. Although no one back then connected salty sweat with pancreatic dysfunction and malnutrition, the shrewd observation about sweat in essence recognized that there was a fundamental metabolic problem, one that would not be decoded until the 1950s!

Throughout the 19th and early 20th centuries, CF was considered to be a fatal gastrointestinal disease of early childhood. The cardinal features were foul, fatty stools and severe malnourishment (without the help of pancreatic enzymes, he or she could not digest most food) that set the child up for death from infection. But childhood malnutrition was common and there were many reasons why it might arise and persist. Identification of the specific disorders that lay hidden within the more general diagnostic category began in the 1930s. The first great students of CF of the pancreas were two women physicians who worked on different continents—Margaret Harper and Dorothy H. Andersen.

Margaret Harper (1879–1964) spent her career as a pediatrician at Royal Alexandra Hospital in Sydney, Australia. Unlike other women who were among the first to break the gender barrier in medical schools, Harper seems not to have been unduly constrained by male prejudice. By 1914 at the young age of 35 she earned a title equivalent today to that of full professor of pediatrics. Harper was intensely interested in gastrointestinal illnesses in children at a time when physicians tended to lump them into the broad diagnosis of celiac syndrome. Among other insights, Harper believed that by studying the stools of affected children she could gain insights into the underlying pathology.

In 1930 in the *Medical Journal of Australia,* Dr. Harper published a paper entitled, "Two cases of congenital pancreatic steatorrhea due to pancreatic defect," in effect placing the two babies in a new diagnostic niche. Note her use of the term congenital; Harper had an inkling that the cases were in some sense hereditary. In 1938 she published another paper discussing eight more patients with the same disorder and the same anatomic abnormalities of the pancreas. She also noted that eight of the 10 had died with pneumonia. A decade later she published findings on 42 patients, this time more strongly asserting that a congenital (present at birth) disorder of the pancreas caused this form of malabsorption syndrome. During these years Harper devoted much effort to developing the best possible diet for these infants (settling finally on bananas and cottage cheese as key constituents), work that introduced the key role that nutrition would play in extending the lives of children with CF over the subsequent decades.

The other great early student of CF, Dorothy H. Andersen, was born to Danish parents in 1901 in Asheville, North Carolina. She graduated from Mount Holyoke College and then attended the Johns Hopkins University School of Medicine, graduating in 1926. She did her internship in surgery at the University of Rochester, but her career was redirected by gender discrimination. After being denied a surgical residency at Rochester, she began to do research in pathology. In 1930 she moved to what was then called Babies Hospital at the Royal College of Physicians and Surgeons of Columbia Medical School in New York where she obtained a doctoral degree and did immensely important work in understanding congenital malformations of the heart. Andersen began her lifelong interest in CF when she performed an autopsy on a child thought to have died of celiac disease and realized that the anatomical abnormalities of the pancreas did not fit that diagnosis. This led her to search hundreds of autopsy reports and to examine hundreds of children who came to clinic with malabsorption syndromes.

At a scientific meeting on May 5, 1938, Andersen presented her exhaustive studies of 49 children who had died with malnourishment and who all had similar abnormalities in pancreatic tissue. It was on this day that the age-old disease finally got a new name—"cystic fibrosis of the pancreas." But despite the major advance that she had made in understanding the pancreatic disorder, no one yet realized that the disease was a single-gene disorder. Indeed, in 1943, Dr. Sidney Farber, who would later become a legend for his groundbreaking work in childhood cancer, offered a competing name for the disease, "mucoviscidosis," an abstruse term that calls attention to the fact that all known elements of the disease in question can be attributed to thickened mucus. In a sense this term was a better descriptor, for it emphasized that many organs were affected. Mucoviscidosis was commonly used for more than 20 years in the United States, and it is still often used in Europe.

The exigencies of World War II redirected most scientists away from pediatrics, but shortly after its end, research in CF grew rapidly. In 1946, Dr. Andersen published a detailed study of families with CF children, and provided the first convincing evidence that it was a single-gene disorder. She, however, remained convinced that an environmental factor (perhaps vitamin A deficiency) drove the disease. In a sense she was correct. Today children with CF die of respiratory failure secondary to repeated bacterial infections, which are environmental factors. Given her deeply held suspicion that still unknown environmental agents could cause severe lung disease, it is ironic that Andersen was a heavy smoker who died of lung cancer.

The next big advance came in 1953 when Dr. Paul di Sant'Agnese, the scion of a rich Italian family who was working in New York with Dr. Andersen, began to wonder if the unusually salty sweat could point him to the underlying cause of the disease. He and others were soon able to show that the salt concentration in the sweat of affected children was fivefold higher than in unaffected children. This enabled them to develop a relatively simple, but highly discriminatory, diagnostic test to determine if infants with pancreatic problems had CF or some other form of malnutrition. The observations of centuries could now be quantified by chemistry. The sweat test quickly became the definitive diagnostic test for CF. But sweat testing is relatively expensive and time-consuming to perform, and the results are not always accurate in the first days of life, so it could not be used to screen all babies at birth for CF.

During the late 1970s, scientists, who had discovered that it was elevated in the blood of infants with CF, developed an accurate, low-cost test to quantify the concentration of a pancreatic protein called trypsinogen in the blood samples that were collected on all babies shortly after birth as part of newborn genetic screening programs. By studying the distribution of this protein among all children, they were able to show that those with levels above the 99th percentile were highly likely to have CF. CF advocates proposed that the immunoreactive trypsinogen test (IRT) should be used to screen all babies and that those with IRT results that were above the 99th percentile should then undergo sweat testing, the diagnostic "gold standard."

In 1982 Colorado became the first state to incorporate the IRT assay into its newborn screening program. Over the next 20 years about 10 other states followed Colorado's lead, but it was only after 2004 when a group of experts convened by the Center for Disease Control and Prevention (CDC) recommended that all states add CF screening to their newborn programs that it became universal in the United States.

Procedures to confirm that an elevated IRT really is due to CF vary among the states. Until about 2005, a sweat test conducted about 2 weeks after birth was considered diagnostic, but now most state programs follow up with DNA testing. One problem with DNA testing is that more than 1000 different CF mutations—some exceedingly rare—have been found, and it is too expensive (for now) to test for all of them. A panel that looks for the 30 or so most common mutations will occasionally, albeit rarely, fail to diagnose an affected child. From a clinical point of view sweat testing is still more sensitive.

In 1949 the median life expectancy of a child born with CF was 13 months; by 1954 it had increased to 4 years. This was a horrible fact to share with parents of newly diagnosed children, but it still constituted a remarkable leap forward. The reason for the big improvement was the advent of antibiotics. During the early 40s a handful of doctors were able to cage a few doses of precious penicillin from the United States military, which then controlled the national supply. When the doctors used penicillin to treat lung infections in young children who had the symptoms of CF, they marveled at the rapid response. After the end of World War II, as antibiotics became more easily available, doctors for the first time could offer parents a glimmer of hope.

Another hero in the struggle to treat children with CF is LeRoy Matthews, a pediatrician who worked in Cleveland. In 1954 he became one of the first physicians to develop a clinic that specialized in the *aggressive* treatment of children with this disease. He surmised that by going to extraordinary lengths to improve nutrition, to reduce airway obstruction due to excess mucus (CF kids have very sticky mucus that is hard to cough up), and to fight every infection, that CF life expectancy could be greatly improved. The following year saw the birth of the Cystic Fibrosis Foundation (CFF), a group that soon became international. This is without question one of the most important developments in the history of treating any genetic disorder. Over the years the foundation became a powerful and effective advocate, raising and allocating scores of millions of dollars into research to improve the care of affected children. Its core activity was to develop and support CF specialty clinics. The results were gratifying.

From 1954 through 1980, largely because of improvements in the quality of care generated by these clinics and the tireless efforts of parents who everyday spent much time in performing physical therapy to maximize lung function, median life expectancy for children born with CF increased steadily. Earlier diagnosis, vastly improved nutritional care, vigorous efforts to reduce the risk of getting pneumonia and an ever-growing array of antibiotics were the key contributors. The United Kingdom, in which about one in 22 whites is born with one copy of a mutated gene, has long compiled nationwide survival data on persons with CF. In 1960, 40% of children born with CF died before their first birthday, in 1970 only 16% died in infancy, in 1980 the figure fell to 4%, and in 1990 it was only 1%. This fine progress heralded a major change: CF ceased to be an acute illness of infancy; it became a chronic illness of young adults. In the 30 years from 1960 to 1990 in the United Kingdom, median life expectancy rose from less than 10 years to an estimate of 40 years.

Parents of children with CF became versed in how the enzymes secreted by the pancreas digested food. Following in the footsteps of Margaret Harper, they adhered tenaciously to low-fat diets. To protect their children from pneumonia, each day they gently pounded their children on their backs to help them mobilize and get rid of the thick mucus in their lungs. Doctors placed the children on preventive antibiotics and responded vigorously to any signs of infection. First in England and then throughout the world, new CF centers focused on offering comprehensive care by experts to the children.

But the painful fact remains that during the 1980s parents of new babies with CF quickly realized that they would be locked in an unending struggle to keep their children alive, a heavy emotional burden indeed. One day in 1981 when I was a first-year medical student, I went on rounds with a professor of pediatrics. He looked uncharacteristically grim. When I asked him what was wrong, he told me that one of his patients, a 12-year-old girl with CF, had died a few hours earlier, and that her distraught father had attempted suicide by trying to jump off the hospital roof. Guards had caught and restrained him.

During the 1990s the increases in life expectancy slowed. A key reason was that as antibiotics conquered the more common bacterial infections, more exotic, often drug-resistant, strains colonized the patient's lungs. The most notorious was *Pseudomonas aeruginosa*, which has taken many lives of adults with CF. Although the quality of life for a person with CF and his family greatly improved over the last four decades, it still requires great physical and emotional stamina to fight the disorder. The typical patient may spend 3 or more hours a day in a highly orchestrated series of maneuvers (postural drainage) to clear airways of accumulating mucus. Pneumonia is a constant threat and patients are frequently hospitalized.

The most heroic effort in the quest to conquer CF has been the use of lung transplants. The first lung transplant for a patient with CF was performed at the University of Pittsburgh in 1983. Since then, surgical techniques and the control of organ rejection have steadily improved. In 2008 there were 1221 lung transplant operations in the United States; nearly 15% for patients with CF. Many centers in Europe are also highly skilled at lung transplant surgery. There too the procedure is most commonly performed for CF patients. Described as the "ultimate treatment" for CF, controversy has long simmered as to whether this hugely expensive, extremely difficult intervention conferred enough benefit to the patient to be justified. In the first decade of use the 1-year mortality rate and problems with organ rejections were disturbingly high. But since 2000 the data offer more hope. Only ~5%–10% of patients now die within a year of transplant, and ~80% are living 5 years after the procedure. A recent report by the Zurich Lung Transplant Program on 100 patients with CF treated from 1992 to 2009 showed remarkable improvements in survival over time. Cases done before 2000 had a 1-year survival of 85% and a 5-year survival of 60%, whereas cases done since 2000 had a 1-year survival of 96% and 5-year survival of 78%.

Unfortunately, lung transplant does not cure CF. In the most successful interventions the new lungs will give the patient much improved lung function for a few years. But patients who undergo lung transplant face ongoing challenges that compromise their daily lives. The biggest limitation on lung transplant is the same one that hampers all solid organ transplant programs—there are many fewer organ donors than there are patients in need. In the United Kingdom in 1990, 40 CF patients received lung transplants. As the size of the adult population with CF in the United Kingdom continues to grow, current projections are that unless some new therapy emerges, about 100–120 CF patients will need lung transplants each year. They will compete for these organs with other patients with other diseases. It is almost certain that donor lungs will be available to less than one-half of the CF patients who need them. The rest will die as they wait for this ultimate gift.

In the United States the Organ Procurement and Transplantation Network (OPTN) oversees allocation of donated organs. One of its allocation rules is that children who are 12 or younger who need lung transplants must receive the donated lung from a deceased child who is also 12 or younger. This is in large part because adult lungs are often too big to transplant safely into the chests of children. Many of the children who are on the transplant list for lungs are afflicted with CF. Some people believe that this rule operates to reduce the chance that a child with CF will get the lung transplant he or she needs and that, in effect, it hastens his or her death. Recently, the parents of a 10-year-old critically ill child with CF challenged the "12 and under rule" in court. Before a judicial decision was reached, the child received two lung transplants from two adult donors. Recognizing that age is an imperfect proxy for size (some children may be large enough to accept an adult lung or a partially resected one), the OPTN is currently reviewing its policy. In 2013 only 30 children under 12 were on the waiting list for lung transplants; the low number reflects the great strides in managing CF in childhood.

In 1989 a group led by Francis Collins, now the Director of the NIH, cloned the gene mutations that cause CF, opening a new era in research on the disease. It was quickly discovered that ~70% of all the mutations were owing to a single specific change called ΔF508, signifying a deletion of only three base pairs of DNA, which coded for just a single amino acid of phenylalanine, but which ultimately caused the protein in question to have an improper shape and position in the epithelial cells of the pancreas and the lung. It was quickly determined that the protein, defects in which caused the disease, played a key role in regulating the transport of chloride ions. Suddenly, the salty sweat made sense!

During the early 1990s it became possible to use DNA testing to confirm the diagnosis of CF in newborns, offer carrier testing to adults, and perform prenatal diagnosis. As these tests became more widespread, physicians and others wondered what impact that would have on reducing the number of children born with the disease. Thus far, after more than a decade of newborn screening for CF and of the availability of prenatal diagnosis, studies of changes in incidence of live births of affected children give contradictory results. Massachusetts reports a decline, but Colorado does not. One region of Italy reported a 24% decline, whereas another reported a decline of only 4% (within the margin of error). The trend seems to suggest that newborn screening has fostered about a 15% decline in incidence of live births, probably by alerting parents to the one in four recurrence risk they face with each pregnancy. In most cases at-risk parents do not roll the dice with subsequent pregnancies. Usually, they just forego more pregnancies. Prenatal diagnosis coupled with abortion of affected fetuses is rare. Among the major reasons are that the parents are profoundly attached to the affected child for whom they are caring so they cannot countenance a therapeutic abortion of a future affected fetus. In addition, CF does not cause brain injury, and there is good reason to hope that it will become an ever more treatable disease. The CF Foundation has never condoned prenatal diagnosis and selective abortion of affected fetuses.

The cloning of the CF gene in 1989 stimulated much interest in the pharmaceutical and biotech industries to develop new drugs to treat the underlying problem—the defect in chloride ion transport. During the 1990s the NIH and many biotech and pharmaceutical companies spent tens of millions of dollars on a variety of novel therapeutic approaches to ameliorate or (in the case of gene therapy) cure CF—to no avail. It was during this era that the now influential CF Foundation, under the leadership of Dr. Robert Beall, a former biochemist at the NIH who became CEO in 1994, redoubled its efforts to drive for a cure. Impressed by new technologies that permitted researchers to screen literally hundreds of thousands of different molecules to assay if they might improve the flow of chloride ions across, in, and out of cells from CF patients, Beall approached future Nobel laureate Roger Tsien and who had helped to start a company called Aurora Biosciences in San Diego to work on it. In 1998, Beall arranged for the CF Foundation to give a grant of $2 million to fund CF research at Aurora. But he knew the costs of finding and developing a drug for CF would be far greater than that. In 1999, Beall approached William Gates, Sr., then the president of the Gates Foundation, and shared his dream of developing a small molecule that could be taken orally and that could reach all the cells in the body of a child with CF and restore normal ion channel function. A few weeks later the Gates Foundation committed $20 million to the Aurora CF project. The CF Foundation contributed an additional $17 million. In addition it embarked on an unprecedented fund-raising effort, which since then has garnered more than $300 million for research.

Even with excellent funding, Aurora faced an immense challenge. There are literally hundreds of different mutations in the *CFTR* gene that can (in any combination of two) cause CF. The initial challenge was to gain a deep understanding of the impact of the various mutations. In time, scientists agreed that the various mutations could generally be grouped into about three broad categories, depending on the effect of the mutation on the chloride channel. This helped them to chart a chemical approach to searching for new drugs. They began to call compounds that enhanced a defective gating function "potentiators," whereas they called compounds that might help the defective protein travel through the ER and reach the part of the cell where they belong "correctors." The thought of having to develop mutation-specific drugs was truly daunting. Aurora screened hundreds of thousands of compounds and was able to identify some that looked promising, but screening was only the first of many steps in drug development. Each molecule with a hint of promise then needed close study by medicinal chemists, the wizards who add a methyl group here or close a ring there to modify and improve the "hits" identified through screening.

In 2001 a biotech company in Cambridge named Vertex Pharmaceuticals acquired Aurora Biosciences and expanded the research effort. In a 10-year effort, Vertex scientists meticulously studied the impact of various mutations on the shape of the dysfunctional proteins. One aspect of this included studying the three-dimensional structure of the crystallized protein (the same technique, much refined, that helped Watson and Crick deduce the double helical structure of DNA in 1953). They then created and or examined thousands of different small molecules in an effort to find one that would interact with the protein in a manner that compensated for the defect in it.

By 2007 the Vertex research team had greatly advanced a compound called VX-770. Extensive preclinical studies showed the molecule was not toxic in two animal species, and it appeared safe in a small early human trial in adults with CF. In 2010 scientists who had used VX-770 in a Phase I trial reported their results in *The New England Journal of Medicine.* Working in several different centers, the principal investigators had enrolled 39 adults with CF who carried at least one G551D mutation (the one that seemed to be most responsive to VX-770) in a short-term placebo-controlled trial. The trial was conducted in two stages. In stage one the 39 subjects were divided into smaller groups to receive in a blinded fashion either one of three different doses of the new drug or a placebo for 14 days. Next the group was given one of two higher doses of the drug or placebo for another 14 days.

This was primarily a safety study, and the research team collected extensive clinical and biochemical data on each subject. Overall the group tolerated the new molecule well; no one dropped out of the study because of adverse effects. In addition to the safety study, the research team studied a number of "secondary" end points that might give some hint as to whether the drug showed a promise of clinical efficacy. Among other things, they looked at chloride ion flow in nasal epithelial cells and the forced expiratory volume (a practical measure of lung function) that each subject could generate. Although the treated groups did not show results that were statistically different from the placebo control group, there were encouraging signs of improvement over time among particular subjects. Vertex pushed forward with a much larger and longer study intended to more thoroughly assess the efficacy of VX-770.

In November of 2011, the principal investigators in the STRIVE study reported the results of testing the drug on 161 patients with CF who were 12 or older and who had at least one so-called G551D mutation. This seemingly abstruse terminology is just biochemical shorthand. Scientists have assigned 20 letters to represent the 20 essential amino acids that make up our proteins. G stands for glycine and D stands for aspartic acid. The designation G551D signifies that, unlike the normal protein, this one has a single amino acid change at position 551 of the peptide chain that constitutes the protein product of the chloride transmembrane function regulator (*CTFR*) gene in patients with CF.

The 48-week study yielded impressive results. The investigators reported a net *gain* in lung function (as measured by FEV1) of 10.5%, an extraordinary result given that CF patients on average *lose* 2% of lung function each year. Except for the few patients who have received lung transplants, improvement in lung function had never before been observed. It will take many years to determine the effect of this improvement on the life span and quality of life among patients with the G551D mutation, but it is plausible that it will modestly improve both.

The effects of the drug, now called Kalydeco, bordered on magical. The CFTR protein is 1480 amino acids long. Yet, Kalydeco exerted a positive change on the function of the chloride channel that was due to a single acid change (at 551). How can such a small change cause such devastating results? To grasp the answer, one must remember that proteins are not linear. To do their biochemical chores, they must fold into exquisitely organized three-dimensional structures that often include *pockets* where local interactions between the protein and other proteins or small molecules take place. If the protein is even slightly "misfolded," its ability to do its work may be severely compromised.

It took more than a decade and about $1 billion before Vertex generated its first dollar of product revenue from treating CF. Vertex is now a fully integrated pharmaceutical company that can claim to have created the first drug to correct a fundamental molecular defect in a protein. Even though the G551D mutation is only found in ~4% of kids with CF, the CFF and CF families everywhere are ecstatic. If you could find a molecule to fix one mutation they reasoned, the same algorithm should be able to find molecules to fix other mutations.

One of the many challenges in developing therapies for orphan genetic disorders is that in most there are scores or hundreds of distinct mutations that can cause the disease. CF is different in that ~70% of all the mutant alleles (each patient has to have two—one from each parent) are accounted for by a three base pair deletion called ΔF508. The rest of the 2000 documented mutations are rare or very rare. Scientists call them familial mutations because throughout the world they have only been found in one or a few families. The G551D mutation is not so very rare; it accounts for 4% of all CF alleles and is the runner-up to ΔF508.

Kalydeco became available to patients in the United States in the second quarter of 2012; soon thereafter regulators in Europe also approved it for sale. No one yet knows how patients who take the drug will fare over time. But it appears that the scientists at Vertex have opened up a new chapter not only in the treatment of CF, but also in the approach to treating many other orphan disorders. They have shown that it is possible to develop a small molecule that can be administered orally and that will reach the surface of particular cells where it can interact with a misshapen protein and make it work better!

The development of Kalydeco is extraordinary in another way. Vertex developed the drug in partnership with the CFF, one of the world's largest and most sophisticated not-for-profit orphan disease groups. Over 15 years, the CFF contributed about $150 million to the development effort. In exchange, it received a royalty right on profits from the sale of the drug, a right that it well knew might never have any value. On November 19, 2014 (less than 2 years after the FDA approved Kalydeco), a front-page story in *The New York Times* reported that the foundation had sold its rights to those royalties for $3.3 billion! Over time, the foundation might have received considerably more than that amount, but it wanted the money at once. The reasoning was simple; the foundation wanted to put as much money to work as soon as possible to help the many thousands of CF patients who could not benefit from Kalydeco. This sum instantly catapulted the CFF into a level of wealth far above that of any other foundation primarily concerned with single-gene disorders.

The early clinical results achieved with Kalydeco are so impressive that in his 2015 State of the Union message, President Obama specifically mentioned the cystic fibrosis story, using it as an example of why he would ask Congress to allocate hundreds of millions of new dollars to research in order to expand the field of precision or molecular medicine.

In addition to acting as the most powerful driver of research to help patients with CF, the CF Foundation has pioneered the concept of venture philanthropy. In the past patient foundations operated as charities. Those groups able to award grants to investigators typically made small, no-strings-attached gifts to academic scientists. The CF Foundation and a few others are increasingly putting some of their financial resources into partnerships with both large and small biotech companies. In so doing they both advance the area of research that most concerns them and open up the possibility that success will cause dollars to flow back to them, thus further enabling their activities.

The beneficial effects provided by Kalydeco to persons with CF due to the G551D mutation suggested that Vertex might be able to accomplish the same feat for patients with the much more common F508 mutation (causing the loss of one single amino acid from the CFTR protein). From data generated in a Phase II (dose ranging study), Vertex found preliminary evidence that, when given with Kalydeco, a second compound that it calls lumacaftor improves lung function in patients with the 508 mutation by ~8% after 56 days. This is too short a period from which to derive a good sense of lumacaftor's efficacy, but it encouraged the company to conduct clinical trials to investigate the efficacy of the two drugs in patients with the most common mutation in CF.

In June of 2014 Vertex excited the biotech world by announcing results to date from two trials (called TRANSPORT and TRAFFIC) that involved more than 1000 CF patients age 12 and older who were born with two copies of the 508 mutation. When compared with placebo, the two drugs improved lung function by a small but statistically significant (2%–3%) amount and, perhaps more important, met the trial's goal of achieving a statistically significant reduction in the need for hospital care. Although Kalydeco was designed to help only about 2000 patients in the world, the two-drug regimen, if approved by the FDA, could be used by many thousands, which could immensely increase revenues. On the day of the announcement, Vertex, which bet the company's future on these trials, saw its stock soar 40%, a multibillion increase in value.

There are several exciting avenues of research that complement the work that led to the development of Kalydeco and lumacaftor. A few scientific groups are trying to develop "correctors" that act on the mutated CFTR protein to eke out a bit more activity. One small molecule (dubbed Corr-4a) did improve function in cells with the V232D mutation, one that happens to be common among CF patients in Spain. Because there are so many different CF mutations, we may ultimately need mutation-specific drugs, a daunting challenge. However, it is likely that mutations may be clustered; that is, that the defects they cause fall into groups in which all members might be addressed with the same drug. To do this with small molecules would be really hard, but new gene editing techniques (that I discuss in the next chapter) might offer an efficient strategy. Yet another approach is to find small molecules (including the well-studied aminoglycosides, some of which are used as antibiotics) that will overcome "stop" mutations that halt the manufacture of CFTR protein by masking the mutation and tricking the RNA transcriptional machinery to keep reading the DNA instructions.

The exciting news from Vertex that its two-drug combination could slow the progress of CF and extend life will likely elicit intense debate over the cost of therapy. Although it is likely that in the United States, the FDA will approve these drugs and that most insurers will pay about $300,000 a year for the cost of the drug for each patient, it is possible that other nations will balk at the price tag. It is possible, for example, that NICE, the agency in the United Kingdom that is charged with drug approval, might decide that there is not yet sufficient long-term clinical data to show convincingly that the drugs measurably improve quality of life or extend survival.

12

What Is Next
Emerging Therapies

SO FAR, PHYSICIAN-SCIENTISTS HAVE IDENTIFIED about 7000 phenotypes that are clearly associated with a variation within a particular human gene. However, many of them are really just recognizable, but benign, variants. Brachydactyly type D, for example, the nosological term geneticists use to describe short thumbs, is caused by a variation in the *HOXD1* gene and is transmitted as an autosomal dominant. Some of the other phenotypes cause suffering only if the person with the genetic variant (such as with *G6PD* deficiency) experiences a particular environmental trigger.

A significant percentage of the hundreds of single-gene disorders may be so rare that current models of commercial drug development may be expensive to permit drug companies to develop therapies for them. For example, thanks to the great success of screening programs to alert carrier couples to their risk of bearing an affected child, only about 10 children with Tay–Sachs disease are born each year in the United States. Although studies of gene therapy in a cat model of this disease are encouraging, it will be challenging to build an economically viable model to justify a drug discovery effort around such a small number of patients. We need new financial and regulatory approaches to accommodate efforts to treat persons with ultrarare genetic conditions (a topic to which I will return in the final chapter). One possible solution will be to create novel partnerships between patient groups and biotech companies in which a not-for-profit foundation funds a specific company project in exchange for a downstream royalty right.

Despite the economic challenges, there is today an unprecedented interest in the biotech industry, as well as in the much larger pharmaceutical industry, in developing new drugs for the small (sometimes exceedingly small) groups of patients with rare, heretofore untreatable, disorders. The extraordinary power of genome sequencing technology and bioinformatics tools today make it relatively easy to find the proverbial needle in the haystack (the mutation that causes the disease) for a large number of disorders, diseases, and syndromes about which, until recently, no one had any idea as to the molecular cause. For that reason, the number of disorders for which we can know with certainty arise from mutations in genes will continue to increase. The correlation of a syndrome (a collection of physical findings) with particular changes in DNA is in a particularly fruitful period. Researchers are quickly resolving ancient clinical mysteries. A couple of examples will suffice.

In the 1980s, when I led the clinical team in charge of the care of about 800 adults with severe intellectual disability, we were often at a loss to understand the cause of their neurological problems. Even though my team was aided by consultation with experts from Boston teaching hospitals, it was possible to make a diagnosis in considerably less than one-half of the patients. Even when we did, we were doing little more than applying a label. I recall a young woman in her 30s with a very small head, flattened cheekbones, epicanthal folds around her eyes, and unusual hands and feet with broad thumbs and big first toes. She was severely mentally handicapped and had never learned to speak. She matched the description of a group of patients who had first been characterized by two physicians—Jack Herbert Rubinstein and Hooshang Taybi—in 1963. Because her physical abnormalities were so similar to those in the group of patients that they had described, we so labeled her. Physicians and patients (or in this case their families) take comfort in labeling. Yet, no one had the faintest idea as to what caused Rubinstein–Taybi syndrome until 2006, when an enterprising research group in Europe collected blood from a group of such persons and showed that one-half of them had mutations in a region on the short arm of chromosome 16 that includes the gene coding for a protein called CBP. Today we know that mutations in that gene do cause about one-half the disease; more recently, mutations in another gene do have been shown to cause most of the remaining cases.

Even when astute physicians can use physical examination, laboratory tests, and imaging technologies to make a genetic diagnosis, the actual molecular cause has until recently remained unknown. Hereditary deafness and retinitis pigmentosa are two examples of relatively common genetic diseases that are caused by mutations in many different genes that result in similar presentations. Over the last few years DNA sequencing has matured to permit one to precisely identify the causative mutation. With the increase of "whole exome sequencing," physician-researchers are rapidly refining our molecular diagnostic skills. This is, of course, but the first step toward developing a therapy.

A paradigmatic example of the application of the new tools of genetic analysis to understand a disease at the molecular level is the work performed by Jim Lupski, an accomplished clinical geneticist at Baylor College of Medicine in Houston. In 2010 he used these tools to solve a clinical problem in his own family. Dr. Lupski had long known that he had a form of Charcot–Marie–Tooth disease (named for the three physicians who independently first described it), a hereditary disease of the peripheral nerves that can over time severely compromise the ability to walk. When he and others reviewed his family history across three generations, they found a confusing picture. His paternal grandmother had lived a long life with a mild neuropathy in her wrists. His father had a somewhat more extensive, but still quite mild, "patchy" neuropathy. But, among eight adult children, four had no evidence of disease and the other four, including Dr. Lupski, had physical problems. Lupski and his colleagues at Baylor used sequencing tools to determine that not one, but *two* different mutations in a gene called *SH3TC2* on chromosome 5 were traveling in the Lupski family, each of which had different clinical effects. A single copy of one—*Y169H*—could cause a mild disorder, whereas a single copy of the other—*R954X*—did not by itself cause any abnormalities. Dr. Lupski's father and paternal grandmother carried a single copy of *Y169H* that behaved as a dominant disorder, whereas a single copy of *R954X* manifested as a benign variant. In combination, the two mutations caused a more significant disability that a clinician might label as an autosomal recessive disorder.

Another recent example of how advances in use of the new sequencing technologies have revolutionized diagnostics is evident in a paper from a group at Harvard Medical School led by Vamsi Mootha, one of the world's experts on mitochondria. It reported on the molecular analysis of 102 patients with rare disorders that by clinical presentation seemed most likely to be diseases of mitochondrial dysfunction. The mitochondrion is a subcellular organelle with its own small genome that plays a critically important role in making energy available to the cell. Among other features, mitochondrial disorders in children often manifest with muscle weakness and fatigue. For each patient, Mootha's group sequenced the mitochondrial genome and about 1000 nuclear genes known to code for proteins related to mitochondrial function. Using DNA sequencing, they were able to find mutations in 17 of 18 children who had been given a diagnosis of being affected with a mitochondrial disorder. More impressively, the team also discovered the molecular basis for disease in five other, theretofore undiagnosed patients. Essentially, in one go, the scientists molecularly defined *several* rare disorders!

Deep sequencing of individual genomes turns clinical "lumpers" into "splitters." One can be certain that in the coming years the list of single-gene disorders will grow. Those interested in developing new therapies will confront a burgeoning number of disorders, most of which have been diagnosed in exceedingly few (often less than 100 in the world) patients. Technologies that will enable new therapeutic approaches to help people with ultrarare disorders will continue to emerge. In this chapter I provide a brief overview of several of the more exciting ones: adeno-associated virus (AAV) gene therapy (which I have also discussed in Chapter 7), induced pluripotent stem cells, RNA interference, and gene editing. I will also return to the rapidly growing reach of DNA-based prenatal diagnostics.

AAV Gene Therapy

There are many problems to overcome if we hope to ameliorate genetic disorders by delivering a *normal* copy of a gene to a person who was born with one or two abnormal copies in each cell. Among the most crucial challenges are to develop viral vectors (think of them as magically small vehicles carrying within them even smaller payloads) that can be given safely to people, that can reach the key tissues that must be corrected, that can avoid attack by our immune systems, and that after penetrating the target cells will produce the needed protein at a safe and effective level. This is indeed a daunting task.

After four decades of investigation, the relatively small community of scientists committed to developing gene therapy is now focused intently on using the AAV as the delivery vehicle. Since it was discovered about five decades ago, scientists have developed a deep understanding of AAV, and determined that it does not cause disease in humans. AAV, a member of the Parvoviridae family whose genome is composed of single-stranded DNA, has a deceptively simple structure. It has an outer capsid that is formed by three proteins (called VP1, 2, and 3) in repeat arrays and a hollow core that can be used to carry about 4700 base pairs of DNA (the payload). Its relatively small carrying capacity means that some genes are too large for it. On the plus side, one of the most important aspects of AAV is that unlike human immunodeficiency virus (HIV) and many other viruses, it does not integrate into the host's genome. This greatly reduces the risk that it could disrupt the work of some key gene and pose a risk of cancer (as happened a decade ago when French researchers used lentiviruses to treat children with immune deficiency).

Since the effort to work with AAV in gene therapy began in the early 1980s, the vast majority of the animal and most of the early human studies have used a serotype called AAV2. It is experience with this serotype that has largely reassured the experts that AAV has a good safety profile. But AAV2 is not particularly good at transducing certain kinds of cells (liver and muscle) so its potential clinical utility is limited.

Much of the work to refine our understanding of AAV was performed by a team led by Jim Wilson at the University of Pennsylvania School of Medicine and by his colleague Guangping Gao, now a professor at the University of Massachusetts Medical School. They did important research to define different serotypes (the term refers to slight differences in the outer structure of the virus called the capsid that have a major influence on which cells the particular version can enter), and to understand the pros and cons of using various serotypes to achieve different goals in animals (usually mice). Today, some AAV serotypes are free for anyone to use. Other vectors, such as AAV9, are protected by patents, so efforts to develop AAV gene therapy to treat disease sometimes depend on navigating a complex and expensive licensing landscape. Because AAV is widespread in the environment, many humans have unknowingly been infected by it. This means that our immune systems have generated antibodies to the virus, a fact that complicates, but does not necessarily negate, efforts to use it as a vector to deliver a corrective gene.

Clinical trials using AAV vectors to treat severe human diseases began more than two decades ago. Different groups have used it to attempt to ameliorate cystic fibrosis, muscular dystrophy, hemophilia, Parkinson disease, and a variety of other disorders. Two early clinical trials involved heroic efforts to treat children with severe fatal genetic disorders called Batten disease and Canavan disease. In 2002, Dr. Ron Crystal, a pulmonologist who had been studying the possibility of using gene therapy to cure cystic fibrosis, α-1-antitrypsin deficiency, and other lung disorders, was urged by the parents of children with a form of Batten disease called late-infantile neuronal ceroid lipofuscinosis (LINCL), a uniformly fatal genetic storage disorder, to attempt to deliver a normal copy of the mutated gene directly into the brain. Children with LINCL have a storage disorder; because the cells in their brains do not make enough of an enzyme called TTP-1, their brain becomes swollen, distended, and miscolored. The affected children typically develop severe neurological problems early in childhood and often sink into a vegetative state and die before age 12. The parents hoped that by supplying a new gene to many cells throughout the brain of an affected child (which would enable them to make the needed enzyme) that researchers could halt the relentless course of the disease.

In 2004 Crystal initiated an 11-patient study to assess the safety and efficacy of gene therapy using AAV2 as the vector to carry the needed gene. The plan required that neurosurgeons create multiple burr holes in the children's skulls, and then use tiny catheters to deliver about 10 billion copies of the genetically engineered virus to various parts of the brain. The protocol required that the first five patients be severely affected, whereas the next six patients could be less severely compromised. The children tolerated the surgical procedure well, and the viral vectors did not seem to cause serious side effects. The major problem was that there was little likelihood that the virus would spread widely enough and transduce enough cells to make a therapeutic difference. In 2008 Crystal reported that the intervention seemed to have slowed disease progression, but the trial was small and the trajectory of decline varies for each patient, so the evidence is slim.

In 2012 the research team used a newer vector called Rh10 (the Rh stands for rhesus monkey, the animal from which the virus was isolated) characterized by James Wilson's group at the University of Pennsylvania to determine if in animal models this vector would distribute more widely once it was injected into the brains. In studies of rats and African green monkeys he found improved distribution. Because storage disorders (and many other single-gene brain diseases) generally affect all areas of the brain, gene therapy is unlikely to be of much value if the vector does not travel widely once injected.

One of the scientists at the forefront of the effort to develop gene therapy to treat orphan brain diseases is Guangping Gao. A small man with a big smile and an even bigger intellect, Guangping, whose remarkable life includes hard years during the Cultural Revolution in Communist China, focuses on finding and studying new forms of adeno-associated viruses. He is deeply interested in helping to increase the array of vectors to carry genes to cells in people with monogenic disorders.

Nearly 20 years ago, Guangping and his colleagues cloned the gene for a fatal neurogenetic disorder known as Canavan disease (CD). CD is an autosomal recessive orphan brain disease caused by a deficiency in an enzyme known as aspartoacyclase (ASPA), which is responsible for breaking down a compound called *N*-acetylaspartic acid (NAA). The most severe form causes symptoms as early as the first month of life. Affected children have poor muscle control; they are floppy and have trouble keeping the head upright.

They also develop unusually large heads during the first year life. Tragically, the disease has a rapid, downhill course; most patients die in childhood. On autopsy the brain shows massive destruction of cells and much water accumulation. Although CD can appear in any human group, it is unusually common among Ashkenazi Jews. Among them, about one in 40 persons carries a mutation; the a priori risk of an untested Ashkenazi Jewish couple giving birth to an affected child is about one in 6400 with each pregnancy. Among the Ashkenazim, just two mutations account for 98% of all carriers.

This devastating disorder is named in honor of Myrtelle Moore Canavan, one of the first women to achieve prominence in pathology, and one of the most important students of neuropathology in the first half of the 20th century. After graduating from the Women's Medical College of Pennsylvania and marrying fellow physician, James F. Canavan, she won an appointment to study bacteriology at Danvers State Hospital in Massachusetts. There she met Dr. Elmer Ernest Southard, a professor of neuropathology at Harvard who invited her to work with him. In 1910 she began a four-decade career as a pathologist at various state hospitals that housed persons with severe neurological disorders. In 1920 after Southard died, Dr. Canavan was named "acting" director of the laboratories at the Boston Psychopathic Hospital (a title that almost certainly indicated a reluctance to give the full appointment to a woman). About 1921, she published a monograph entitled, "The First One Thousand Autopsies of the Pathological service of the Massachusetts Commission of Mental Diseases 1914–1919," a work that is a classic in the history of neuropathology. Dr. Canavan served as an *associate* professor at Boston University School of Medicine for 25 years, and was the "*assistant* curator" of the famous Warren Anatomical Museum at Harvard for nearly as long. Harvard never named her as the full curator or appointed her to its faculty.

Dr. Canavan worked at a time when there was much speculation that criminal behavior was heavily influenced by genetic factors and or abnormal brain pathology. Thus, she studied the brains of hundreds of mentally retarded or institutionalized mentally ill persons, and sometimes published articles that seem bizarre today, such as "The Brains of 50 Insane Criminals," as well as monographs in the series "Waverly Researches in the Pathology of the Feeble-minded." The word, Waverly, refers to the area in Waltham, Massachusetts, which is home to the Fernald State School. Beginning about 1930, prominent neuropathologists from Harvard, including Dr. Canavan, performed hundreds of "brain cuttings," as they were called, on those who died at Fernald, in an attempt to understand the anatomical correlates of mental retardation. A small red brick building on the campus was set aside for this activity and, in time, a massive library of microscope slides of various brain tissues was built up. I remember visiting the deserted red brick building and rubbing my hands across the stone tables on which the autopsies were performed. As fate would have it, in the mid-1990s, I was involved in overseeing the transfer of the slide collection of brain tissues taken during these autopsies to the Armed Forces Institute of Pathology in Washington, DC (and, thus, saving them from almost certain destruction). I may well have handled slides that Myrtelle Canavan prepared 60 years earlier.

For all her extraordinary importance in the field of neuropathology—in 1959 someone calculated that she had been involved in training about half the practicing neurosurgeons in the United States—today Dr. Canavan's fame rests on a single case report. In 1931 she prepared a detailed analysis of the brain of a 16-month-old child who had died after a year of neurological decline. This was the first published report of a description of the spongy degeneration in the brain of a child with the disorder now named after her.

About a decade ago a team of neurologists and gene therapists, including Dr. Gao, conducted a gene therapy trial in children with CD. They used essentially the same protocol as that chosen by Dr. Crystal for Batten disease—delivery of an AAV vector (AAV2) carrying a normal copy of the gene to the brain parenchyma via multiple burr holes prepared by neurosurgeons. The surgeons performed the operation on 11 affected children. Long-term follow-up of them has confirmed that the vector is safe. It has been much harder to assess whether the intervention helped. Compared with a control group, the treated children seemed to have had some reduction in the frequency of seizures, some lessening in the rate of brain shrinkage, and a definite reduction in the amount of storage material in the brain. But clinically they have fared poorly.

Some of Dr. Gao's more recent research may provide reason for hope. He has been working with a "knockout" mouse model of CD. The mouse typically lives a short unhappy life, dying at about 4 weeks of life. On necropsy, its brain is already mush, the result of the massive amount of sphingolipids that accumulated during gestation. In 2012, Dr. Gao reported dramatically positive results from treating the affected mice. He injected each animal intravenously with a single large dose of a new vector carrying the missing gene. He showed that, as long as he injected early in life, this treatment *reversed* signs and symptoms. In some cases the Canavan mice treated with gene therapy have lived for 2 years. On necropsy, their brain tissue looks essentially normal. He is now planning to treat them at birth to see if he can prevent the disease from manifesting at all.

In many genetic disorders, even though the mutations are in every cell, most of the manifestations of the disease arise because of a failure in a specific cell type. Designing viral variants that have "tropism" for (prefer to transduce or infect) a particular organ, was until recently more art than science. If researchers could modify an AAV to carry a therapeutic gene of interest only to the cells that most need it, there might be a dramatic improvement in treatment. This more refined delivery would also reduce concerns about harmful off-target effects, thus improving the safety profile, reduce overall costs, and provide general support to the scientific outlook for gene therapy.

Although researchers have been working in the field of AAV gene therapy for 30 years, no drug based on this technology has yet been approved for use in the United States. In November of 2012 the European Medicines Agency (EMA)—the equivalent of the Food and Drug Administration (FDA)—and the European Commission approved the first gene therapy drug in the West (several years before that authorities in China approved a gene therapy vector to treat patients with brain cancer). The EMA's decision was the culmination of a decade-long process initiated by Amsterdam Molecular Therapeutics (AMT). This novel drug is composed of an AAV1 virus that carries the gene coding for lipoprotein lipase. It is used to treat persons with an ultrarare life-threatening genetic disorder of fat metabolism called lipoprotein lipase deficiency (LPLD). In the absence of this enzyme, extremely high levels of triglycerides build up in the body. The drug is administered once as a series of small injections into the leg muscles. This is because muscles normally produce a lot of lipoprotein lipase and AAV1 preferentially enters muscle cells.

The road to regulatory approval is long. The drug (which is marketed under the name Glybera) was developed in 2003. It took about 2 years of experimenting with a knockout mouse (one in which scientists have been able to cause the disease by removing the gene) and with cats with a feline version of the disease to determine the dose with which to treat. In addition, the researchers had to treat normal animals to explore safety issues. For example, they gave normal mice a 100-fold dose of the amount they had given to the animals with LPLD. The only adverse finding from this extreme overdose was a reduction of 30% of the normally expected weight gain over the 90-day experiment.

The first human trial of Glybera was initiated in 2005. Eight adult subjects with LPLD underwent 40 to 60 injections into their muscles. The primary efficacy end point was to achieve a 40% reduction in serum triglyceride levels at 12 weeks, a goal that was met in four of the eight patients. Unfortunately, the benefits faded in subsequent weeks, in part because of immunological reactions to the enzyme. In August 2007, AMT launched a second, larger trial that included higher doses and immunosuppressant drugs. This time seven of 14 patients met the agreed on end point of a reduction in serum triglycerides at 12 weeks, but again the effect wore off by 26 weeks. A third trial was launched early in 2009. Treatment of five patients with a stronger immunosuppressive dose led to sustained reduction in the levels of circulated chylomicrons. The study did not, however, meet its major end point—a reduction in the bouts of pancreatitis that required hospitalization—so regulatory officials demanded further studies.

The company submitted its application for approval to market Glybera in 2009. The reviewing agency (known as the Committee for Advanced Technologies) raised relatively few issues about safety or manufacturing. It did, however, have significant concerns about long-term efficacy. This was the first of four regulatory reviews that the Glybera application would face. While it was struggling to develop the drug, AMT became insolvent, but was recapitalized under the name uniQure. Late in 2012 the European Commission approved the Glybera application. The approval came despite the fact that an important minority of the agency's scientific reviewers were unconvinced of the drug's value. To address this, the Commission demanded a postmarketing pharmacovigilance plan, biannual safety reports, and that the company create a registry and follow every patient treated.

A strong candidate to be the first gene therapy drug to be approved in the United States is being developed to treat a rare genetic blindness known as Leber congenital amaurosis. As discussed in Chapter 7, in the 1990s researchers discovered that a blindness essentially the same as LCA2 in humans arose naturally in some Briard terriers. In 2001, a team at the Cornell University College of Veterinary Medicine led by Dr. Gregory Acland and Gustavo Aguirre reported that they had successfully used an AAV vector to deliver the *RPE65* gene via subretinal injection in affected dogs, and that the vision of the three animals had measurably and substantially improved. In 2008 three research teams, one at the Children's Hospital of Philadelphia, one at University College London, and a third in Italy, reported in companion articles in *The New England Journal of Medicine* on similar experiments that they had conducted in patients with LCA2. In all three trials, the subretinal injections were well tolerated; the only appreciable side effect was a small retinal tear (a surgical risk of the procedure) in one patient.

In the ensuing years the team in Philadelphia has moved forward with experiments in both dogs and people. In 2009 it enlarged its clinical trial to 12 patients. The outcome of these treatments was so reassuring that they began to treat the second affected eye in each person. Patients reported visual improvement (increased light sensitivity) in just a few weeks, and several of them recaptured enough vision that they were no longer legally blind! In the fall of 2013 the researchers at the Children's Hospital of Philadelphia (CHOP) launched a Phase III (also called a pivotal) clinical trial of AAV gene therapy to treat LCA2. The development of new therapies is never smooth. In May of 2015 a research group in England reported on the long-term (>4 year) follow-up of 12 patients with RPE 65 deficiency who had been treated with gene therapy. The results were mixed; although some persons definitely showed some evidence of improvement in vision, overall the underlying disease continued to worsen.

Researchers have also recently reported stunning success in using AAV therapy to treat hemophilia. In December of 2011 a group at University College London published its study of six adult men with hemophilia B who were treated by a single systemic injection of a form of AAV, which carried the normal version of the gene for Factor IX, the protein missing in the disease. After 6–16 months of follow-up, all men were producing an increased amount of Factor IX, four no longer needed prophylactic administration of Factor IX to avoid bleeding episodes, and the other two needed less than they had needed. Thus far, the safety profile of the drug looks good.

Success with attempting to use AAV gene therapy to ameliorate one or two rare disorders will be an important milestone, but many challenges remain. Among the most important are (1) scaling up a production system that will satisfy FDA regulations and make enough vector at a reasonable cost, (2) overcoming immunological problems that arise because many people have been infected with AAV through naturally occurring infections, (3) engineering new vectors with much more limited infectivity (currently vectors may infect a far greater range of cell types in the body than is needed to treat most diseases), (4) determining and controlling the proper dose of a drug that will likely be given only *once*, and, ultimately, (5) creating regulatable systems (methods to increase or decrease production of the protein by the vector). This last point is particularly worrisome in that it is possible that a vector that causes some cells to make a *much larger than normal amount of a natural* protein may be harmful. In addition to all these concerns is the more pedestrian, but quite important, issue of how to fit a one-time therapy into the complicated, ever changing health care reimbursement system.

In 2013–2014 the creation and funding of new AAV gene therapy companies proceeded at an unprecedented pace. In the fall of 2013 the biggest name in gene therapy was the European giant, Sanofi, which when it acquired Genzyme in 2011 gained control of a gene therapy franchise that is more than 20 years old and that has done especially important work on a fatal disorder of childhood called SMA1. In 2013 Dr. Katherine High and her colleagues at CHOP teamed up with their parent hospital to launch Spark Therapeutics, which will initially specialize in developing gene therapies for eye and liver disorders. Given the impressive quality of the scientific founders and their colleagues, Spark quickly raised capital and went public early in 2015. Over the last 2 years about a dozen well-funded AAV gene therapy companies have either grown dramatically or debuted, including Avalanche, Audentes, AGTC, and Dimension Therapeutics. All of them will use a similar viral delivery system and all will face similar clinical challenges.

During 2012–2014 a TRV team built an AAV therapy company intended to focus initially on orphan disorders that harm the central nervous system. As part of that exercise, we recruited world leaders in vector manufacturing, neurosurgery, gene therapy, and vector engineering, analyzed about 100 potential target disorders, interviewed dozens of clinical experts and scientists working to develop next-generation vectors, met with the leaders of many patient disease foundations, and hired several leading scientists to design preliminary experiments. Voyager Therapeutics (named in honor of the Voyager I spacecraft, the first man-made object to leave the solar system, and now about 20 billion miles from earth) will be among the first companies to develop large-scale vector manufacturing operations.

The weight of the evidence is that after three decades of struggle, gene therapy is coming of age. Within a decade, children affected with severe genetic disorders and adults suffering from neurodegenerative disorders such as Parkinson disease and some forms of amyotrophic lateral sclerosis may receive substantial clinical benefit thanks to our ability to precisely deliver in a single treatment vast numbers of viral vectors carrying genes to ameliorate these diseases.

Induced Pluripotent Stem Cells

As we age, our bodies, like the ancient carriage made famous in the 19th century poem by Dr. Oliver Wendell Holmes, "The One Hoss Shay," simply wear out. Even if we avoid heart attacks and cancer, sooner or later degenerative diseases, whether of the heart or the immune system or the brain, will cause our demise. But what if it were possible to reprogram old cells to make them young again? This is an ancient dream, but it is also the brash goal of the science of *regenerative* medicine.

The current flowering of this new field can be traced to the work of British scientist and 2012 Nobel Laureate, Dr. John B. Gurdon, who in the early 1960s showed that it was possible to create a frog *embryo* from a single *cell* taken from a tadpole. To achieve this stunning feat, Gurdon collected frog eggs, removed the nucleus from each, and then transferred a nucleus from *an intestinal* cell of the tadpole into each of the eggs. In so doing, he created an egg with a diploid set of genes (both sets of genes came from the cell of the tadpole), in effect bypassing fertilization and creating a genetic clone of the animal from which the cell was obtained. His work ignited much interest, and hopes for cloning mammals ran high. But work with mammals floundered until 1997 when a Scottish physiologist named Ian Wilmut shocked and delighted the world by using DNA testing to prove that he had cloned a sheep that he named Dolly, and which became a media star. But the success rate in cloning sheep was very low. Only ~2% of the embryos created by transferring the somatic nuclei from older animals survived. Work in other species was also very challenging.

Soon after the cloning of Dolly, Dr. James Thomson, a developmental biologist at the University of Wisconsin, and his team published the extraordinary finding that they had succeeded in creating specific types of cell lines derived from early human embryos (which had been aborted). After months in culture these cells retained the ability to differentiate into any of the major cell lineages. This work provided part of the foundation for the rapidly growing field of regenerative medicine. Thompson showed that it was possible under certain conditions for human cells committed to one lineage to be "reprogrammed" to act as universal stem cells.

The catalytic effect of Thompson's publication in *Science* was quickly dampened by a vitriolic national debate. Anti-abortion forces argued that it was unethical to use cells derived from human embryos for research. During the administration of George W. Bush, public debate drove the decision by the National Institutes of Health (NIH) to limit its funding of embryonic stem cell research to just the handful of already existing (and in many cases poor quality) human embryonic cell lines already in its possession. This decision sharply curtailed federally funded research, and even led several prominent researchers to move to other countries that did not have such restrictions. Fortunately, several states, notably California, provided major pools of funding for scientists who wished to work in this arena. In the midst of the political battle over funding research with cells derived from human embryos came a dramatic discovery from a researcher in Japan.

In 2006, Dr. Shinya Yamanaka reported in *Cell* that his laboratory had succeeded in reprogramming already differentiated cells from mouse fibroblasts (skin cells) to return to an embryonic-like state by transferring their nuclei into enucleated egg cells or by fusion with embryonic stem (ES) cells. Astoundingly, he accomplished this feat by introducing merely *four* transcription factors (called oct3/4, sox2, c-myc, and klf4) into the cultures. These transformed cells, which are known as *induced pluripotent stem cells* (IPSCs), show the morphology and growth properties of embryonic cells, and they express the expected cell surface markers. When he transplanted IPSCs into nude mice, Yamanaka created tumors containing a variety of tissues from all three embryonic germ layers. Injecting these cells into very early mouse embryos, he proved that they contributed to embryonic development. These data show that he generated IPSCs from fibroblast cultures by the addition of only a few defined factors, and then controlled their fate. The *Cell* paper earned Dr. Yamanaka a Nobel Prize in Physiology or Medicine in 2012.

Yamanaka's paper ignited an explosion of research in IPSCs. In 2007, Thompson's laboratory in Wisconsin and Yamanka's laboratory in Tokyo reported in *Science* that they had successfully recapitulated the work performed in mice in human cells. Since 2007, dozens of top research laboratories have been working furiously to extend the research. Of special interest are the hundreds of publications showing that IPS cell lines can be created from patients with particular genetic disorders, cell lines that constitute a powerful new resource for searching for potential new medicines to treat those diseases.

In January of 2014 a research team led by a woman named Haruko Obokata at the Riken Center in Kobe, Japan published two remarkable papers reporting even simpler methods to redirect differentiated cells to become pluripotent. Obokata reported that by merely exposing white blood cells from newborn mice to a solution at low pH for a short time that a fraction of those cells lost their "white cell-ness" and developed surface markers indicating that they had become pluripotent. In a second paper, she reported that these cells (named STAP cells) could integrate into early mouse embryos and develop, and that under certain conditions in cell culture they became indistinguishable from ES cells. If this could be done with human cells, it would be a huge advance in the development of stem cell therapy. Unfortunately, within months the research on STAP cells was engulfed in scandal, and Obokata was forced to withdraw her papers amidst allegations of fraud. Nevertheless, research in regenerative medicine is expanding at a prodigious pace.

The ability to create IPSC lines has given great impetus to the field of regenerative medicine. The holy grail of this field is to repair damaged and aging organs through the proliferation of youthful cells. These new, but still early, therapeutic approaches include such challenges as providing new heart cells after heart attacks or injecting programmed cells into the right areas of the brain to treat Parkinson disease. Currently, there are more than a dozen stem cell companies in the United States; in addition, scores of larger companies have stem cell programs, and there are many privately held start-ups. The companies have in aggregate a market capitalization of a few billion dollars. Most of the stocks are trading below the price of their initial public offering. This is largely because the road from basic science (no matter how exciting) discovery to an approved drug usually takes more than a decade and is fraught with danger. Still, clinical research with IPS is starting to advance. In the United States a small company called Advanced Cell Technology (ACT) is sponsoring a clinical trial in Los Angeles in which patients with certain diseases caused by retinal degeneration are being treated by injecting IPS-derived retinal cells into the backs of their eyes. In Japan the government-supported Riken Institute expects to begin treating patients with similar problems in 2014.

RNAi

Simply speaking, one can divide most single-gene disorders into two categories. Most autosomal recessive disorders arise because of mutations that cause partial or complete loss of function of a particular protein. Literally hundreds of such mutations can arise across any gene, and these can result in a range of injuries to the protein, often by causing it to fail to take its proper three-dimensional shape, which is essential for it to do its biochemical work. These are "loss-of-function" diseases. Some disorders that are dominantly acting, that is, caused by a de novo mutation (one that newly arose in the sperm or egg) or by inheriting a mutated gene from an affected parent (often true for late-onset disorders), are "gain-of-function" mutations. In such disorders the mutated protein interferes with the work of the normal version of the same protein produced by the sister gene. Thus, if one could develop a method to block or scale back the production of that harmful protein, one might ameliorate the disease.

The central dogma of molecular biology teaches that DNA provides the template from which RNA is transcribed, and that in the cytoplasm messenger RNA (mRNA) provides the protein-manufacturing unit with the design instructions. Starting about two decades ago, molecular biologists began attempting to develop ways to block the mRNA to constrain production of unwanted proteins.

In the early 1990s several scientists noticed that under certain conditions RNA molecules injected into cells seemed to suppress gene function. About the mid-1990s, Andrew Fire at the Carnegie Institution for Science and Craig Mello, a professor at the University of Massachusetts Medical School, were investigating the fundamentals of gene transcription in the tiny worm known as *Caenorhabditis elegans*, which is widely used in the laboratory and about which a great deal of knowledge has been generated. Spurred on by puzzling findings in other laboratories, they set out to deduce the mechanism by which RNA silencing took place. Their key discovery was that one could design and use short double-stranded RNA molecules of a specific sequence to suppress production of proteins by a gene with an analogous sequence. Their paper, which was published in *Nature* in 1998, earned them a Nobel Prize in 2006.

Over the last decade, thousands of scientists and a few daring start-up companies have rushed to drive RNA interference (RNAi) toward clinical utility—so far with only modest success. However, in 2013 hope that RNAi could someday play a major role in treating genetic and other diseases advanced on the report published in *The New England Journal of Medicine* that the method had been used to sharply reduce production of a harmful protein in a rare disorder called transthyretin amyloidosis (in which patients suffer because a molecule made in the liver is deposited in excess amounts in many of their tissues, including the heart and peripheral nerves). The research team developed two small interfering RNAi models with a nucleotide sequence that matched that of a section of the transthyretin molecule. They placed these molecules in tiny hollow lipid nanoparticles and injected them in different doses into two groups of people. One molecule was given to 32 persons with the disease, whereas the other was given to 17 normal controls. The researchers found that in both groups of human subjects there was a rapid, dose-dependent lowering of transthyretin in the blood. The team is now conducting a study on the clinical efficacy of these drugs. News of the success in suppressing production of transthyretin was enormously positive for *Alnylam Pharmaceuticals*, one of the few companies working to commercialize this field. On the strength of positive results in its effort to develop RNAi therapy for this and several other single-gene disorders, in the summer of 2014, the company, once low on money, had a market capitalization of more than $5 billion.

Gene Editing

Even more exciting than RNAi is the new field of gene editing. Given the many billions of cells in our bodies, the notion that it might be possible to develop and deliver molecules to each cell to physically *correct* the mutations so that those cells could then make normal copies of the protein in question seems like magical thinking. Yet, over the last 20 years scientists have discovered at least three different methods for doing so, each more powerful than the one that preceded it. These three technologies are known as zinc finger nucleases (ZFNs), transcription-activator-like effector nucleases (TALENs), and clustered regularly interspersed short palindromic repeat (CRISPR) elements. This is highly technical, so I will merely provide a brief overview (which readers may guiltlessly skip!).

Zinc finger nucleases are man-made enzymes created by fusing a part of a protein that binds DNA with a part of a protein that cuts DNA. Scientists can design them to have multiple elements, each targeting a unique set of three DNA base pairs, with spacer regions that extend their reach along a gene. After the ZFN has excised a target DNA sequence, other naturally occurring, cellular enzymes will patch in a normal sequence. For the last decade Sangamo BioSciences has led the charge to develop this technology. It has focused on trying to determine if ZNFs can be used to stop HIV infections. HIV uses a cell receptor called CCR5 to enter white cells. Since 2009 Sangamo scientists have been working with academic collaborators at the University of Pennsylvania to see if they can remove CD34 cells from patients with HIV, use ZFNs to delete the ability to make the CCR5 receptor, and return them to the patients where they will repopulate the immune system with cells resistant to HIV infection. Thus far, the study is still in its safety phase, and no sign of clinical efficacy has been reported.

TALENs are another man-made system. They are created by combining a DNA recognition system found in certain species of plants (called TALEs) with a DNA cutting enzyme called an endonuclease (N). Beginning about 2009, a number of research laboratories began focusing on how to exploit this inherent gene-targeting feature to study gene function. It soon became apparent that TALENs had certain features that made them somewhat easier to work with than ZFNs, and that they too could become a therapy that cuts out deleterious mutations and then relies on the *cell's* enzymes to repair the gene after the targeted sequence is excised. No one has yet initiated a clinical trial using TALENS to target disease genes.

Yet another gene editing system, called CRISPR-cas9, has recently taken the field of genetic engineering by storm. The acronym CRISPR refers to a molecular defense system that evolved eons ago in bacteria that uses small RNA molecules to defend against foreign DNA (such as from invading viruses). It can be thought of as a primitive form of innate immunity. A small group of microbiologists has studied CRISPR for many years, but interest in the field exploded in 2012. In August of that year a scientific team led by French-born Emmanuelle Charpentier and University of California, Berkeley professor Jennifer Doudna published a paper in *Science* demonstrating that, when combined with a small protein called cas9 that has two DNA cutting domains, the system can be easily programmed to selectively alter any DNA sequence of interest. The discovery, which immediately drew immense interest, suggests a simple, yet radical, new approach to altering DNA, and perhaps curing many orphan diseases. In 2013 a group led by Feng Zhang at MIT and the Broad Institute demonstrated that they could use the system to edit DNA in human cells. Many other labs, notably that of Harvard's George Church, were rapidly refining our understanding of this new tool. By late 2013 the editors at *Science* were referring to the "CRISPR craze" in molecular biology, applications of the system were generating hundreds of papers, venture capital firms were building new companies, debates over priority of discovery were becoming tense, and some intellectual property attorneys were getting a lot of new business. It is a good bet that some of those who worked in this field will win a Nobel Prize in the near future. As with the other gene-targeting technologies, the two major challenges that developers of CRISPR will face are (1) devising methods to deliver this novel form of therapy to enough cells and (2) to avoid off-target effects that might cause unintended harm. No one has initiated a clinical trial that attempts to treat genetic disease with a CRISPR-based approach. That milestone is probably still a few years away.

The Therapeutic Horizon: Down Syndrome

What therapeutic interventions might we dream about? Perhaps gene therapy will someday provide a means to reverse plaque buildup in the arteries, greatly diminishing the incidence of heart attacks and strokes. Possibly, each of us (who can afford it?) will rely on banked IPS cells to generate a new supply of healthy dopaminergic neurons to avert Parkinson disease. Scientists might be able to use AAV vectors to deliver antibodies against the tau proteins that are likely a cause of Alzheimer's disease. The decision by some parents to bank cord blood harvested at the births of their children, a resource that could help the child survive certain uncommon disorders such as childhood leukemias (and which will be eventually used to help less than one in 1000 of those children) will be remembered as a forerunner of cell therapy. Advances in our ability to make and deliver large proteins may permit us to ward off certain birth defects in utero. We might be able to harness RNA interference to halt the progression of some forms of amyotrophic lateral sclerosis (Lou Gehrig's disease). Should we solve the significant challenge of figuring out how to deliver the CRISPR gene-editing technology to the right cells at the right time, we might eventually be able to ameliorate or cure almost any single-gene disorder.

The genetic disorders for which it will be most difficult to develop effective therapies would be those in which an infant is *born* with serious brain impairment. Among the hundreds of rare disorders that fall into this domain those that are caused by the presence of an entire extra chromosome—the trisomies—seem particularly intractable because they involve the presence in each cell of an extra copy of hundreds of different genes. Putting aside the trisomies involving an extra X or Y chromosome (which are comparatively mild disorders), nature permits humans to live with extra copies of three of the 22 autosomes: chromosome 13, chromosome 18, and chromosome 21. A human embryo burdened with any other trisomy dies (usually during pregnancy). It is quite unusual for infants born alive with trisomy 13 or trisomy 18 to live for more than a few months. But trisomy 21, better known as Down syndrome (named for John Langdon Down, a British physician who in 1866 provided an early description of the condition) is different.

The 22 pairs of autosomes (those other than the X and Y) are numbered in order of size. Number 1 is the largest; number 22 should be the smallest, but the numbering system (which was agreed to more than 50 years ago at a conference in Paris) miscalled the smallest pair. (The short arm of a chromosome is called "p" for petit, a nod to the Paris meeting). Number 21 is slightly smaller than 22. It makes sense that if any individuals with a trisomy should survive, it should be those with an extra copy of a small chromosome because it would have fewer extra genes, which in turn would create a smaller overall burden of excess proteins that might cause birth defects (a phenomenon called gene dosage effect).

In the United States today there are about 100,000 persons living with Down syndrome, and about 4000 affected babies are born each year. We see them enrolled in regular classes in grammar schools, sitting quietly during church services, and working in local grocery markets. A few have even become successful television actors. This is vastly different from the life of such persons in 1963, the year I first met an individual with Down syndrome. My family had just moved to a new town, and I enrolled in a new school. One September afternoon a classmate invited me to his home, where I was surprised to meet his older brother who, quite literally, lived in the attic and who, I learned, was rarely allowed outside. This may seem cruel, but in the era in which he was born (late 1930s), many families sent infants with Down syndrome to live in "state schools," after which most had only perfunctory contact with their parents. In 1963 folks with Down syndrome were still frequently referred to as "Mongolian imbeciles." This term entered medical parlance in the 19th century because persons with an extra chromosome 21 have downward slanted palpebral fissures, which gives their eyes an Asian look. As late as 1949, the major textbook in the field, *Principles of Human Genetics,* still used the words. It was only after 1958, the year that French geneticist, Jérôme LeJeune, discovered that such persons have an extra chromosme 21, that people slowly abandoned this racist label. The stigmatization that families of children with Down syndrome felt back then is—from today's vantage point—almost incomprehensible, but that does not make it less true.

Despite their many clinical problems, including congenital heart defects that require surgery, an increased risk for pneumonia, and a much increased lifetime risk for leukemia, there have been remarkable gains in maintaining the health of patients with Down syndrome. Since 1950 the median age of survival has increased from about 15 to about 60. Intensive efforts to maximize good health combined with vigorous efforts to educate children with Down syndrome have yielded great results. Today, the intelligence testing scores of such children are higher than those of the mid-20th century (yet, still nearly 3 standard deviations below the population mean with a typical IQ of about 60). Unfortunately, improved longevity has uncovered another cruel feature of the syndrome: by about age 50 or so essentially *all* persons with Down syndrome develop Alzheimer's disease. This is because there is a gene on chromosome 21, an extra copy of which greatly increases susceptibility to this neurodegenerative disorder.

The parents of children with Down syndrome have over the years built a powerful organization—the Down Syndrome Congress—to advocate for their children. Understandably, most parents of children with serious disabilities worry about what will happen to their child after their deaths. Will they be employable? Where will they live? Who will look after them? Until recently, though, most did not even dream of treatments that would compensate for or correct their cellular defects. But steady advances in the understanding of how genes are turned on or off, a line of research stretching back more than 30 years, suggests that it might someday be possible to develop therapies that reduce the impact of trisomies on the human brain.

Scientists successfully created the first trisomy 16 mouse in 1984, and by the late 1990s they had developed an animal that was trisomic only for the so-called "critical region"—the chromosome segment most associated with the features of Down syndrome in humans. This is possible because evolution is remarkably conservative. Over the eons, genes have evolved on specific locations on chromosomes. On an evolutionary scale, mice are closer to humans than they are to most other organisms on the planet, so it should not be surprising to learn that the mouse versions of the genes strung along human chromosome 21 are aligned in a similar order on mouse chromosome 16. The differing chromosome numbers reflect the fact that over comparatively short periods of evolutionary time large chunks of chromosomes break off and recombine. Mice with segmental trisomy 16 provide a key model used by scientists to study the cell biology of Down syndrome.

One important line of research has been to use tests, such as the Morris water maze test, to study the problem-solving abilities of mice. In many studies involving several different test systems, scientists have convincingly shown that trisomy 16 mice are not as clever as their normal brethren. Thus, they provide a system in which one can study interventions that might improve cognition.

Roger Reeves, a neurophysiologist at Johns Hopkins University, has been attacking the problem of how to improve cognition in Down syndrome for 20 years. His team has demonstrated that giving a drug to the trisomic mouse that up-regulates a gene called "sonic hedgehog" improves its cognitive abilities. Even more dramatic is other work showing that in another mouse model a small molecule appears to increase the growth of an area of the brain called the hippocampus that plays a major role in memory. That growth is not merely in volume; there is also a marked increase in the number of cells per unit volume. So dramatic were the results that a major pharmaceutical company is now conducting Phase I (safety trial) of this molecule in humans. It will be years before we know if this drug can measurably improve cognition in humans.

An important area of research targeting methods to improve cognition in children with Down syndrome involves efforts to inactivate the extra chromosome 21 in each cell, the condition that is the underlying cause of the syndrome. Nature has taught us that it is possible to inactivate whole chromosomes. In each cell of every woman, one X chromosome is active and one is inactive (a condition that is programmed by the time the embryo consists of eight cells). Scientists have studied the mechanism by which this occurs for decades, and we have some understanding of *XIST*, the gene that is responsible for this event. Might it be possible to target and inactivate the extra chromosome 21? The road to achieve this seems very long and the journey may not lead to success, but scientists like Jeanne Lawrence at the University of Massachusetts Medical School have embarked on it.

Another area of research that holds increasing promise to provide breakthrough therapies is epigenetics, a field that focuses on understanding and manipulating the biological "marks" on chromatin (the complex arrangement of DNA and several types of histone proteins that wrap around it to determine if a gene or genes will be active or quiescent). Some orphan disorders arise as a consequence of mutations in single genes that cause inappropriate silencing of others. Geneticists have identified more than 40 such disorders (each quite rare), nearly all of which profoundly harm neurological development. Of special interest is research by Hans Bjornsson at Johns Hopkins with a mouse model of Kabuki syndrome (which affects about one in 30,000 infants). He has shown that by countering the epigenetic errors that afflict the mouse early in life, he can greatly improve their behavior and performance on various tests, a surrogate for mouse intelligence. Given the plasticity of brain development in early life such work holds out hope for treating single-gene disorders of the "epigenome."

Fetal DNA Sequencing

In a few years women who wish to do so will likely have access to whole fetal exome sequencing to screen for genetic disorders in the fetus. *This could be the biggest development in the history of prenatal diagnosis, perhaps in the history of reproductive medicine.* Once widely adopted, which may take less than a decade in the United States, a significant fraction of those fetuses destined to be born with rare, severe single-gene disorders will be aborted early in the second trimester of pregnancy. Of course, for religious or moral reasons some women (current evidence suggests at least one-quarter) will not use this technology.

Routine use of prenatal diagnosis arose in the early 1970s once it became possible to perform chromosome analysis on fetal cells obtained by amniocentesis. In 1973 a large study showed that among pregnant women who were 35 or older, the risk of miscarriage posed by that procedure was equal to or less than the risk of bearing a child with Down syndrome (about 1:300). Because the risk of carrying a fetus with Down syndrome rises with maternal age, it quickly became the clinical (and legal) standard of care to offer fetal karyotyping to women who would deliver by their 35th birthday. In the ensuing 40 years, screening of fetuses for serious birth defects has evolved relatively slowly. The development of biochemical assays that could identify abnormal levels of several chemicals that are associated with an increased risk for carrying a fetus with a chromosomal disorder or a neural tube defect constituted important advances, but the screening tools probed for only a tiny fraction of all possible fetal disorders.

An important step on the road to comprehensive prenatal screening occurred in 2012 when researchers reported the results of a large study (funded by the NIH) to compare a technology called "chromosomal microarray analysis" with standard fetal karyotyping. Microarray tools are good at finding small (i.e., not visible under the light microscope) deletions and/or duplications in a human genome—a growing number of which have been strongly associated with serious neurobehavioral disorders. Dr. Ron Wapner, a professor of obstetrics at Columbia University, led the large team that studied the amniotic fluid samples of more than 4000 women who were undergoing prenatal diagnosis for standard reasons. They split each sample into two aliquots and conducted both tests. The results showed that the microarray test is as good as fetal karyotyping in detecting full chromosome abnormalities, *and* that it is *superior* in detecting smaller abnormalities that are associated with severe developmental delay. The test uncovered one of those more subtle abnormalities in 1.7% of pregnancies in women undergoing testing because of advanced maternal age or because of having scored positive on a biochemical screening test.

In November of 2013, the American College of Obstetricians and Gynecologists formally recommended to its members that chromosomal microarray analysis (CMA) should be offered as "the first line genetic test" in pregnancies in which ultrasound has detected a structural abnormality in the fetus, a first step in the adoption of this test for much wider use. Since then a growing number of obstetricians have guided their patients who seek amniocentesis to screen for fetal defects away from standard karyotyping in favor of microarray testing. But, despite its larger scope, microarray testing, which does not probe individual genes, still only assesses a tiny portion of the potential disorders with which a fetus could be affected. To perform a much broader analysis, a laboratory must have access to and the ability to analyze the entire fetal exome, a far more comprehensive method of screening. Such access is imminent.

It has long been known that small numbers of nucleated fetal cells circulate in the maternal blood. In 1990 Dr. Diana Bianchi demonstrated that fetal DNA could be isolated from nucleated red cells that crossed the placenta into the maternal circulation. During the subsequent decade, several labs were able to show that it was possible to capture such cells and test them to see whether an at-risk fetus was destined to be born with a particular genetic disease. But the technical challenges limited the undertaking to highly experienced research labs. During the 1990s, Bianchi and others made herculean efforts to improve the rate of identification and capture of circulating nucleated fetal cells. Over time, the capture rate improved, but it was still far from being ready for routine use for prenatal diagnosis. During the period 2000–2005, the group led by Bianchi and other groups were able to show that they could capture large amounts of cell-free fetal DNA from amniotic fluid and accurately identify chromosome abnormalities such as Down syndrome and Turner syndrome without needing to culture fetal cells (thus compressing the wait for a diagnosis). Research groups also began to devote intense efforts to capturing cell-free fetal DNA that also can be found in low amounts in maternal blood. In 2005 a group in Bari, Italy published a remarkable report in which it showed that among 32 couples known to be at risk for bearing a baby with β-thalassemia that they had been able to identify the paternal mutation in fetal DNA circulating in the mother's blood. But despite immense efforts, scientists were for many years unable to develop methods to capture enough fetal DNA to form the basis for a highly reliable clinical test.

Several years ago, a company called Sequenom, using technology developed by a Stanford scientist, successfully showed that by capturing small amounts of *fetal* DNA in a blood sample from the *expectant mother* it could identify fetuses with the more common chromosomal disorders (trisomy 21, 18, and 13 and the abnormalities in number of the X and Y). The technical reach of its test (and those of a growing number of competitors) is growing;

it now is able to detect many relatively rare large chromosomal defects that are associated with severe congenital malformations. The ability to look for the common fetal chromosomal abnormalities by a *noninvasive* method such as taking a sample of maternal blood has altered the practice of prenatal diagnosis. This approach is often called NIPT (noninvasive prenatal testing).

In 2013, Sequenom performed more than 125,000 tests (~3% of all pregnancies) on women in the United States. This was about one-third of all prenatal testing for fetal risk for Down syndrome. In February of 2014, Dr. Bianchi, now a professor at Tufts School of Medicine, and her colleagues published their findings on using a similar technology to screen for the more common chromosomal abnormalities among a typical (not especially high-risk) community population of women. They found that by sequencing the cell-free *fetal* DNA in blood samples obtained from pregnant women, they were able to identify *all* fetuses that were burdened with an abnormal number of chromosomes and, in comparison with current indirect testing methods, reduce the false positive rate by ~90%! This noninvasive screening test is nearly perfect in its ability to rule out Down syndrome and the other, less common, trisomies (13 and 18) in fetuses.

These tests should soon greatly reduce the *number of amniocenteses* performed each year to screen for aneuploidies, certainly welcome news to pregnant women. But note that I said the test was "nearly" perfect. Cell-free fetal DNA actually is captured from the apoptosis (death) of cells (called cytotrophoblasts and syncytiotrophoblasts) in the early placenta. Rarely, there is discordance between the chromosomal constitution of these cells and those of the actual fetus, a condition called mosaicism. In such cases, NIPT can generate both false positive and false negative results. For the next few years, an NIPT that identifies a trisomy is likely to be confirmed by amniocentesis to confirm the fetal karyotype.

The ultimate goal in prenatal testing is to be able to noninvasively capture intact fetal cells circulating in maternal blood, extract and amplify the DNA, and screen for hundreds of severe monogenic disorders. Currently the rate-limiting step (in addition to cost) to this approach is the efficiency with which the fetal cells can be captured. However, several start-up companies and some academically based laboratoies are diligently working on this technology. I expect significant advances over the next several years. But developing methods to capture the fetal cells and screen the DNA for a host of conditions is only part of the challenge. At first, it will also be extremely difficult to interpret the data and to provide informed counseling to the woman about the massive amounts of information that will be generated. Over time, as we become more familiar with the *millions* of variations in the human genome and their association with disease risk, counseling will become more accurate. This will naturally follow from the immense efforts that are under way to gather in and study vast amounts of DNA information.

The largest gene sequencing center in the world, a Chinese company called BGI, that employs more than 4000 scientists and technicians, was founded in the fall of 1999, and is now located in Shenzhen, China, a city of about 11 million people. BGI has industrialized the task of whole genome sequencing. By the end of 2013 it had sequenced the genomes of about 60,000 humans. Operating nearly 200 sequencing machines around the clock, on a daily basis BGI sometimes compiles more genomic data than that generated in total by the laboratories in the rest of the world combined. Its ambitions are impressive. BGI has launched a "The Million Human Genomes Project," "The Million Animal and Plant Genomes Project," and "The Million Microecosystems Genomes Project." The company has bet—and it is a good bet indeed—that generating this massive amount of data will provide the basis for developing products ranging from drought-resistant cereals to new drugs for complex diseases like diabetes, new algorithms for manipulating the microbiomes in our digestive tracts to avert inflammatory bowel disease, and, of course, the ultimate fetal screening tool.

Although it is a quixotic endeavor, BGI also has launched a "Cognitive Genomes" project, essentially a brute force effort to parse and study all the genetic variants that help shape human intellectual capacity. Working with Robert Plomin, a behavioral geneticist at King's College in London, who has been studying intelligence in human twin pairs for decades, BGI plans to sequence the genomes of several thousand people with IQs that measure above 150. Another project will analyze the genomes of more than 1000 mathematically gifted persons. Statistical analysis of vast amounts of genomic data gathered from such special populations will advance our understanding of the role of gene variants in setting the range of human intelligence (a poorly defined concept, to be sure), but I struggle to come up with a practical application. First, almost certainly there are thousands of gene variants—all with very small effects—that set the boundaries of intellectual capacity. Second, because the maternal–fetal environment, parenting skills, and hundreds of other environmental factors interact with human genome, it will, except for a few rare mutations with large effects, be almost impossible to capitalize clinically on relevant genetic variations. One gene variant (for which there is evidence, but not proof) that might be of interest to recognize at birth is the ability to recognize pitch in music; the capacity for "perfect pitch" seems to fade if not developed. One could, of course, use whole fetal genome sequencing to identify fetuses destined for certain severe intellectual disabilities, and that will be done. However, the idea of using the same technology to select for fetuses with high intellectual potential is just shy of delusional.

The advent of fetal genomic DNA analysis is likely over time to cause a significant *reduction* in the number of births of infants with severe monogenic disorders. A likely example of this is the neuromuscular disorder of infancy known as spinal muscular atrophy type 1 (SMA1). About 400 children with this autosomal recessive disorder are born each year in the United States. Currently, there is no approved treatment, but supportive (including placement on a ventilator) care may extend life a few years. Studies suggest that about two-thirds of parents who give birth to affected infants opt *against* aggressive medical care, in which case the infants often die before their first birthday. Most parents who learn through prenatal testing (triggered by family history) that the fetus is affected with SMA1 terminate the pregnancy. I anticipate that a decade hence the incidence (births) of children with SMA1 may be less than half of what is today. This will translate into a much smaller prevalence of patients with the disorder, and in turn make it a less attractive target for drug development. It is possible that population-wide noninvasive whole-genome sequencing of fetuses will in time sharply reduce the birth incidence of scores of serious monogenic disorders.

Archibald Edward Garrod, British physician, was the father of "inborn errors of metabolism" (1908). Portrait, Wellcome Library, London.

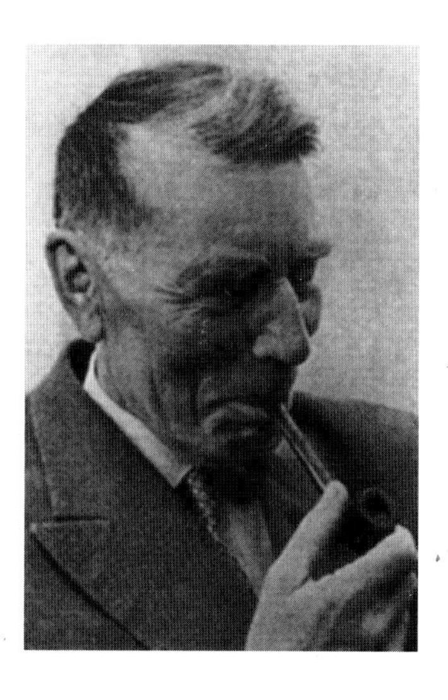

Ivar Asbjørn Følling, Norwegian physician, was the discoverer of PKU as a distinct cause of mental retardation (1934). Image reproduced from Folling, I. 1994. *Acta Paediatrica,* 407 (Suppl), 4-10, with permission from John R. Wiley and Sons.

Robert Guthrie, American physician–scientist, developed newborn screening for PKU (late 1950s–1960s). Photograph courtesy of the Museum of disABILITY History.

James V. Neel, American physician and prominent early human geneticist (University of Michigan), established that sickle cell anemia was an inherited disorder (1940s). Photograph, with permission, from News and Information Services Collection, Bentley Historical Library, University of Michigan.

Victor McKusick, American physician and prominent early human geneticist (The Johns Hopkins University), contributed greatly to our understanding of Marfan syndrome (1950s). Photograph courtesy of National Library of Medicine.

David Weatherall, British physician–scientist, has devoted his life to understanding β-thalassemia and helping patients in less developed countries. Photograph courtesy of Sir David Weatherall, Weatherall Institute of Molecular Medicine, University of Oxford, John Radcliffe Hospital.

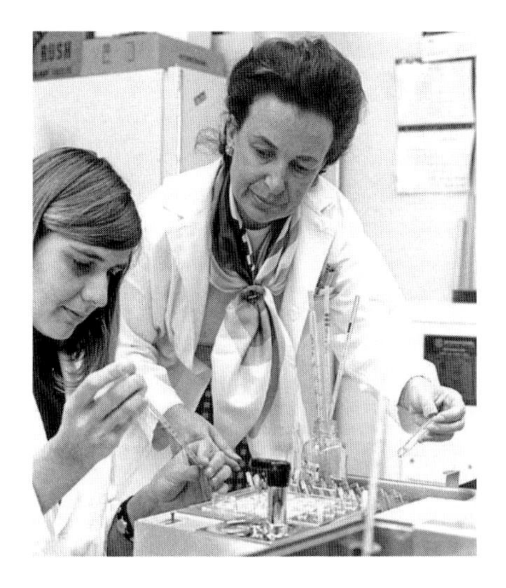

Judith Pool, American scientist, was the discoverer of cryoprecipitate, which greatly improved the lives of many thousands of patients with hemophilia. With permission from Stanford Medical History Center.

Michael Kaback, American geneticist and physician (University of California, San Diego, School of Medicine), played a leading role during the 1970s in developing carrier screening programs which helped to greatly reduce the number of children born with Tay–Sachs disease. Photograph courtesy of Michael Kaback.

Antonio Cao, Italian physician, played a key role in developing screening programs to help couples avoid bearing children with β-thalassemia (1978). Photograph courtesy of University of Cagliari Archives.

Henry Nadler, American physician, did key work in making amniocentesis for prenatal diagnosis for chromosomal disorders routinely available for women (1970). Reprinted, with permission, from Shulman ST. 2014. *Children's Memorial Hospital of Chicago.* Arcadia Publishing, ©2014 by Stanford T. Shulman, MD.

Robert Good, American physician–scientist (University of Minnesota), led the team that performed the first bone marrow transplant for an orphan genetic disorder (1968). Photograph, with permission, from University of Minnesota Archives, University of Minnesota–Twin Cities.

Joseph Murray, American surgeon and Nobel laureate, performed the first kidney transplants (1954). Photograph, with permission, from Harvard Medical School.

William Krivit, American physician–scientist (University of Minnesota), pioneered bone marrow transplantation to treat genetic disorders (1970s–1990s). Photograph courtesy of the Krivit family.

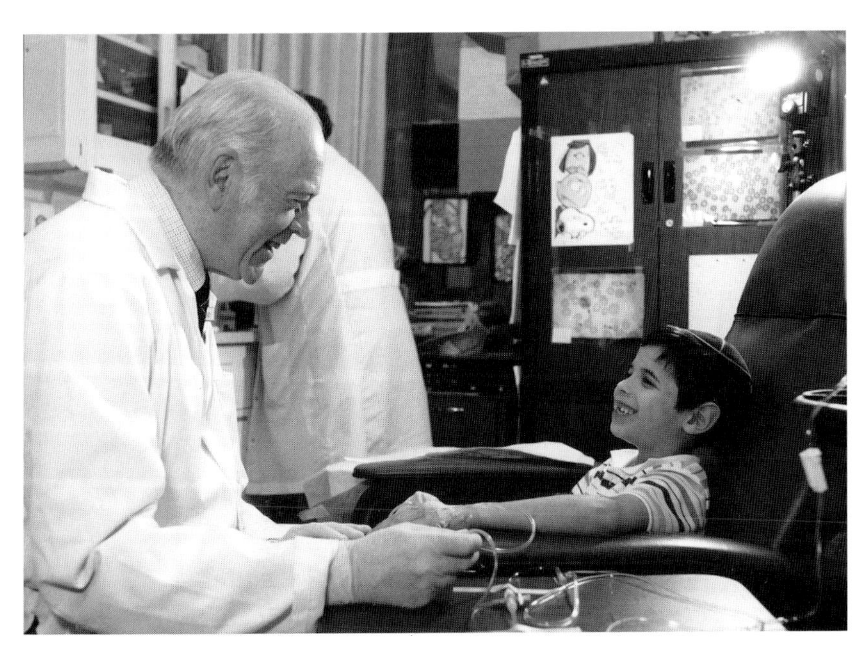

Roscoe Brady, American physician–scientist and Lasker Award winner, pioneered development of enzyme replacement therapy for lysosomal disorders (1960s–1990s). Photograph courtesy of Roscoe Brady.

Henri Termeer, CEO of Genzyme for 30 years, led an immense effort to develop enzyme replacement therapy, thus saving the lives of thousands of patients with orphan diseases. Photograph courtesy of Henri Termeer.

Katherine High, American physician and prominent hematologist and gene therapist (Children's Hospital of Philadelphia), did pioneering work in gene therapy for hemophilia. Photograph courtesy of Katherine High.

Luigi Naldini, Italian geneticist, among the most prominent gene therapy researchers in the world, has done especially important work on metachromatic leukodystrophy. Photograph courtesy of Luigi Naldini.

Antoine Marfan, French physician, first described Marfan syndrome (1885).

Nikolaus Friedreich, German physician, first well described Friedreich's ataxia (1876). Wikimedia Commons.

Theodore Puck, American scientist and Lasker Award winner (1958), did pioneering work in culturing mammalian cells (1950s). Photograph courtesy of the Albert and Mary Lasker Foundation.

Mei Chen, University of Southern California, her research in the biology of epidermolysis collagen VII provided key knowledge for a drug development effort in rDEB. Photograph courtesy Mei Chen.

David T. Woodley, University of Southern California, made fundamental contributions over two decades to understanding the biology of rDEB. Photograph courtesy of David Woodley.

Mary Kaye Richter, founded and for 30 years led the National Foundation for Ectodermal Dysplasia. Photograph courtesy Mary Kaye Richter, NFED.

Pascal Schneider, biochemist (University of Lausanne), has made major advances in understanding the problem that causes XLHED. Photograph courtesy of Pascal Schneider.

Dorothy H. Anderson, accepting an award for her research in cystic fibrosis from the Cystic Fibrosis Foundation in 1958. Library of Congress, Prints and Photographs Division, New York World Telegram & Sun Collection.

Guillaume Duchenne, French physician, among the first to describe muscular dystrophy (1868). Wikimedia.

Robert J. Beall, American scientist, as its Executive Director has built the Cystic Fibrosis Foundation into a highly influential and well-funded advocacy group. Photograph courtesy of the Cystic Fibrosis Foundation.

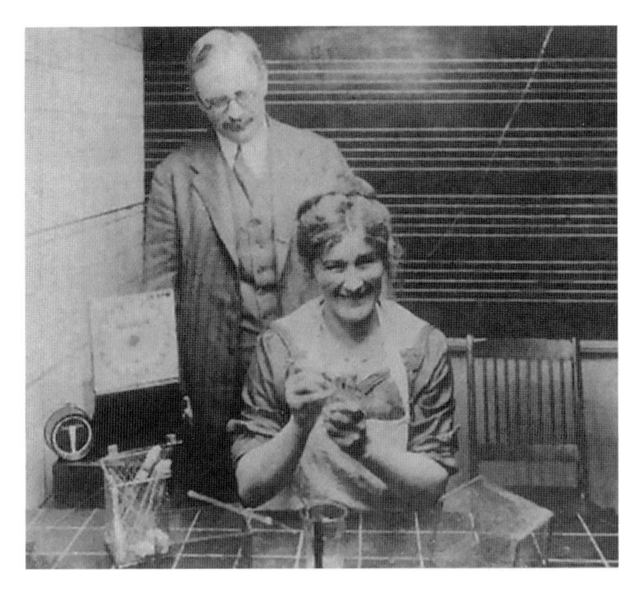

Myrtle Moore Canavan, a prominent neuropathologist, who in the mid-1900s trained many neurosurgeons, first described Canavan's disease. Wikimedia Commons.

Arvid Carlsson, American scientist and Nobel laureate (2000) for work done in the 1960s on the value of L-dopa as a drug for Parkinson disease. Photograph by Göran Olofsson/GU, provided courtesy of The Sahlgrenska Academy, University of Gothenburg.

Shinya Yamanaka, winner of the Nobel Prize (2012) for his work in stem cell biology. Wikipedia.

Monica Coenraads and her daughter Chelsea. Monica Coenraads is the Executive Director of the Rett Syndrome Research Trust. Her work over the last decade is a prime example of how determined parents can drive research forward. Photograph by Paul Fetters, HHMI, courtesy of Monica Coenraads.

Matt and Kristen Wilsey with daughter, Grace. Matt Wilsey, Parent Advocate and President of the Grace Wilsey Foundation, is changing the way rare diseases are discovered, researched, and treated. Carrie Chen Photography LLC.

Leo Kanner, one of the fathers of pediatric neurology, described in 1943 the first collection of patients who would come to be known as having "autism." Wikimedia.

13

We Are All Orphans

Lessons for Common Diseases

Progress So Far

THE MARCH OF POWERFUL NEW TECHNOLOGIES SUCH AS DNA sequencing, the large-scale manufacture of immensely complex proteins, new systems that permit the production at industrial scale of viral vectors for gene therapies, methods (such as magnetic resonance imaging [MRI] guided stereotactic surgery) for delivering those vectors, RNA knockdown, exon skipping, and gene editing, suggest that in the next decade we will see impressive progress in developing new drugs to treat heretofore untreatable rare single-gene disorders. Yet, because it may take a decade to move a drug from idea to approval, temporal and quantitative predictions of success are fraught with uncertainty. Ultimately, the key metric of progress in drug development is the number of drugs that win approval from the Food and Drug Administration (FDA) and its sister agencies in other nations each year, and that then prove to make a meaningful difference in the lives of patients. For orphan genetic disorders, one may gain some reassurance about the future by assessing recent drug approvals.

Over the last two decades regulatory agencies in the United States and Europe have approved about 10 recombinantly produced coagulation factors to treat rare single-gene clotting disorders such as the hemophilias. As of 2014, the FDA has approved seven enzyme replacement therapy drugs to treat some of the many lysosomal disorders, as well as three to treat a rare disease called hereditary angioedema or HAE (a life-threatening disorder caused by inadequate amounts of a protein called C-1 esterase inhibitor, which impairs the ability to control fluids in cells and can lead to sudden massive swelling in the throat). European authorities have approved one gene therapy drug (Glybera, to treat a rare lipid disorder), but the FDA has not yet approved a single-gene therapy product (although I expect several approvals in the next few years).

But excluding drugs that have been approved because they may give *symptomatic* relief or that are used to treat *secondary* aspects of a disease (such as chelating agents to remove excess iron from patients with β-thalassemia), the list of approved drugs for orphan genetic diseases is still distressingly short. In 2011 (if an iron chelator called Ferroxipone is excluded) the FDA approved only one drug for an orphan genetic disorder—Firazyr for hereditary angioedema. In 2012 the FDA approved 30 new drugs of which only three were for monogenic diseases (including Kalydeco for persons with cystic fibrosis owing to a particular mutation in the *CFTR* gene that afflicts ~4% of those patients). One of the three was not particularly innovative, as it is a modification of an established approach to treating Gaucher disease, for which other therapies exist. In 2013 the Center for Drug Evaluation and Research (CDER) at FDA approved 27 new drugs, of which only one targeted an orphan genetic disorder. The year 2013 witnessed a setback in the effort to develop *exon-skipping* therapies to treat boys with Duchenne muscular dystrophy, as the FDA informed one of the companies working on this approach that its data would not support approval and another that it was not willing to recognize increases in the level of dystrophin (lack of which is the root cause of the disease) as an "approvable end point." In general (and appropriately), the FDA prefers using functional end points such as the widely used "6-minute walk test" for muscle disorders over indirect biochemical measures.

This might seem to constitute a rather meager report, but it does not reflect the true situation in drug development for orphan disorders today. New approvals of orphan drugs over the last decade should be viewed as a harbinger of the future. As recently as 2008, few large pharmaceutical companies were seriously engaged in developing new drugs for rare genetic disorders. Over the last few years many of the major pharmaceutical companies that had been ignoring the rare genetic disease market as being too small for their commercial maw have either launched internal programs to develop such drugs or are strongly pursuing partnerships with small companies that were founded to do so. As their internal research budgets have shrunk or remained flat, a growing number of the larger players have altered their business strategy to include buying biotech companies that are working on an orphan drug of interest to them.

No doubt the impressive valuations that more than a dozen orphan disease companies, such as Alexion and Synageva, have achieved after the approval of a single new drug for a rare disorder have catalyzed this trend. In December of 2011 Alexion set the biotech industry on fire when it announced it was buying a tiny company called Enobia that was well along in developing a drug to treat a rare genetic bone disease called X-linked hypophosphatasia, for up to $1.1 billion (if it met all its drug development milestones). Reports of new companies raising large sums to develop therapies for rare genetic disorders now appear regularly. One celebrated example is a California-based company called Ultragenyx, founded by Dr. Emil Kakkis, who once led the clinical development work at BioMarin. Started in 2010, the company soon raised $100 million in venture capital. It quickly in-licensed several early stage clinical projects aimed at orphan genetic diseases, and went public in 2014. Although the company is still several years from generating revenue based on approval of these drugs, investors have valued it in excess of $1 billion.

It has taken some of them a few years to come around, but the big players in the pharmaceutical industry now realize that the industry really is entering the long-anticipated age of *personalized* (also called *precision*) *medicine*. They know that in future years there will be few, if any, blockbuster new drugs like Lipitor (which at its height earned more than $10 billion in revenues a year), and that it will be ever more common for the use of new drugs to be prescribed *only* if a genetic test shows that the patient has a particular genetic variant that the drug was designed to treat. This is already the case with new cancer drugs targeted to certain mutations in tumor cells. For example, the FDA recently approved both Tafinlar (dabrafenib) and Mekinist (trametinib) to treat those patients with metastatic melanoma whose tumors harbored a particular mutation in the *BRAF* gene known as V600E. There is little likelihood that commercial interest in developing orphan drugs for cancer will soon abate. Indeed, molecular biology now so efficiently subtypes cancers that the number of drugs that are launched with *orphan* status is proliferating. Approvals of "orphan" cancer drugs substantially exceed those for rare single-gene disorders.

Perhaps the most persuasive evidence that development of new drugs to treat rare genetic disorders has entered a significant growth phase is the emergence of about a dozen new gene therapy companies during 2013 and 2014. Fidelity Biosciences, a venture group in Boston, backed a new company called Audentes to focus on liver diseases; Children's Hospital in Philadelphia (CHOP) launched a new company called Spark Therapeutics, of which three of the founders were women who are top gene therapy scientists, that will focus on eye and liver disorders. In 2014 Third Rock Ventures launched Voyager Therapeutics Corporation, which will initially focus on using adeno-associated viral (AAV) vectors to treat disorders of the central nervous system. Also in 2014, a California-based company called Avalanche Biotechnologies that focuses on eye diseases had a highly successful initial public offering (IPO). Several other small start-ups (including AveXis and AAVLife) have secured capital and are initiating clinical trials. Most are using AAV vectors in an attempt to deliver a normal copy of a gene that is defective in patients with the hope that the vectors will deliver enough of the corrective genes to relevant tissues to ameliorate or even reverse the disease. A growing number of commercially conservative pharmaceutical companies have a deepening interest in AAV gene therapy.

Another way to assess efforts in developing new drugs to treat genetic disorders is to count the clinical trials underway to address them. The government-sponsored website Clinicaltrials.gov lists the vast majority of drug trials under way in the *world*. At this writing about 170,000 are listed. Of these, more than 1700 are focused on using various methods to treat rare genetic disorders. In June of 2014, a query for gene therapy (which includes studies that are focused on cancer) listed 3369 entries. There were 258 trials categorized as being focused on gene therapy for single-gene disorders. These numbers are much larger than similar listings were a decade ago.

Perhaps an even better way to estimate activity in developing drugs to treat rare genetic disorders is to estimate the number of compounds in Phase I, II, or III clinical trials that are listed on the websites of biotech and pharmaceutical companies. Early in 2014, I identified more than 400 such *trials*, a cohort that covered more than 100 different *disorders* (some of the more common genetic disorders, such as cystic fibrosis and Duchenne muscular dystrophy, are the subject of many research projects). A significant fraction of these drug development efforts will fail, but if even a tenth are successful, we should expect to see 40 new drugs in this space over the next 5–7 years. Because the biology of single-gene disorders is better understood than that of cancer and heart disease, the number of approvals that flow out of these trials has been shown to be significantly higher than is the case with small molecules (which are much more likely to have off-target adverse effects).

Yet another simple way to assess the status of drug development for orphan genetic disorders is to investigate the number of projects for which the FDA has agreed to grant "orphan drug status." This category, created by legislation in 1983 to promote new therapies for neglected diseases, is intended in part to incentivize biotech companies to do research on diseases for which there are only small markets. In the United States to qualify as an "orphan" disorder there must be fewer than 200,000 persons (about one in

1500 persons) affected with the disease. If a drug with "orphan" status wins FDA approval, it gains a commercial monopoly of 7 years. During that time no other company can market a closely similar drug designed for a similar purpose. However, a company can attempt to develop a different type of drug for the same disorder. Europe has similar legislation to accomplish similar ends that offers 10 years of protection.

Since it became law in 1983, the Orphan Drug Act has been invoked ever more often by drug developers, and this trend is accelerating. This is in no small part because as they learn more about the genetic component of many disorders, especially somatic cancers, researchers are splitting diseases into more discrete entities that have less than 200,000 patients. How many chemical compounds or biologics (as recombinant proteins are called) targeting genetic disorders have been awarded orphan status by the FDA? In the decade before 1983, when the law was enacted, just 10 drugs that would qualify as "orphan" were approved. Since 1983 more than 400 drugs with this protection have been approved. Most of these are for disorders other than monogenic diseases. But an important fraction has been for single-gene disorders.

The substantial growth in interest of the biotech and pharmaceutical industry in developing drugs to treat rare genetic disorders is driven in part by the growing power and advocacy of *nonprofit patient groups*, almost always entities started by parents of affected children who band together to generate support for each other and to drive research. Understandably, such groups have little tolerance for the slow time lines that have traditionally characterized drug development. Perhaps the most sophisticated and successful of these groups is the Cystic Fibrosis Foundation that under the able and passionate leadership of Dr. Robert Beall has over three decades raised and allocated hundreds of millions of dollars for research on this disorder. Today there are hundreds of similar (albeit much smaller) groups in the United States and Europe. They range from tiny organizations that have little more than a website and an annual meeting of concerned parents to scientifically sophisticated groups (often run by parents who have become disease experts) that focus with determination on driving research forward. One inspiring example is the Rett Syndrome Research Trust, directed by a parent named Monica Coenraads who over the last decade has become a well-known figure among neuroscientists who are trying to better understand this complex developmental disorder.

It is difficult to pinpoint the emergence of family groups that have chosen to focus predominantly on being directly involved in catalyzing research for specific disorders. Certainly the work of the Odone family in the 1980s to develop a diet modifying therapy for children with X-linked adrenoleukodystrophy (made famous in the movie *Lorenzo's Oil*), the efforts of John Crowley stimulated by the birth of a child with Pompe disease, and the work of Brad Margus, the father of two affected boys, to drive research into ataxia telangiectasia constitute three of the most inspiring examples of parents who became "players" in driving and *shaping* research into ultrarare disorders. Thanks to the rise of whole genome sequencing, which has sharply truncated the multiyear diagnostic odyssey that so many families once endured, one can now find a starting point for new research—the causative mutation—much earlier. In recent years I have met several families who after the birth of a child with an ultrarare disorder became passionately engaged in trying to ignite research into its molecular cause to an unprecedented level. At the risk of slighting such inspiring parents, I will say a few more words about one of them.

In January of 2014, while at the J.P. Morgan Healthcare Conference, the big annual biotech industry meeting in San Francisco, I met Matt Wilsey, for whom the birth of a child with an ultraorphan genetic disorder has also been life changing. Grace Wilsey was only the second child in the United States to be diagnosed with NGLY1 deficiency. Children with a mutation in the *NGLY1* gene do not produce sufficient amounts of an enzyme to degrade a protein, resulting in severe damage to cells. Grace's diagnosis came after an exhaustive search at several top medical centers and included whole-genome sequencing at Stanford University, Baylor College of Medicine, and the Broad Institute in Cambridge. Since Grace was diagnosed, Matt and his family have researched, recruited, and assembled a global team of some 40 researchers at 12 centers on three continents. The family also started the Grace Wilsey Foundation (GWF) to fund research and to raise awareness about this disorder. The foundation is the central hub managing multiple research projects.

What is unique about this story is the model. Matt Wilsey is a Silicon Valley entrepreneur. He has used lessons he learned from founding and managing technology start-ups to begin to unlock the mysteries of (and someday develop treatments for) an ultrarare genetic disease. The Wilseys did this by leveraging contacts, building relationships, and employing targeted cold calls and e-mails. Scientists were interested in joining the team because the parents were well informed and passionate to proceed. Additional recruitment tools included an early team of top talent, private funding, the lure of a complex medical challenge, and the group structure the family established. Much like the start-ups Mr. Wilsey has built, the NGLY1 team is run with a flat structure with no hierarchy. The team's core consists of open and fast sharing of information, trust, and collaboration. The global team operates as a business with objectives, milestones, and quarterly check-ins. The speed and collaborative methods with which the foundation operates may be unparalleled.

Mr. Wilsey strongly believes that the combination of dedicated families in concert with academic researchers and pharmaceutical companies is critical to developing rare disease treatments. He built his relationships with researchers and companies "one by one." He found a trusted source and asked for other experts. An interconnected network developed. The family foundation continuously monitors gaps in the scientific puzzle and searches out the best researchers to address those holes. Mr. Wilsey cold calls or sends concise, sincere e-mail requests and finds this generates responses. Of course, the work requires constant follow-up. Each time he adds a new researcher to the team, he circles back to existing teammates to ensure there is little to no overlap and that new researchers are well integrated. The GWF specifically looks for researchers who are willing to collaborate, and the foundation plays a critical role in encouraging collaboration, sharing, and follow-up. Mr. Wilsey has found that patient advocacy efforts and private financing catalyzes academia and industry researchers because they feel they are on a heroic mission, which they are. NGLY1 deficiency is far from having a treatment, but the team the Wilseys assembled is worth noting. Their effort exemplifies a new strategy that can be used for faster, cheaper, and more comprehensive disease understanding and treatment.

Both venture capital groups interested in rare disorders and biotech companies working to develop drugs for them regard patient groups as essential partners in their commercial strategies. A patient group that represents a particular orphan disease can be of great assistance in speeding enrollment in clinical trials. There are hundreds of extraordinary persons like Monica Coenraads, Brad Margus, and Matt Wilsey for whom the lightning strike of a terrible genetic disease in a child have redirected their lives and spurred them to set goals that they might have once thought to be impossible.

Still, these are long journeys. There is, undoubtedly, good reason to be optimistic about the ever-growing pipeline of new drugs to treat orphan genetic disorders. But even the most optimistic forecaster would likely agree that over the next decade (an eternity for parents of children suffering with these disorders) we are likely to witness approval for breakthrough therapies for only about 25 orphan genetic disorders. The sheer number of disorders, the paucity of our physiological knowledge, and the inherent uncertainties of clinical trials are rate-limiting factors. However, if we make significant progress with some platform technologies—in particular, gene delivery, gene knockdown, and gene editing—there may be more approvals for new drugs to treat orphan disorders than I anticipate. At the very least, a decade hence the annual number of approvals could be significantly higher.

Beyond Single-Gene Disorders

One of the most important insights that we have gained from efforts to develop therapies for rare genetic disorders (and one that is sure to foster much more investment) is the growing realization that *every one of us* is some way burdened by genetic variations that will eventually threaten our health. We now know that asthma, cancer, heart disease, diabetes, Alzheimer's disease, and virtually every other condition that we think of as *common* evolves and becomes manifest *in part* because of the inheritance of variations in genes that code for proteins that somehow influence the threshold of disease risk.

Although our odyssey of discovery through the human genome is only beginning, we are learning a great deal about the genetics of common disorders. Within each of the two haploid sets of genes with which we start life there is immense *variation*. A comparison of the DNA sequences of any two persons (excluding identical twins) would reveal several *million* spots where the nucleotides (DNA letters) differ. At a higher level of resolution—the protein sequence for which the genes code—there is also substantial variation among humans. Much of this variation is owing to differences in just one or a few of the several hundred amino acids (the median number among 30,000 human proteins is 384) in a protein, many of which have no detectable effect on function and appear (at our current level of resolution) to be inconsequential to human health. But statistical calculations based on comparing genetic variations in large groups of patients with risk for various disorders (efforts called GWAS for genome-wide association studies) convincingly show that many common variations in the human genome have small, but measurable effects on risk for disease. Since about 2005, scientists have conducted hundreds of GWAS, most of which have discovered many new genes in which variations affect risk for common disorders. For example, researchers at Johns Hopkins have identified 150 genes that influence the electrical properties of the heart. Dozens of genetic variations have been linked to stature; scores of others have been linked to increased risk for diabetes, obesity, osteoarthritis, asthma, and a host of other disorders. Each newly discovered linkage between a gene and risk for a disorder constitutes a potential research pathway that could lead someday to new therapies.

There are even a *few* variants that confer a surprisingly *large* risk for common complex disorders. A prime example is a variant of the apolipoprotein E (APOE) gene called *APOE4*, which (in the 2%–3% of Caucasians who are born having inherited a copy from *each* parent) confers a *substantial* risk for developing Alzheimer's disease about *10–15 years earlier* than would otherwise be expected. We do not know why some people with two copies of *APOE4* develop this disease whereas others do not. But we have amassed much evidence indicating that each of us on this planet has hidden in his or her genomes many variants—perhaps 20 to 40—that confer increased risk for serious disorders heretofore not viewed as genetic. In effect, *all diseases arise in part for genetic reasons.* This remarkable fact is changing how scientists, physicians, and patients conceive of illnesses and of drug development. Ultimately, it could constitute as fundamental a conceptual change as that wrought by the development of germ theory in the second half of the 19th century. Admittedly, many years will pass before the search for genetic variants yields much actionable information that becomes part of preventative health care (say, like checking blood glucose levels). But there are some important common disorders for which it might happen soon. I will briefly discuss two of them.

Parkinson Disease

A relatively common neurodegenerative disorder that afflicts about one *million* Americans, the more common signs and symptoms that comprise Parkinson disease (PD), were familiar to the ancient Egyptian and Greek physicians. Solid epidemiological data suggest that the lifetime risk for developing PD is 4%, and that it rises steadily for men until the age of 89 when it begins to decline! James Parkinson, an English physician, compiled the first full modern description in *Essay on the Shaking Palsy*, published in 1817. During the 19th century, a number of eminent European physicians, especially Jean-Martin Charcot in Paris, closely studied this condition, and more than a century ago physicians were already familiar with the classical signs of tremor, slowness of movement, muscle rigidity, and diminished affect. In the early 20th century a research neurologist who studied the brains of deceased patients found that certain brain cells contained abnormal material, now called "Lewy bodies." Today we know that PD is in large part a consequence of the selective dying off of cells in the portion of the brain called the substantia nigra, which produces much of the neurotransmitter (chemical messenger) known as *dopamine.*

The great therapeutic advance in treating PD came in the late 1950s growing out of work by the Swedish chemist Arvid Carlsson who showed that a compound called L-dopa (which is able to cross the blood–brain barrier) ameliorated symptoms in animals with induced PD. A few years later, a team led by a Harvard neurologist named George Cotzias showed that high doses of L-dopa were efficacious in humans with the disease. In 1969 Cotzias won a Lasker Award for the work he did to determine how to administer the drug to humans. Fittingly, the development of L-dopa, which once it crosses the blood–brain barrier, provides a substrate for those brain cells that are still alive to make dopamine, garnered a Nobel Prize for Carlsson in 2000.

L-dopa confers immense benefit to patients with PD by providing the chemical that is in short supply due to cell death (the causes of which are not yet well understood). But with time, as cell die-off continues, L-dopa provides fewer benefits and causes more serious side effects (such as the severe uncontrollable movement of the arms and legs that afflict America's best known patient, Michael J. Fox). The basic treatment for PD—supplying L-dopa to eke out as much benefit from residual enzyme as possible—has not greatly changed in four decades. For the most severely affected, a relatively new technique called deep brain stimulation (DBS), which involves the permanent surgical placement of electrodes in the brain, can offer substantial relief, but many people fear the neurosurgery, and only a small fraction (~10%) of eligible patients choose this route.

For some years, pioneers in gene therapy have been trying to develop novel treatments for PD. Some have used stereotactic surgery to deliver vectors containing various *neurotrophic* factors to the substantia nigra and its projecting tracts (a strategy that seeks to protect existing dopaminergic neurons with the hope that it will slow the course of the disease). Others have focused on delivering a vector containing the gene that codes for the enzyme that converts L-dopa to dopamine, in effect supplying the key enzyme that is low in concentration due to cell death. Thus far, the clinical trials with neurotrophic factors have not met their clinical goals.

In June of 2013 I talked with some of the scientific staff at the Michael J. Fox Foundation (MJFF) in part to learn what they thought about the growing evidence for genetic influences in risk for this disorder that afflicts so many Americans. In a little over a decade the likeable actor who was diagnosed with PD at 36—two decades earlier than is the case for most people—has built one of the world's most successful disease advocacy groups. Since 2000, the foundation has raised more than $300 million and channeled much of it to translational research—scientific work that is focused on developing new treatments or cures. Guided by its in-house experts, The MJFF is committed to supporting high-quality innovative research to find new treatments for PD. It supports exploring gene therapy to treat idiopathic (nonmonogenic) PD. The reason? Much evidence derived from the treatment of animals with induced PD suggests that it is possible to deliver the gene that codes for the key enzyme, aryl aromatic decarboxylase (AADC) that converts L-dopa to dopamine, to a small area of the brain. In Taiwan, scientists have used this gene therapy to treat an exceedingly rare *infantile* form of PD with some success in improving motor function. A gene therapy trial in which a vector carrying the AADC gene is surgically delivered under MRI guidance by a fine needle deep into the brain is under way. Perhaps the most crucial determinant of success will be the degree to which the surgeon can deliver the viral vector throughout the area known as the *putamen*. Gene therapy for PD is of great interest in part because it is *not* a single-gene disorder. In this case the viral vector is used as a vehicle to provide an enzyme that is lacking because of selective cell death. A tiny percentage of patients with PD do have a rare genetic form of the disorder (rare mutations in about 10 different genes are responsible).

Despite being mulitfactoral, risk for idiopathic PD is genetically influenced. Over the last few years, perhaps the most high-profile PD philanthropist has been Sergey Brin (the cofounder of Google) who has a family history of the disorder. Brin is a financial backer of 23andMe, one of the world's first "direct to consumer" genetic testing companies, which was cofounded by his wife, Anne Wojcicki. Several years ago, the company began to use the power of social media to create a registry of people with a family history of PD. By 2014 they had collected DNA from more than 7500 affected persons, creating the largest database concerned with that disease in the world. This will surely facilitate research. For over the last decade or so, scientists have found ever more evidence that variations in about a dozen genes increase the risk of developing PD. In most cases, we do not yet know how or why, but this research has provided scientists who develop drugs with new targets to study. Hopefully, lessons learned from studying the natural history of the disease in large cohorts will not only lead to new treatments for patients with the variants, but to therapies that will help others as well.

Of the several genes being studied one (called *LRRK2*) stands out. Population studies suggest that a common variant in this gene more than doubles the lifetime risk for PD. The discovery of the risk for PD associated with *LRRK2* and others has created a challenge for the MJFF and the many thousands of families that are counting on it to drive research forward. As one staffer at MJFF said to me, "we are undergoing a paradigm shift in regard to the disease. We have got to educate our families about genetics, and we may need to urge them to provide DNA for testing for research purposes. We have never done that before." The discovery of important genetic risk factors for Alzheimer disease, PD, and virtually all other common disorders offers the hope that each gene creates a new research opportunity that could lead to new therapies.

The idea of using gene therapy to treat advanced PD raises interesting and complicated questions about the economics of this new field of drug development. If 10% of patients with PD underwent gene therapy, that would require about 100,000 neurosurgical procedures a year to deliver a drug that might be priced at around $100,000. This scenario envisions new therapeutic costs to the health care system of about $10 billion a year in the United States alone. That sounds like an extremely large sum, until one begins to calculate the costs that it would likely *offset*. Currently, the care of persons with advanced PD is immensely expensive, even though largely custodial. The Parkinson Disease Foundation estimates that each year in the United States alone more than 60,000 persons are diagnosed with PD. The foundation also estimates that the combined direct and indirect annual cost of caring for all affected patients in the United States exceeds $25 billion. Given those credible numbers, it is at least possible that a one-time gene therapy that greatly slowed the steady descent of PD patients into the need for custodial care could actually *reduce* health care costs. Of course, such an analysis must wait until the FDA decides that a new therapy for PD is safe and effective, an event that may be at least 5 years in the future.

Autism

One of the most serious common disorders for which deep genomic analysis will capture new and actionable molecular information is *autism*, a developmental disability that is currently said to affect more than one in 100 children in the United States. Unfortunately, autism is a broad term for a collection of signs and symptoms that arise from many different kinds of genetic and/or environmental injury to the brain. Despite decades of research, experts still have great difficulty defining its clinical boundaries, which have been steadily broadened over the years.

During the 1990s, I oversaw a clinic that evaluated children with developmental problems, many of whom were suspected of having autism. The staff consisted of about a dozen people, including experts in audiology, psychology, occupational therapy, physical therapy, pediatric neurology, speech therapy, and nursing. They worked in teams to evaluate children from throughout Massachusetts who were experiencing serious developmental delay and/or behavioral problems.

The clinic's mission included educating trainees. Each Friday for nearly a decade I spent several hours with Dr. Raymond D. Adams, one of the fathers of modern neurology, as he taught residents to evaluate individuals with neurological disorders. Tall, slightly stooped, ruddy faced, and patrician in nature, Adams had a mildly condescending attitude toward nonphysicians, a trait that did not endear him to the members of the multidisciplinary teams. Nevertheless, his encyclopedic knowledge of the brain made for memorable educational sessions.

Adams (who died at 93 in 2008) spent much of his career at the Massachusetts General Hospital and Harvard Medical School, where he made extraordinary contributions in neuropathology. Long before there were MRI scans, neurologists dissected the brain at autopsy in an attempt to correlate the findings that burdened the patient during life with lesions ranging from stroke to Down syndrome. When I was a resident at Boston City Hospital in the early 1980s trying to learn the rudiments of neuropathology, I sometimes attended "brain cutting" sessions. After one medical resident summarized the history of the patient who had died, the others opined as to whether we would find a discernible lesion in the brain, and if so, where it would be located and what it would look like. A mark of a fine neurologist was his or her ability to assess the clinical history and physical findings and from them infer the nature of the brain lesion. After we had guessed, the senior physician would quite literally slice the brain, just as one might carefully slice a loaf of bread, looking for abnormalities.

One of the disorders that most perplexed Adams was autism, a condition that he had been thinking about since soon after an Austrian physician named Leo Kanner (who immigrated to the United States in the 1930s to work at Johns Hopkins, where he established one of the nation's first child psychiatry clinics) described it in 1943. From the start, one challenging aspect of autism has been that when the brains of the few children who died young with nonsyndromic autism (i.e., not part of some other disorder) were dissected, neuropathologists could not (except in rare cases) find any obvious abnormalities. This was especially curious because in the mid-1980s, the notion of "autism spectrum disorders" (ASD) was not so well accepted as it is today, and the few brains that were studied were usually obtained from patients with severe disease, a situation in which researchers expected that the brains should have discernible abnormalities. Based on his experience with other severe neurological disorders, Adams was puzzled that even under the microscope the brain tissue of autistic children did not yield good clues concerning the cause of their disability.

I well remember how he described the syndrome to the medical students and house officers. Children with autism, he said, were disconnected from the social fabric of the world. They seemed to construct an inner, private world within the boundaries of which they were content. They feared intrusion and when stressed found physical release in highly repetitive behaviors as though riding a neural loop that gave comfort. Most were not mentally retarded, but many were cognitively disorganized. Adams particularly enjoyed discussing the case of a young autistic man who lived at Fernald. Unable to read, with limited ability to speak, and severely constrained in his ability to interact socially with other humans, the man was, nevertheless, a superb, self-taught pianist who had memorized several hundred songs. Even more intriguing was his talent for calculating for any given year the day of the week on which one's birthday would fall! How could one's brain operate with such constraints in some areas and function so magnificently in others? Adams viewed autism as a collection of *orphan* disorders most likely due to errors in one or more genes affecting the biochemistry of the neuron. He attributed the fact that it was diagnosed about four times more often in boys than girls (one in 1500 compared to one in 6000) to mutations in genes on the X-chromosome.

Today, the dominant narrative about autism is quite different. I know of no other disease or disorder about which thinking has changed so profoundly and so quickly (but not necessarily correctly) in the last two decades. Consider that because of uncertainty about how to define the disorder, the American Psychiatric Association did not even accord autism an official diagnostic status until 1980. As recently as 1988, experts in the field taught that autism affected only about 4 to 5 children in 10,000. Yet in 2014 in cities across the United States one could see messages on billboards soliciting funds to help cure a disorder that is claimed to affect as many as *one in 88* children.

Although the diagnosis of autism is still made more often in boys, the upsurge in diagnosing girls with autism is such that they now account for ~30% of new diagnoses (an increase of nearly 50%). When the prevalence of a disorder in a population of several hundred million people increases 20-fold over a generation, one might reasonably infer that it is because of the emergence of a novel infectious agent or some new environmental toxin, not because of changes in our genome. The frequency of a DNA variant in the population changes over much longer time periods. Yet, despite immense efforts by many scientists, none have been able to provide an environmental explanation for the dramatic increase in the incidence of autism. This mystery has, not surprisingly, stimulated wildly misguided and often harmful suggestions, the most notorious of which is the persistent (and demonstrably untrue) argument that vaccines or some chemical in them are the cause.

What is going on? A substantial reason for the vast increase in diagnoses is that we have *broadened the definition* of the disorder. We no longer speak of autism; we speak of ASD. In some states, if a pediatrician diagnoses a child (who is clearly burdened with some developmental problem) as having an ASD, a cascade of otherwise unavailable social services flows to the child and his family. Does the decision by medical groups to enlarge the domain of ASD reflect greater sensitivity to the need to help kids with behavioral problems earlier and more intensively? Perhaps. Are clinicians today more adept at recognizing behavioral and cognitive problems in young children than those of a generation ago? I doubt this. Training in the early recognition and management of subtle childhood behavioral problems has never commanded great emphasis in primary care pediatrics. It has always been the case that general pediatricians, aware of the wide range in which children develop, engage in watchful waiting, and, only when they develop a high level of suspicion that there is a problem, do they refer the child to a neurologist.

A full discussion of the diagnostic challenges in autism is beyond the scope of this book. I will focus instead on the rapid and steady increase in our *knowledge* of the *genetic* causes of autism. These diagnostic advances result in part from the wide availability of new analytical tools that can discern small changes (smaller than can be seen with the most powerful light microscope) in the stretches of DNA that reside on each chromosome.

One of the earliest indications that we would someday be able to attribute much of autism to a large number of as yet to be described single-gene conditions arose out of a chance discovery in 1969 by a clinical geneticist named Herbert Lubs (which I described more fully in Chapter 8). He discovered that if one cultured white cells from men with a certain, fairly common type of mental retardation in a medium that lacked folic acid, and then broke those cells open and examined them under a light microscope, one could see a region on the long arm of the X chromosome that appeared unraveled or broken. Lubs' discovery was quickly replicated in other laboratories. About a year later, a geneticist named Fred Hecht coined the term Fragile X syndrome, and the name stuck.

Because most causes of mental retardation were largely unknown, researchers were quick to screen large populations of mentally retarded men to determine if any had this newly defined molecular disorder. They were surprised and gratified to learn that ~2%–4% of such men had Fragile X syndrome. Even if there is no treatment to offer, a new diagnosis corrects prior errors, helps to focus research efforts, and assists families in understanding and coping with the problem. I recall one middle-aged man who I diagnosed with Fragile X. In doing so, I destroyed the family narrative that had been told and retold for four decades—that he was mentally retarded because of a fall from a crib in his infancy, an event about which his still living parents felt immense guilt. When I told his mother that the man's problems did not arise from the fall, but were due to a genetic error, she told me that the news had lifted a huge burden of guilt from her mind.

Another diagnostic advance that occurred about the same time was that cytogeneticists (chromosome experts) learned how to improve on a technique called "chromosome banding." By using certain dyes, they could under the microscope characterize in a highly reproducible way each set of human chromosomes with about 850 light and dark bands. If a band was missing, it meant that the underlying DNA was not there. Over the last two decades microdeletion analysis has taught us that there are several spots on human chromosomes that if deleted are associated with, and almost certainly cause, autism or other forms of intellectual disability.

The next major advance in studying microdeletions came with the development of gene arrays. Arrays consist of tens of thousands of short strands of DNA representing specific addresses on the full complement of human chromosomes. One can now draw blood from an individual, harvest and purify his or her DNA, chemically digest it into little pieces, and layer those pieces over these arrays. The probe DNA should blanket the array. If one of the diagnostic strands is not covered, it indicates that the patient is missing a stretch of DNA in that region. Gene arrays can detect deletions that are *much smaller* than the ones that are detectable with chromosome banding. Not surprisingly, by greatly increasing the sensitivity of the test, scientists are finding many more clinically significant microdeletions. Some are so small that they have not yet been definitely associated with any abnormalities. However, a steadily growing number are being found in people with autism or other, largely psychiatric, problems. Currently microdeletion syndromes collectively account for ~5%–7% of autism.

Today doctors could attribute the causes of ~10% of *all* cases of autism to specific genetic errors. This may not seem like much, but it constitutes a major advance—for a positive test provides a precise explanation. The result halts the diagnostic odyssey that parents embark on when they first confront autism. It also often allows the clinicians to offer the parents a much clearer picture of recurrence risks in future pregnancies. When a microdeletion is found in the child, the next step is to test the parents. If neither have the microdeletion, it is a new mutation that occurred in either the egg or sperm that initiated the last pregnancy, making the risk in future children vanishingly small. It is possible, however, that one of the parents also has the mutation, in which case the recurrence risk is high. How is it possible to have a microdeletion, and not be affected? We do not yet know enough to answer this question in detail. Sometimes, on closer study, physicians recognize that the parent with the same microdeletion, although less severely affected and able to function independently, has some psychological problems that are probably due to the same genetic defect. In other cases the parent appears completely normal, and we are left to ponder whether some insidious environmental factor interacted with the genetic error to cause the child's problems.

In November of 2014 two large consortia of scientists published the results of their efforts to use whole exome sequencing (scanning the DNA that provided the code for making every protein in the body) to discover evidence of genes in which variations increase the risk of developing autism. One group studied 2500 families with two normal parents, one affected child, and one normal child. They identified about *400* genes in which new variations (that is, DNA changes in the affected child that are not present in other family members) were associated with developing autism. The second group, taking a somewhat different approach, found that among genes known to be important to brain development, variations in more than 100 were associated with an increased risk for autism. Such large numbers of genes should be no surprise. As the human brain is the most complicated entity on the planet, one should expect that many genes are involved in its development and that there are many ways for things to go wrong.

Each time a researcher correlates a newly discovered microdeletion (the most prevalent is located on the short arm of chromosome 16, which affects about one in 3500 people) with autism, he or she opens up a new avenue of research. What genes are missing? What roles do the proteins for which those genes code play in the development of cognition? To what environmental forces might they be particularly sensitive? Might these microdeletions be common enough that all children should be tested early in life so that we could in some way intervene on behalf of those who are discovered to have them to try to ameliorate potential injury? What do the answers to these questions imply for research into new therapies?

Today, researchers are using refined molecular diagnostic techniques to uncover the many different molecular pathways that lead to the condition we call autism. The term "autism spectrum disorder," anticipated what is now clear. Just as cancer is a shorthand term for scores of disorders, so is autism. It is highly likely that none of these microdeletion disorders will account for >1% of cases. It is still possible, but very unlikely, that some as yet unidentified environmental factor will be found to play a major role in the astounding increase in the prevalence of ASD over the last 30 years. At the moment I am aware of only one investigative study of a single-gene defect that may someday be shown to explain >1% of cases. I briefly summarize it because it reveals some of the complexities that clinical detectives face.

In 2011 Arthur Beaudet, then Chairman of the Department of Human Genetics at Baylor School of Medicine in Houston and one of the nation's leading human geneticists, began investigating a possible new suspect in autism. A highly regarded scientist, Beaudet is typically at work well before the sun is up. He has a skeptical temperament and an obsession with detail. A long-time reviewer and editor, Art demands that each element of any scientific hypothesis be supported by a convincing set of experimental evidence.

In the course of studying some children, researchers in Beaudet's laboratory had discovered a microdeletion on the X chromosome that had not before been associated with having autism. When he and colleagues in The Netherlands investigated what genes were deleted, they found that the region included a gene that was essential for the synthesis of an amino acid called carnitine. When they looked for the microdeletion in the normal population, they found it in about one in 500 boys. If that finding holds up in larger studies, Art will have discovered one the world's *most common* genetic disorders of metabolism! That is itself extraordinary, but the key point is the possible connection between carnitine deficiency and an increased risk for autism. Carnitine is essential for the development of the brain. People get most of what they need from diet (red meat is loaded with carnitine), but humans also have a metabolic pathway to make it from scratch. In boys born with this microdeletion, the endogenous carnitine production pathway is blocked. But, because of the abundance of carnitine in the average diet, the genetic defect should be a benign condition. Yet, in his initial screening for the deletion in boys *with autism,* Art had found that the deletion was *more* common than among boys in the general population.

Beaudet suspected that if boys who were born with the microdeletion that prevented them from *making* carnitine were—for whatever reason—deprived of *dietary* carnitine for even a short period of time (a few days) in the first 2 or 3 years of life, that this might cause or increase the risk for autism. But why might they be deprived? As all parents know, when a little child develops a febrile illness, he or she often refuses to eat. As instructed to, the parents keep the children hydrated, but they usually do not try to force them to eat, waiting instead until the children are ready. So, if the boys born with the deletion for carnitine synthesis are hit with a febrile illness and do not eat much for several days, they could enter into a state of relative carnitine deficiency. Art hypothesized that among young boys with the metabolic disorder, transient carnitine deficiency could harm the developing brain. The fact that the carnitine deficiency microdeletion was on the X chromosome also fit nicely into the autism puzzle; X-linked disorders primarily affect boys, and boys are diagnosed with autism much more often than are girls.

The discovery of this new genetic disorder—trimethyllysine hydroxylase epsilon (TMLHE) deficiency—is an early step in what will be a long, complicated investigation. If one in 500 boys is born with TMLHE, but only a small percentage of them develop autism, the genetic error might not explain many cases of the developmental disability at all. It could be merely a coincidence that a boy had both TMLHE and autism. But it is also possible that TMLHE deficiency significantly increases the susceptibility to developing autism if some other environmental factor—dietary carnitine deficiency—comes into play.

Beaudet scoured the medical literature on carnitine. Of interest was that some parents of children with autism had long claimed that supplementing the diet with carnitine (which can be purchased at low cost from many health food chains) significantly improved behavior and educability. This was of limited scientific value, as many other parents made similar claims about the similar value of other food supplements. Of greater interest was a paper published years ago by a biochemical geneticist in Egypt. He studied carnitine levels in 30 boys who clearly met the diagnostic criteria for autism, and found that most of them had *low* blood levels of carnitine. Perhaps because he published his work in an obscure journal, it had not elicited any follow-up. But his work fit with the TMLHE hypothesis. In Egypt poor children are much more likely to eat a diet that has little meat and that is therefore relatively deficient in carnitine.

Beaudet and his colleagues are continuing their study of mutations in the *TMLHE* gene and risk for autism. Early data suggest that the microdeletion that disrupts TMLHE is *not* a risk factor for autism in families with one affected son, but that it may be a risk factor in families with *two or more* affected sons (a group of families that appear more likely to have affected sons because of a specific factor as opposed to multiple factors). Of greater interest are studies suggesting that if one supplements the diet of affected boys with carnitine that they experience improvements in behavior. It is not yet possible to know if TMLHE deficiency will explain even 1% of autism, but this finding might turn out to be an important contribution in the study of an immensely complex clinical problem.

Precision Medicine

If one thinks about one's extended family, perhaps starting with hazy stories about great-grandparents one never knew, then moving to grandparents about whom one has warm, but misty, memories, then to aunts and uncles and cousins, one often finds that some disease seems to recur in the extended family more often than might be suggested by chance. This is why medical students have been taught for generations to include the "family history" in the evaluation of any new patient.

Looking over three generations of family history on my mother's side, several common elements of health history emerge. The most impressive is that several of my ancestors were afflicted with colon cancer. Depending on how far I cast the family net, I can identify seven or eight. Some of them died of their colon cancer; some did not. A great-uncle died of colon cancer before the age of 50 and my grandmother died of it at 70. Two of my aunts (her daughters) underwent surgery for colon cancer before they were 65, but one died of heart disease in her 80s and the other at 93. My family history suggests that there is a dominantly acting allele that predisposes to colon (and perhaps other) cancers traveling through the generations. In contrast, another theme of my maternal family history is longevity despite living comparatively unhealthy lifestyles. One great-aunt was obese, smoked heavily, and enjoyed several glasses of wine a day well into her 90s! My dear

Aunt Lorraine who was in good shape until she died of a stroke just shy of 93 might have lived a few years longer had she allowed doctors to treat her chronic atrial fibrillation!

Most families are like mine. Just a little sleuthing usually uncovers some medical problem that runs in the family. We will in the next decade become able to understand the molecular basis for such stories and, in many cases, identify options to reduce or manage the genetic risk. The great diseases that dominate the developed world in the early 21st century will be parsed into many well-delineated subtypes. In time we will learn that there is an optimal way to treat each subtype. That is, we will know a lot more about the particular journey that our cells take as a particular disease unfolds. This will eventually allow us to develop precisely tailored therapies that will target specific molecular profiles that characterize the subtypes.

Right now this evolution is proceeding rapidly in cancer medicine. Molecular analysis has made it clear that the century-old paradigm of thinking about cancer according to which organ it affects (lung, breast, prostate) is inadequate. Cancer is a disease in which over time a number of mutations accumulate in a cell and its progeny. Each mutation deranges a certain metabolic pathway, in turn creating a new biological profile of the developing tumor. The goal of cancer therapeutics has long been to use drugs to kill cancer cells while sparing healthy tissues, a goal that despite seven decades of effort we have failed to achieve. But a new day has dawned. By performing DNA sequencing on tissue removed from a cancer, laboratories can compose a molecular mug shot of the cancer, and point oncologists to a therapy that they might not otherwise have considered. Much research is needed to develop drugs that precisely target newly discovered cancer mutations. But this is almost certainly an easier challenge than it has been to develop a drug to fight cancer when the molecular target is *not* known. If one of the most striking early successes in precision medicine—the impact of a drug called imatinib (marketed as Gleevec by Novartis) on chronic myelogenous leukemia (CML, a cancer in which the body massively overproduces white blood cells) is a harbinger, the results for cancer patients are likely to be transformative.

Scientists mastered techniques that allowed them to study human chromosomes in the 1950s, but it was not until 1956 that scientists confirmed that the human karytope contains 46 chromosomes—22 pairs of autosomes and one pair of sex chromosomes. The availability of these new cytogenetic techniques led many to search for cellular abnormalities that might explain diseases. In 1960, Peter Nowell and David Hungerford applied these techniques to cancer cells and found that virtually all patients with CML had an odd-shaped chromosome, which soon became known as the *Philadelphia* chromosome for the city in which it was discovered. In 1973 Janet Rowley, a human geneticist at the University of Chicago who studied this cytogenetic abnormality for more than a decade, used chromosome dyes to show that the abnormal chromosome was due to a reciprocal translocation. That is, it is derived from breaks in two chromosomes after which the broken segments reattach in a different place. The Philadelphia chromosome is a translocation involving the long ends of chromosome 9 and chromosome 22. It took another decade of research before scientists at the National Institutes of Health (NIH) and in The Netherlands showed that the reciprocal translocation results in the accidental creation of a new protein that is a *fusion* of parts of two others. This new stretch of DNA is in effect a novel aberrant protein of the family known as tyrosine kinases. In its normal state, this protein regulates the production of white blood cells. The aberrant form, known as bcr-abl, operates longer and with higher power, vastly overproducing those cells. It is a key driver of CML in nearly all patients.

In the mid-1990s, Dr. Brian Druker, an oncologist who was working in Boston and who now heads a cancer institute in Oregon, speculated that a molecule that inhibited bcr-abl would be an effective treatment against CML started searching for such molecules. His effort led him to Ciba-Geigy (the forerunner of Novartis), which had a library of tyrosine kinase inhibitors. There he teamed up with Nick Lydon, a chemist who specialized in early drug development. By 1996, the research teams had identified a molecule that reduced the production of white blood cells by ~98% in cell culture.

Druker pushed hard for clinical trials. Even in the first small safety study, the drug (known then as ST571) showed remarkable efficacy. In 2001 the first large clinical trial to test the *efficacy* of the tyrosine kinase inhibitor called imatinib that the teams had chosen as the lead compound, halted the CML in its tracks. Druker's group reported a "complete hematopoietic remission" in 53 out of 54 patients, an absolutely astounding result. Equally impressive, in most cases Gleevec conferred *long-term* benefits. Before the development of imatinib, only 30% of patients with CML lived for 5 years from diagnosis. But after 5 years of treatment with Gleevec nearly 90% of the patients were alive, again an astounding advance in a field in which gains in survival are often measured by several months.

In effect, Drucker and his colleagues had developed a drug that drove a spike into the heart of a cancer. In 2009 Druker, Lydon, and Dr. Charles Sawyer, an oncologist who had worked closely with Druker on CML, won the

Lasker Award for their research. Over the last decade Gleevec has been approved for treating 10 other cancers in which cells have tyrosine kinase abnormalities; it has been shown to work especially well in treating a rare cancer of the gastrointestinal tract, and it is effective in a disorder called hypereosinophilic syndrome (HES). By the end of 1914, Gleevec has greatly extended the lives of more than 150,000 persons with CML throughout the world. It is now routine for ~80% of patients to live for more than a decade after being diagnosed with CML.

Among the most impressive advances in precision medicine are those that have used tumor genome analysis to develop new drugs to restrain highly aggressive metastatic melanoma. In 2011 a paper in *The New England Journal of Medicine* reported that in a large study of patients with a so-called *BRAF* mutation, a drug (vemurafenib) that specifically targeted that mutation so outperformed the standard of care (dacarbazine) after 6 months that the data safety monitoring board terminated the study and recommended that patients receiving dacarbazine be switched to vemurafenib. Such decisions are quite rare in clinical trials. Marketed under Zelboraf, vemurafenib became the first mutation-specific drug that the FDA approved to treat metatstatic melanoma. Of the 17 drugs that the FDA has approved for treatment of melanoma, four have already been developed to treat specific *BRAF* mutations.

Over the next few years, genomic analysis of tumors (known as tumor profiling) will become a *routine* part of cancer care. Mutational analysis of tumors will not always suggest a choice of therapy, but the practice will greatly extend our understanding of the natural history of tumors. It is highly likely that a decade or so from now that the organ of origin will be viewed as secondary to two other questions: What is the genomic profile of the tumor? What is the molecular target that we should assault?

Advances in molecular profiling of cancer is stimulating similar work in other major disease categories, especially heart disease. For decades cardiologists have crudely divided cardiomyopathies into two major categories—hypertrophic cardiomyopathy (HCM) and dilated cardiomyopathy (DCM). The former is characterized by unusual thickening of the heart muscles, which can lead to sudden death (due to restriction on the blood vessels that feed the heart muscle); the latter is characterized by a thin-walled, floppy heart that lacks the ability to squeeze out enough blood with each power stroke. Over the last decade the discovery of mutations in various genes responsible for cardiac structure or function (work pioneered by Dr. Christine Seidman and Dr. John Seidman at Harvard) have led to the realization that there are many forms of late-onset, single-gene, dominantly inherited HCM. It is likely that much late-onset heart disease is influenced by mildly deleterious mutations (often called hypomorphs) in the same genes. Drugs developed for uncommon genetic forms of cardiomyopathy may well offer new treatment strategies for thousands of other patients as well. A similar approach to developing new drugs for heart disease is also unfolding in regard to dilated cardiomyopathy. Armed with DNA sequencing tests, cardiologists will soon be subtyping patients with this disorder to look for predisposing germline variants. In turn, they will use these to make helpful genotype-phenotype correlations. These will eventually inform future drug trials that will analyze new drugs by stratifying responsiveness to the differing genetic profiles of the patients.

Another important consequence of low-cost, highly accurate DNA sequencing is that we will be able to determine with certainty whether or not a child with a disease of unknown etiology in fact has a new single-gene disorder. An early demonstration that this was possible came in 2010 when a team led by Richard Lifton, the Sterling Professor of Genetics at Yale Medical School, was asked to study the DNA of an infant in Turkey who was thought to have a rare kidney disease. In just a few days the team ruled out the suspected gene defect, and discovered a causative mutation in a different gene! For many years the NIH has supported a team to act as the diagnosticians of last resort for patients with truly mysterious disorders. Armed with DNA sequencing tools, they are now able to discover if the disorder is due to a defect in a single gene with relative ease.

Perhaps even more important, by trolling deeply among human populations, we may find gene variants that *protect* against common diseases and that will open important new areas for drug research. In 2014 a group that included scientific teams from academe and from Pfizer reported that they had found a gene mutation that *protected* against developing diabetes! They began their search among about 400 elderly people who had all the bad habits one associates with type 2 diabetes, but who *did not* have the disease. Two had mutations in a gene called *ZnT8*. Expanding the search to 18,000 people in Sweden, they found 31 persons with similar mutations who seemed to be protected from diabetes. Dr. David Altshuler (then at Harvard Medical School) worked with scientists at deCODE genetics in Iceland to query its extensive genetic and clinical records of that population. In an *hour* they were able to determine that *ZnT8* mutations were six times more common in nondiabetics than in diabetics! The data were so counterintuitive that their report was at first rejected for publication. They then repeated the study in another 13,000 persons and got the same results. Also in 2014, a research team reported that in a large population study in which they sought to find gene variants that increased risk for heart disease they had instead found several that *protected* against it. Of special interest is one that is strongly associated with very low levels of triglycerides. Such findings often suggest new avenues for drug development.

Footing the Bill

It will be no surprise that the price of mutation-specific drugs for the care of cancer patients is high. Standard therapy with Gleevec costs about $75,000. The cost of treatment with vemurafenib (Zelboraf) for a year is more than $300,000. This figure is particularly disconcerting because we do not yet know how many years (or months) of survival vemurafenib will on average deliver to patients with metastatic melanoma. Early findings suggest that it may extend median progression-free survival by about 4–5 months. In the case of malignant melanoma this is an impressive improvement, but does it justify the expense? Will government programs continue to cover the costs of this drug, which is about 10-fold greater than that of dacarbazine (the standard therapy)?

So far, new drugs to treat orphan genetic disorders and those that target mutation-specific tumors constitute a tiny portion of the nation's overall pharmaceutical bill (about $325 billion in 2012). In the case of orphan genetic diseases, that is likely to remain the case for the next few decades (partly because under the best of circumstances it takes at least 7 years to develop a drug to the point of approvability). Also, the cancer market is vast and the resources being mobilized to develop new drugs far eclipse those dedicated to inherited single-gene disorders. Sooner or later, the cost of drugs targeting subtypes of cancer will raise difficult ethical questions about allocation of resources. The most likely outcome of debates over cost will be to develop algorithms to examine drug efficacy. In the language of the policy wonks, the question will be: What is the projected cost of a quality adjusted life year (QALY) for the patient? Already in Great Britain, an agency called the National Institute for Health and Care Excellence (NICE) sometimes rejects efforts to market new drugs in that nation based on its analysis that a small potential gain in survival does not justify the expense involved.

Several years ago, as part of an effort to assess how those who pay the bills might react to very expensive drugs to treat rare genetic disorders, I interviewed an executive of a major health care insurer in Florida. I asked him: "How much could a company charge for a drug that had been approved by the FDA as safe and effective in treating a life-threatening genetic disorder in a child? He gave a terse response: "It can charge any amount it wants." Because the drug had been approved by the FDA and constituted the only available treatment for a severe disease, his company would have no choice but to cover the cost. But then he surprised me. "We might even brag about paying so much to help save a child," he said, "but don't ask me to increase the amount I pay for drugs for diabetes." He later explained that given how rare the disease is there would likely only be a few children that his company covered who would need it. Thus, it would have little impact on his business.

Since that conversation some resistance to the high cost of new drugs to treat children with rare genetic disorders has recently emerged from an expected quarter. Since 2008, many state-based Medicaid programs (which get about half their funding from the federal government)—which nationally cover health care costs for 70 million Americans—have experienced unrelenting financial stress. In 2014 the Arkansas Medicaid program balked at paying for Kalydeco, the new cystic fibrosis drug (which costs about $300,000 a year), for those few patients in the state who carry the particular mutations for which it is clinically indicated. The program's refusal was couched in a technical argument—that it would pay for the drug only after other the clinical benefits of other cystic fibrosis (CF) drugs were exhausted. This is a specious argument, for Kalydeco was created for a precisely defined subgroup of patients for whom there are no other really efficacious drugs to use first. Indeed, the molecular logic of Kalydeco argues that patients should begin using it early in life, when they are still quite healthy. Shortly after the Arkansas Medicaid program's recalcitrance became known, the patients sued, and the state backed down. The settlement removed additional drug approval restrictions that the state agency had demanded. Now patient eligibility for the drug merely requires that they clinically satisfy criteria recognized by the FDA.

Despite the events in Arkansas, I do not believe that there will be substantial societal resistance to paying (indirectly through overall increases in insurance premiums or directly through increased taxes) for new, highly expensive therapies for people burdened with rare genetic disorders. First, unlike most cancers, which are diseases of the elderly, most monogenic disorders affect children. There are many congenital disorders for which our society routinely absorbs the cost of many hundreds of thousands of dollars in care during the first year or two of life. Examples include liver transplants, multiple operations to repair severe malformations of the cranial bones, and severe hemophilia. In addition to our natural inclination to help children, the fact is that effective new drugs generate far more QALYs than do interventions in the lives of old people. Further, efficacious drugs targeting pediatric age groups might be less expensive than the current cost of delivering mainly symptomatic care. Successful therapies will convert many economically unproductive lives into productive ones. Of special interest is that in several of the most promising new therapeutic approaches, including gene therapy, a *single* treatment may cure or ameliorate the disease for life. For many genetic disorders, a therapy costing as much as $1 to $2 million could still be cost-effective. Some existing chronic therapies—such as the treatment of severe hemophilia with Factor VIII therapy—sometimes exceed that level in just a few years. Arguments to refuse to cover or limit coverage of therapies for genetic disorders will be especially difficult to make if they are highly effective, which many will be.

Challenges Ahead

The 20th century saw immense strides in the struggle against infectious diseases. In 1900 tuberculosis—the white plague—was the leading cause of death in the United States and Europe. In 2000 heart disease held that title, and tuberculosis (TB) did not make it into the top 20. That year only three persons in 100,000 in the United States were diagnosed with TB, and most of them were successfully treated. In 1900 children died primarily of pneumonias. The beds in pediatric wards were filled with children struggling to survive acute onset, short-term illnesses that either killed quickly or left the patient in good health with a strengthened immunity. Currently, about one-quarter of the beds in pediatric hospitals are occupied by children whose illnesses are genetic in origin, including many of the diseases I have discussed (like cystic fibrosis and sickle cell anemia), as well as scores of other much more rare disorders. It is likely that over the next few decades the fraction of beds devoted to persons with genetic disorders will continue to increase.

Although all crystal balls are cloudy, by considering our current state of clinical knowledge about rare disorders, the efficacy of the existing therapies that I have discussed, the availability of new technologies to better understand the disorders at the molecular level and to develop efficacious methods to treat them, I can hazard some reasonable *guesses* about the next two decades.

Most current therapies—moderately successful interventions that took decades to develop such as diet for inborn errors of metabolism, protein replacement for hemophilia, blood transfusions for β-thalassemia, and enzyme replacement therapy for lysosomal storage disorders—will gradually be replaced by superior therapies. Some of the sophisticated new approaches that I have discussed—substrate reduction, chelating agents, delivery of oligonucleotides, chaperones, and the creation of small molecules that improve the function of defective enzymes—will provide moderately positive benefit for a some disorders, and will play an important, but *relatively brief*, tenure in the therapeutic arsenal. This moderately skeptical comment is driven by the fact that such approaches generally do *not* address the underlying causes of the diseases.

Two major technological developments will drive the next major advances in combating severe childhood genetic disorders. The first will be *gene therapy*. Drug developers will harness ever more sophisticated viral vectors for three major purposes: (1) to deliver a normal copy of a gene of interest to target tissues, (2) to deliver small RNA molecules to shut down the production of aberrant proteins that are damaging the cells, (3) to deliver gene-editing systems such as the CRISPR system to *cut* mutations out of genes and replace them with the correct nucleotides. All three approaches share one major hurdle.

The three great problems in gene therapy are delivery, delivery, and delivery—getting the right amount of viral vector to the right tissues at the right time (and in a way that ensures that they will make the right amount of protein). Based on the dramatic developments we are beginning to see from systems that focus on transducing human stem cells and returning them to the body to make needed proteins and from others that use AAV vectors to transduce human cells in vivo and deliver normal genes to them, I believe gene delivery will be widely and successfully used to treat many, many disorders over the next two decades. The viral engineers will develop new capsids (the surface of the viral particles) that will circumvent the problem that many people have neutralizing antibodies (generated from subclinical infections sometime in their past) to current vectors. They will make vectors that have high affinities for tissues of interest and that detarget (do not infect) tissues that are not of clinical interest. Why deliver a virus to *all* the organs if the goal is to treat a *liver* disease? By having an array of vectors that are sufficiently distinct, it will be possible to re-treat patients without fear of a dangerous "allergic" reaction. In time we will become much more precise in our ability to control the amount of protein that the vectors express (thus greatly reducing the risk of drug overdosing). In coming years, scientists will use next-generation AAV vectors to deliver CRISPR and other systems that edit mutations right out of DNA molecules and enable them to make normal proteins. Gene surgery may become a reality!

But no matter how exciting the technologies, drug development takes a lot of time and immense amounts of money. Over the next 20 years it is reasonable (but optimistic) to expect that the FDA will approve 50 to 75 drugs to treat orphan monogenic disorders. This would constitute a great boon to patients, but would have only a modest impact in the *universe* of orphan disorders we would like so much to treat. To create new therapies for the hundreds of *ultraorphan* disorders, we will have to develop a radically new, highly streamlined, much less costly, approach that regulatory authorities accept as reasonably balancing safety risks with compelling clinical need.

The second critical technological development that will be crucial to developing new therapies for orphan disorders will be advances in the field of biomarkers—the use of changes in the measurement of some chemical in the body (or in imaging studies) that are acceptable as *surrogate* markers that predict clinical benefits to follow. The development of such biomarkers (likely to be highly dependent on advances in proteomics) is still in its infancy, but one can imagine great advances in coming years. If we are going to treat rare disorders that affect only a few dozen children each year, we must compress the standard three-phase approach to clinical trials in favor of a single study that combines safety and efficacy data, an approach for which gene therapy may be particularly well suited. When one treats with a gene therapy, one may well be treating for life. This blurs the distinction between Phase I safety studies and later efficacy studies. It also raises ethical issues about the fairness of safety studies that knowingly use subtherapeutic dosing in a context in which the patient cannot (for immunological reasons) be re-treated at a later time.

All persons concerned with developing new therapies for children (and adults) with orphan genetic disorders need to ponder how we might streamline the regulatory pathways needed for approval of new therapies for rare diseases. If we do not come up with a more efficient, less time-consuming, and less expensive approach than that currently in use, many disorders will never garner the attention of the experts who could make new therapies become a reality.

Simply put, what might be done to convince biotech companies that specialize in developing new drugs for orphan genetic disorders to take on a project which might ultimately result in a therapy that will be used to treat less than 100 children a year in the United States and Europe? I do not know the answer to this, but I do know that parents of such children are desperate to alter the system to give their children a measure of hope. In the pages of this book, I have called out several of them, and I wish I could pay homage to the many others I have met who do battle every day to find a cure for their children.

Perhaps, ultimately, for disorders for which there are only a few patients born each year, the federal government will develop a financing system that defrays development costs and promises a reasonable return on investment. Taxpayers in the United States spend more than $50 billion each year to underwrite the costs of providing care (hemodialysis and kidney transplants) to tens of thousands of patients with end-stage renal disease. An annual budget that is just 10% of that amount could underwrite the cost of developing 50–100 new drugs for orphan genetic disorders each year.

Given the slow and expensive nature of drug development, and the more rapid pace and lower costs of developing noninvasive prenatal testing (NIPT), it is likely that *avoidance* of the conception or births of infants destined to be affected with severe orphan disorders will become widespread within the next decade. That possibility provokes the question of whether in countries that have health care systems that are able to cover the costs of NIPT, we will experience a new form of *eugenics*. Technologically enabled, demanded by couples, and rationalized by the severity of and lack of adequate treatment for many disorders, widespread use of NIPT could sharply reduce the numbers of children born with the most burdensome genetic illnesses. Although this may be morally repugnant to some, it is a logical extension of current screening of pregnancies to detect fetuses with chromosomal disorders. Of course, the smaller the size of the annual birth cohort for a particular disease, the less persuasive is the case to develop a new drug to treat it.

I think it is likely that by 2025 the majority of young couples who are planning to have children will first undergo comprehensive DNA testing to assess whether they share mutations in genes associated with serious single-gene disorders. That diagnostic information will inform their reproductive planning. For example, it could lead them to preimplantation genetic testing to avoid the birth of a child with a severe disorder. In addition because many genetic disorders arise de novo in the fetus, it will be the standard of care in medicine to offer analysis of the entire fetal genome to search for evidence of risk for severe genetic disorders.

Unfortunately, even sophisticated information will not infrequently leave women and their physicians facing great uncertainties in how to interpret information. Obtaining a definitive diagnosis often does not provide much prognostic power about disease severity. Over the last three decades thousands of women have faced this problem in regard to learning that they are carrying fetuses with Down syndrome or spina bifida. In most cases, it is impossible to predict how affected individuals might do in life.

Perhaps the most disturbing issue for the future will concern how to deal with genetic information that has been associated with a risk for having behavioral disorders or intellectual impairment. In a technologically advanced society in which families typically have only two children, I suspect moderately disturbing news, especially if made available early in pregnancy, will often trigger a pregnancy termination. For nearly a century scientists and ethicists have been speculating if genetic information might someday influence parents to act to maximize the potential of their children. What might parents do if their physician or genetic counselor informs them that the fetus has a 20%–30% chance of developing autism, and a 70%–80% chance of developing normally? I doubt that our society will avoid this dilemma by denying couples access to imperfect tests. In 2015 it is already clear that people believe that access to genetic information about themselves and their children should be part of the legally protected right to privacy. It is highly likely that helpful, but imperfect, predictive tests, will complicate reproductive planning for a couple of decades.

Regardless of the steady expansion of carrier and prenatal testing, there will always be hundreds of thousands of children and adults burdened with hundreds of orphan genetic disorders for which there will be a compelling argument to develop new therapies. Scientific curiosity and human compassion for the suffering of children will constitute a powerful stimulus to the pace of drug development for decades to come.

Bibliography

Chapter 1: Diet

References

American Academy of Pediatrics Committee on Genetics. 1999. Folic acid for prevention of neural tube defects. *Pediatrics* **104:** 325–327.

Beutler E. 1993. Study of glucose-6-phosphate dehydrogenase: History and molecular biology. *Am J Hematol* **42:** 53–58.

Beutler E. 2008. Glucose-6-phosphate deficiency: A historical perspective. *Blood* **111:** 16–24.

Beutler S, Beutler B. 2011. Ernest Beutler: His life and contribution to medical science. *Br J Haematol* **152:** 543–550.

Centers for Disease Control and Prevention. 1991. Use of folic acid for prevention of spina bifida and other neural tube defects: 1983–1991. *MMWR Morb Mortal Wkly Rep* **40:** 513–516.

Centerwall SA, Centerwall WR. 2000. The discovery of phenylketonuria: The story of a young couple, two retarded children, and a scientist. *Pediatrics* **105:** 89–103.

Committee for the Study of Inborn Errors of Metabolism, National Research Council. 1975. *Genetic screening: Programs, principles, and research.* National Academy of Sciences, Washington, DC.

Finger S, Christ SE. 2004. Pearl S. Buck and phenylketonuria (PKU). *J Hist Neurosci* **13:** 44–57.

Guthrie R. 1995. The introduction of newborn screening for phenylketonuria: A personal history. *Eur J Ped* **15** (Suppl 1): S4–S5.

Guthrie R. 1961. Blood screening for phenylketonuria. *J Am Med Assoc* **178:** 863.

Hecker PA, Leopold JA, Gupte SA, Recchia FA, Stanley WC. 2013. Impact of glucose-6-phosphate dehydrogenase deficiency in primary prevention of cardiovascular disease. *Am J Physiol Heart Circ Physiol* **304:** H491–H500.

Hibbard BM, Roberts CJ, Elder GH, Evans KT, Laurence KM. 1985. Can we afford screening for neural tube defects? The South Wales experience. *Br Med J (Clin Res Ed)* **290:** 293–295.

Jay AM, Conway RL, Feldman GL, Nahhas F, Spencer L, Wolf B. 2015. Outcomes of individuals with profound and partial biotinidase deficiency ascertained by newborn screening in Michigan over 25 years. *Genet Med* **17:** 205–209.

Kumar RK, Nagar N, Ranieri E. 2014. Newborn screening for G6PD deficiency—Why is it important for India? *Indian J Pediatr* **81:** 90–91.

Luzzatto Z, Seneca E. 2014. G6PD deficiency: A classic example of pharmacogenetics with on-going clinical implications. *Br J Haematol* **164:** 469–480.

MRC Vitamin Study Research Group. 1991. Prevention of neural tube defects: Results of the MRC vitamin study. *Lancet* **338:** 131–137.

NIH Consensus Statement October 16–18, 2000. Phenylketonuria (PKU): Screening and management **17:** 1–33. NIH Office of the Director.

Oakley GP Jr. 1998. Folic-acid-preventable spina bifida and anencephaly. *Bull World Health Organ* **76** (Suppl 2): 116–117.

Oakley GP Jr. 2010. Folic acid-preventable spina bifida: A good start but much to be done. *Am J Prev Med* **38:** 569–570.

Oakley GP Jr, Adams MJ, Dickinson CM. 1996. More folic acid for everyone, now. *J Nutr* **126:** 751S–755S.

Raghuveer TS, Garg U, Graf WD. 2006. Inborn errors of metabolism in infancy and early childhood: An update. *Am Fam Physician* **73:** 1981–1990.

Stevenson RE, Allen WP, Pai GS, Best R, Seaver LH, Dean J, Thompson S. 2000. Decline in prevalence of neural tube defects in a high-risk region of the United States. *Pediatrics* **106:** 677–683.

Youngblood ME, Williamson R, Bell KN, Johnson Q, Kancherla V, Oakley GP Jr. 2013. 2012 update on global prevention of folic acid preventable spina bifida and anencephaly. *Birth Defects Res A Clin Mol Teratol* **97:** 658–663.

Further Reading

Bailey LB, ed. 2009. *Folate in health and disease,* 2nd ed. CRC, Boca Raton, FL.

Koch JJ. 1997. *Robert Guthrie—The PKU story: Crusade against mental retardation.* Hope, Pasadena, CA.

Özek MM, Cinalli G, Maixner WJ, eds. 2008. *Spina bifida: Management and outcome.* Springer, New York.

Rimoin DL, Connor JM, Pyretz RE, Korf BR, eds. 2007. *Emery and Rimoin's principles and practice of human genetics,* 5th ed. Churchill Livingstone, Philadelphia.

Timmermans S, Buchbinder M. 2013. *Saving babies?: The consequences of newborn genetic screening.* University of Chicago Press, Chicago.

WWW Resources

www.fda.gov/AboutFDA/WhatWeDo/History/ProductRegulation/SelectionsFromFDLIUpdateSeriesonFDAHistory/ucm091883.htm United States Food and Drug Administration, Folic Acid Fortification: Fact and Folly.

www.mchb.hrsa.gov/programs/newbornscreening/ Newborn Screening.

www.ninds.nih.gov/disorders/spina_bifida/detail_spina_bifida.htm National Institute of Neurological Disorders and Stroke, Spina Bifida Fact Sheet.

www.npkua.org National PKU Alliance.

www.spinabifidaassociation.org Spina Bifida Association.

www.ssiem.org Society for the Study of Inborn Errors of Metabolism.

Chapter 2: The Rise of Medical Genetics

References

Chandra HS, Heisterkamp NC, Hungerford A, Morrissette JJ, Nowell PC, Rowley JD, Testa JR. 2011. Philadelphia Chromosome Symposium: Commemoration of the 50th anniversary of the discovery of the Ph chromosome. *Cancer Genet* **204:** 171–179.

Clarke CA. 1972. Genetic Counselling. *Br Med J* **1:** 606–609.

Dice L. 1950. A panel discussion on genetic counseling. *Am J Hum Genet* **1:** 251–258.

Dreifus C. 2008. "A genetics pioneer sees a bright future cautiously." *The New York Times*, April 29, 2008, C1.

Dronamraju KR, Francomano CA (eds.) 2012. *Victor McKusick and the history of medical genetics.* Springer, Seacaucus, NJ.

Francomano CA, McKusick VA, Biesecker LG. 2003. Medical genetic studies in the Amish: Historical perspective. *Am J Med Genet C Semin Med Genet* **121C:** 1–4.

Greifensten C. 2007. Arno Motulsky papers at the American Philosophical Society. *Mendel Newsl* **16:** 3–6.

Heimler A. 1997. An oral history of genetic counseling. *J Gen Counsel* **6:** 315–325.

McKusick VA. 2006. A 60-year tale of spots, maps, and genes. *Annu Rev Genomics Hum Genet* **7:** 1–27.

McKusick VA. 1956. *Heritable disorders of connective tissues.* Mosby, St. Louis.

Mckusick VA. 1989. HUGO news: The Human Genome Organization: History, purposes and membership. *Genomics* **5:** 385–387.

McKusick VA, Egeland JA, Eldridge R, Krusen DE. 1964. Dwarfism in the Amish I: The Ellis–van Creveld syndrome. *Bull Johns Hopkins Hosp* **115:** 306–336.

McKusick VA. 1993. Medical genetics: A forty-year perspective on the evolution of a medical specialty from a basic science. *JAMA* **270:** 2351–2356.

Neel J. 1949. The inheritance of sickle cell anemia. *Science* **110:** 64–66.

Neel J. 1994. *Physician to the gene pool.* John Wiley & Sons, New York.

Neel JV, Salzano FM, Junqueira PC, Keiter F, Maybury-Lewis D. 1964. Studies on the Xavante Indians of the Brazilian Mato Grosso. *Am J Hum Genet* **16:** 52–140.

Olopade OL. 2014. Obituary: Janet Davidson Rowley 1925–2013. *Cell* **156:** 390–391.

Pollack A. 2012. "The ethics of advice: Conflicts seen when genetic counselors work for test companies," *The New York Times*, July 14, 2012, pB1.

Reilly PR. 1991. *The surgical solution: A history of involuntary sterilization in the United States.* Johns Hopkins University, Baltimore.

Witkowski JA, Inglis JR (eds). 2008. *Davenport's dream: 21st century reflections on heredity and eugenics.* Cold Spring Harbor Laboratory Press, Cold Spring Harbor, NY.

Further Reading

Berliner J. 2014. *Ethical dilemmas in genetics and genetic counseling.* Oxford University Press, Oxford.

Bowman JE, Murray RF Jr. 1990. *Genetic variation and disorders in peoples of African origin.* The Johns Hopkins University Press, Baltimore.

Dronamanraju KR, Francomano CA, eds. 2012. *Victor McKusick and the history of medical genetics.* Springer, New York.

Kevles D. 1985. *In the name of eugenics: Genetics and the uses of human heredity.* Alfred A. Knopf, New York.

Neel JV. 1994. *Physician to the gene pool.* Wiley, New York.

Witkowski JA, Inglis JR, eds. 2008. *Davenport's dream: 21st century reflections on heredity and eugenics.* Cold Spring Harbor Laboratory Press, Cold Spring Harbor, NY.

WWW Resources

www.acmg.org American College of Medical Genetics.

www.ashg.org American Society of Human Genetics website.

www.dnalc.org DNA Learning Center.

www.nsgc.org National Society of Genetic Counselors website.

www.ohhgp.pendari.com/Chronology.aspx UCLA Oral History of Human Genetics Project: Timeline of medical genetics 1900–2000.

www.orphanet.net Orphanet.

www.rarediseases.org National Organization for Rare Disorders.

Chapter 3: Blood

References

Bass MH. 1959. In memorium: Reuben Ottenberg, 1882–1959. *J Mt Sinai Hosp NY* **26:** 421–423.

Blundell J. 1828. Observations on transfusion of blood by Dr. Blundell with a description of his gravitator. *Lancet* **II:** 312–324.

Clark RW. 1968. *J.B.S.: The life and work of J.B.S. Haldane.* Oxford University, Oxford.

Darby SC, Kan SW, Spooner RJ, Giangrande PL, Lee CA, Makris M, Sabin CA, Watson HG, Wilde JT, Winter M; UK Haemophilia Centre Doctors' Organisation. 2004. The impact of HIV on mortality rates in the complete UK haemophilia population. *AIDS* **18:** 525–533.

Diamond LK. 1965. History of blood banking in the United States. *JAMA* **193:** 128–136.

Dragsten PR, Hallaway PE, Hanson GJ, Berger AE, Bernard B, Hedlund BE. 2000. First human studies with a high molecular weight iron chelator. *J Lab Clin Med* **135:** 57–65.

Dreifus C. 2013. "A doctor's intimate view of hemophilia." *The New York Times*, Dec. 24, 2013, D5.

Dubin C, Francis D. 2013. Closing the circle: A thirty-year retrospective on the AIDS/blood epidemic. *Transfusion* **53:** 2359–2364.

Fif M, Pelinka LE. 2004. Karl Landsteiner, the discoverer of blood groups. *Rhesus* **63:** 251–254.

Franchini M, Mannucci PM. 2014. The history of hemophilia. *Semin Thromb Hemost* **4:** 571–576.

Giangrande PLF. 2000. Historical review: The history of blood transfusion. *Br J Haematol* **110:** 758–767.

Ingram GIC. 1976. The history of haemophilia. *J Clin Pathol* **29:** 469–479.

Kasper CK. 2012. Judith Graham Pool and the discovery of cryoprecipitate. *Haemophilia* **18:** 833–835.

Massie R, Massie S. 1975. *Journey.* Alfred A. Knopf, New York.

McKusick VA. 1965. The royal hemophilia. *Sci Am* **213:** 88–95.

Murphy SL, High KA. 2008. Gene therapy for haemophilia. *Br J Haematol* **140:** 479–487.

Obituary. 1946. Thomas Benton Cooley, MD 1871–1945. *JAMA Pediatrics* **71:** 77–78.

Pippard MJ, Callender ST. 1985. The management of iron chelation therapy. *Br J Haematol* **54:** 503–507.

Pool JG, Shannon AE. 1965. Production of high potency concentrates of antihaemophilic factor in a closed bag system. *New Engl J Med* **273:** 1443–1444.

Pool JG, Gershgold EJ, Pappenhagen AR. 1964. High-potency antihaemophilic factor concentrate prepared from cryoglobulin precipitate. *Nature* **203:** 312.

Silvestroni E, Bianco I. 1973. Screening for microcytemia in Italy: Analysis of data collected in the past 30 years. *Am J Hum Genet* **27:** 198–212.

Weatherall D. 2010. Thalassemia: The long road from the bedside through the laboratory to the community. *Nat Med* **16:** 1112–1115.

Weatherall D. 2010. *Thalassemia: The biography.* Oxford University Press, Oxford.

Zetterstrom R. 2008. Alfred Nobel's will and the Nobel Prize to Karl Landsteiner for his discovery of human blood groups. *Acta Paediatr* **97:** 396–397.

Further Reading

Hill SA. 2010. *Managing Sickle cell disease in low-income families.* Temple University Press, Philadelphia.

Howard J, Telfer P. 2015. *Sickle cell disease in clinical practice.* Springer, New York.

Jones P. 2002. *Living with haemophilia*, 5th ed. Oxford University Press, Oxford.

Kamal J. 2013. *Thalassemia: My lifelong companion.* Xibris, New York.

Massie R, Massie S. 1975. *Journey.* Knopf, New York.

Resnik S. 1999. *Blood saga: Hemophilia, AIDS, and the survival of a community.* University of California Press, Berkeley.

Weatherall D. 2005. *Thalassemia: The biography.* Oxford University Press, New York.

WWW Resources

www.cdc.gov/ncbddd/sicklecell/map/map-nationalresourcedirectory.html Sickle Cell Disease National Resource Directory.

www.fscdr.org Foundation for Sickle Cell Disease Research.

www.hemophilia.org National Hemophilia Foundation.

www.profiles.nlm.nih The Charles R. Drew Papers at the National Library of Medicine (blood banking).

www.sicklecelldisease.org Sickle Cell Disease Association of America, Inc.

www.thalassemia.org Cooley's Anemia Foundation.

www.wfh.org World Federation of Hemophilia.

Chapter 4: Genetic Testing

Avoiding Disease

References

ACMG Professional Practice and Guidelines Committee. 2004. Second trimester maternal serum screening for fetal open neural tube defects and andeuploidy. www.acmg.net.

Benn PA, Fang M, Egan JF. 2005. Trends in the use of second trimester maternal serum screening from 1991 to 2003. *Genet Med* **7:** 328–331.

Bowman JE. 1983. Is a national program to prevent sickle cell disease possible? *Am J Pediatr Hematol Oncol* **5:** 367–372.

Bowman JE, Murray RF. 1990. *Genetic variation and disorders in peoples of African origin.* Johns Hopkins University Press, Baltimore.

Brock DJ, Scrimgeour JB, Steven J, Barron L, Watt M. 1978. Maternal plasma α-fetoprotein screening for fetal neural tube defects. *Br J Obstet Gynaecol* **85:** 575–581.

Cao A, Rosatelli MC, Galanello R. 1996. Control of β-thalassaemia by carrier screening, genetic counseling and prenatal diagnosis: The Sardinian experience. *Ciba Found Symp* **197:** 137–151.

Cao A. 1994. 1993 William Allan award address, *Am J Hum Genet* **54:** 397–402.

Committee for the Study of Inborn Errors of Metabolism. 1975. *Genetic screening: Procedural guidance and recommendations.* National Academy of Sciences, Washington, DC.

Dreesen J, Destouni A, Kourlaba G, Degn B, Mette WC, Carvalho F, Moutou C, Sengupta S, Dhanjal S, Renwick P, et al. 2013. Evaluation of PCR-based preimplantation genetic diagnosis applied to monogenic diseases: A collaborative ESHRE PGD consortium study. *Eur J Hum Genet* **22:** 1012–1018.

Ferguson-Smith MA. 2008. Cytogenetics and the evolution of medical genetics. *Genet Med* **10:** 553–559.

Ferguson-Smith ME, Ferguson-Smith MA, Nevin NC, Stone M. 1971. Chromosome analysis before birth and its value in genetic counseling. *Br Med J* **4:** 69–74.

Grody WW, Thompson BH, Gregg AR, Bean LH, Monaghan KG, Schneider A, Lebo RV. 2013. ACMG position statement on prenatal/preconception expanded carrier screening. *Genet Med* **15:** 482–483.

Gross SD, Boyle CA, Botkin JR, Comeau AM, Kharrazi M, Rosenfeld M, Wilfond BS. 2004. Newborn screening for cystic fibrosis: Evaluation of benefits and risks and recommendations for state newborn screening programs. *MMWR Morb Mortal Wkly Rep* **53:** 1–36.

Hoyme CH. 1998. Antenatal detection of hereditary disorders, by Henry Nadler, MD, Pediatrics 1968, **42:** 912–918. *Pediatrics* **102** (Suppl 1): 247–249.

Kaback M, Lim-Steele J, Dabholkar D, Brown D, Levy N, Zeiger K. 1993. Tay–Sachs disease—Carrier screening, prenatal diagnosis, and the molecular era. An international perspective, 1970–1993. The International TSD data collection network. *JAMA* **270:** 2307–2315.

Mersy E, Smits LJ, van Winden LA, de Die-Smulders CE, South-East Netherlands NIPT Consortium, Paulussen AD, Macville MV, Coumans AB, Frints SG. 2013. Noninvasive detection of fetal trisomy 21: A systematic review and report of quality and outcomes of diagnostic accuracy studies performed between 1997 and 2012. *Hum Reprod Update* **19:** 318–329.

Morain S, Greene MF, Mello MM. 2013. A new era in noninvasive prenatal testing. *New England J Med* **369:** 499–501.

Nadler HL, Gerbie AB. 1970. Role of amniocentesis in the intrauterine detection of genetic disorders. *New Engl J Med* **282:** 596–599.

National Health Services. 2011. Annual Report: Screening Programmes: Sickle cell and thalassaemia. http://sct.screening.nhs.uk.

Rucknagel DL. 1983. A decade of screening in the hemoglobinopathies: Is a national program to prevent sickle cell anemia possible? *Am J Pediatr Hematol Oncol* **5:** 373–377.

Santesmases MJ. 2014. The human autonomous karyotype and the origins of prenatal testing: Children, pregnant women and early Down's syndrome cytogenetics, Madrid 1962–75. *Stud Hist Philos Biol Biomed Sci* **47:** 142–153.

Scriver CR, Bardanis M, Cartier L, Clow CL, Lancaster GA, Ostrowsky JT. 1984. β-thalassemia disease prevention: Genetic medicine applied. *Am J Hum Genet* **36:** 1024–1038.

Sermon K, Van Steirteghem A, Liebaers I. 2004. Review: Preimplantation genetic diagnosis. *Lancet* **363:** 1633–1641.

Slotnick N, Filly RA, Callen PW, Golbus MS. 1982. Sonography as a procedure complementary to α-fetoprotein testing for neural tube defects. *J Ultrasound Med* **1:** 319–322.

Thein SL, Wainscoat JS, Old JM, Sampietro M, Fiorelli G, Wallace RB, Weatherall DJ. 1985. Feasibility of prenatal diagnosis of β-thalassaemia with synthetic DNA probes in two Mediterranean populations. *Lancet* **2:** 345–347.

Thom H, Campbell AG, Farr V, Fisher PM, Hall MH, Swapp GH, Gray ES. 1985. The impact of maternal serum α-fetoprotein screening on open neural tube defect births in north-east Scotland. *Prenat Diagn* **5:** 15–19.

Wald NJ. 2010. Prenatal screening for open neural tube defects and Down syndrome: Three decades of progress. *Prenat Diagn* **30:** 619–621.

Wald NJ, Brock DJ, Bonnar J. 1974. Prenatal diagnosis of spina bifida and anencephaly by maternal serum-α-fetoprotein measurement: A controlled study. *Lancet* **1:** 765–767.

Wald NJ, Cuckle HS, Boreham J, Brett R, Stirrat GM, Bennett MJ, Turnbull AC, Solymar M, Jones N, Bobrow M, Evans CJ. 1979. Antenatal screening in Oxford for fetal neural tube defects. *Br J Obstet Gynaecol* **86:** 91–100.

Further Reading

Becker AJ. 2013. *What every woman needs to know about prenatal testing: Insight from a mom who has been there.* Patheos Press, Denver.

Desnick RJ, Kaback MM, eds. 2001. *Tay–Sachs disease.* Academic, New York.

Dyson S. 2005. *Ethnicity and screening for Sickle cell/thalassemia: Lessons for practice from the voices of experience.* Churchill Livingston, Philadelphia.

Evans M. 2007. *Prenatal diagnosis.* McGraw-Hill, New York.

Kuliev A. 2013. *Practical pre-implantation genetic diagnosis,* 2nd ed. Springer, London.

NAS Committee for the Study of Inborn Errors of Metabolism. 1975. *Genetic screening: Programs, principles, and research.* National Academy of Sciences Press, Washington, DC.

Sandler A. 2004. *Living with spina bifida.* University of North Carolina, Chapel Hill, NC.

WWW Resources

www.acog.org/Resources_And_Publications/Practice_Bulletins/Committee_on_Practice_Bulletins_Obstetrics/Screening_for_Fetal_Chromosomal_Abnormalities American Congress of Obstetricians and Gynecologists.

www.acog.org/Resources_And_Publications/Committee_Opinions/Committee_on_Genetics/Preconception_and_Prenatal_Carrier_Screening_for_Genetic_Diseases_in_Individuals_of_Eastern_European ACOG Committee Opinion: Preconception and prenatal carrier screening of genetic Diseases in individuals of Eastern European Jewish descent.

www.bmgl.com Baylor Miraca Genetics Laboratories.

www.goodstartgenetics.com Good Start Genetics.

www.hfea.gov.uk Human Fertilisation and Embryology Authority.

www.ntsad.org National Tay-Sachs and Allied Diseases Association.

http://sct.screening.nhs.uk/equality fileid12594 National Health Services. 2011. Annual report: Screening programmes: Sickle cell and thalassaemia.

Chapter 5: Stem Cells
Creating Human Mosaics

References

Anasetti C, Logan BR, Lee SJ, Waller EK, Weisdorf DJ, Wingard JR, Cutler CS, Westervelt P, Woolfrey A, Couban S, et al. 2012. Peripheral-blood stem cells versus bone marrow from unrelated donors. *N Engl J Med* **367:** 1487–1496.

Ballen KK, King RJ, Chitphakdithai P, Bolan CD Jr, Agura E, Hartzman RJ, Kernan NA. 2008. The national marrow donor program 20 years of unrelated donor hematopoietic cell transplantation. *Biol Blood Marrow Transplant* **14** (Suppl 9)**:** 2–7.

Bjoraker KJ, Delaney K, Peters C, Krivit W, Shapiro EG. 2006. Long-term outcomes of adaptive functions for children with mucopolysaccharidosis I (Hurler syndrome) treated with hematopoietic stem cell transplantation. *J Dev Behav Pediatr* **27:** 290–296.

Dean C. 2012. "Joseph E. Murray, Nobel Laureate and transplant surgeon, dies at 93." *The New York Times,* Nov. 28, 2012, A18.

Escolar ML, Poe MD, Provenzale JM, Richards KC, Allison J, Wood S, Wenger DA, Pietryga D, Wall D, Champagne M, et al. 2005. Transplantation of umbilical cord blood in babies with infantile Krabbe's disease. *New Engl J Med* **352:** 269–281.

Hobbs JR. 1981. Bone marrow transplantation for inborn errors. *Lancet* **2:** 735–739.

Johnson FL. 1981. Marrow transplantation in the treatment of acute childhood leukemia. Historical development and current approaches. *Am J Pediatr Hematol Oncol* **3:** 389–395.

Kersey JH. 2007. Blood and marrow transplantation: A perspective from the University of Minnesota. *Immunol Res* **38:** 149–164.

Krivit W. 2002. Stem cell bone marrow transplantation in patients with metabolic storage diseases. *Adv Pediatrics* **49:** 359–378.

Krivit W. 2004. Allogeneic stem cell transplantation for the treatment of lysosomal and peroxisomal metabolic diseases. *Springer Semin Immunopathol* **26:** 119–132.

Krivit W, Whitley CB, Chang PN. 1990. Lysosomal storage diseases treated by bone marrow transplantation: Review of 21 patients. In *Bone marrow transplantation in children* (ed. Johnson FL, Pochedly C), pp. 261–287. Raven, New York.

Krivit W, Peters C, Shapiro EG. 1999. Bone marrow transplantation as an effective treatment of central nervous system disease in globoid cell leukodystrophy, metachromatic leukodystrophy, adrenoleukodystrophy, mannosidosis, fucosidosis, aspartylglucosaminuria, Hurler, Maroteaux–Lamy, and Sly syndromes, and Gaucher disease type III. *Curr Opin Neurol* **12:** 167–176.

Marshall V. 1982. Organ and tissue transplantation: Past, present, future. *Med J Aust* **2:** 411–414.

Mathe G, Schwarzenberg L. 1979. Bone marrow transplantation (1958–1978): Conditioning and graft-versus-host disease, indications in aplasias and leukemias. *Pathol Biol (Paris)* **27:** 337–343.

Miller WP, Rothman SM, Nascene D, Kivisto T, DeFor TE, Ziegler RS, Eisengart J, Leiser K, Raymond G, Lund TC, et al. 2011. Outcomes following allogeneic hematopoietic cell transplantation for childhood cerebral adrenoleukodystrophy: The largest single-institution cohort report. *Blood* **118:** 1971–1978.

Murray JE, Merrill JP, Harrison JH. 1958. Kidney transplantation in seven pairs of identical twins. *Ann Surg* **148:** 343–347.

Murray JE, Lang S, Miller BF. 1955. Observations on the natural history of renal homotransplants in dogs. *Surg Forum* **5:** 241–244.

Neven B, Valayannopoulos V, Quartier P, Blanche S, Prieur AM, Debré M, Rolland MO, Rabier D, Cuisset L, Cavazzana-Calvo M, et al. 2007. Allogeneic bone marrow transplantation in mevalonic aciduria. *New Engl J Med* **356:** 2700–2703.

Parikh SH, Szabolcs P, Prasad VK, Lakshminarayanan S, Martin PL, Driscoll TA, Kurtzberg J. 2007. Correction of chronic granulomatous disease after second umbilical cord blood transplant. *Blood* **49:** 982–984.

Peters C, Charnas LR, Tan Y, Ziegler RS, Shapiro EG, DeFor T, Grewal SS, Orchard PJ, Abel SL, Goldman AI, et al. 2004. Cerebral X-linked adrenoleukodystrophy: The international hematopoietic cell transplantation experience from 1982–1999. *Blood* **104:** 881–888.

Prasad VK, Kurtzberg. 2009. Cord blood and bone marrow transplantation in inherited metabolic diseases: Scientific basis, current status and future directions. *Br. J Haematol* **148:** 356–372.

Sendi H, Schurter M, Letterman G. 1968. The first 68 years of renal transplantation. *J Am Med Women Assoc* **23:** 998–1008.

Shackman R. 1966. The story of kidney transplantation. *Br Med J* **4:** 1379–1383.

Staba SL, Escolar ML, Poe M, Kim Y, Martin PL, Szabolcs P, Allison-Thacker J, Wood S, Wenger DA, Rubinstein P, et al. 2004. Cord-blood transplants from unrelated donors in patients with Hurler's syndrome. *New Engl J Med* **350:** 1960–1969.

Woods WG, Ramsay NK, D'Angio GJ. 2003. The American Society of Pediatric Hematology/Oncology Distinguished Career Award goes to William Krivit. *J Pediatr Hematol Oncol* **25:** 279–281.

Further Reading

Appelbaum FR, Forman SJ, Negrin RS, Blum KG, eds. 2011. *Thomas' hematopoetic cell transplantation,* 4th ed. Wiley-Blackwell, New York.

Brent L. 1997. *A history of transplantation immunology.* Academic, San Diego.

López-Larrea C, López-Vásquez A, Suárez Alvarez B, eds. 2012. *Stem cell transplantation.* Springer-Verlag, New York.

Stewart SK. 1995. *Bone marrow transplants: A book of basics for patients.* BMT Newsletter, Atlanta.

WWW Resources

http://optn.transplant.hrsa.gov Organ Procurement and Transplantation Network.

www.babycenter.com Baby Center.

www.bcbsnc.com Blue Cross Blue Shield of North Carolina, Allogeneic Hematopoetic Transplantation for Genetic Diseases.

www.dukechildrens.org/services/bone_marrow_and_stem_cell_transplantation Duke Children's Hospital and Health Center, Bone and Marrow Transplant Program.

www.parentsguidecordblood.org Parent's Guide to Cord Blood Foundation.

www.peds.umn.edu/bmt University of Minnesota School of Medicine, Division of Blood and Marrow Transplantation.

Chapter 6: Enzyme Replacement Therapy
Genetically Engineered Drugs

References

Anonymous. 1990. Genzyme's Ceredase recommended for approval for treatment of moderate to severe Gaucher's disease; panel concludes. Ceredase's benefits outweigh risks. *The Pink Sheet*, Oct. 29, 1990.

Barton NW, Brady RO, Dambrosia JM, Di Bisceglie AM, Doppelt SH, Hill SC, Mankin HJ, Murray GJ, Parker RI, Argoff CE, et al. 1991. Replacement therapy for inherited enzyme deficiency–macrophage-targeted glucocerebrosidase for Gaucher's disease. *New Engl J Med* **324:** 1464–1470.

Barton NW, Furbish FS, Murray GJ, Garfield M, Brady RO. 1990. Therapeutic response to intravenous infusions of glucocerebrosidase in a patient with Gaucher disease. *Proc Natl Acad Sci* **87:** 1913–1916.

Brady RO. 2010. Benefits from unearthing "a Biochemical Rosetta Stone". *J Biol Chem* **285:** 41216–41221.

Brady RO, Gal AE, Bradley RM, Martensson E, Warshaw AL, Laster L. 1967. Enzymatic defect in Fabry's disease: Ceramidetrihexosidase deficiency. *New Engl J Med* **296:** 1163–1167.

Brusilow SW, Valle DL, Batshaw ML. 1979. New pathway of nitrogen excretion in inborn errors of urea synthesis. *Lancet* **3:** 452–455.

Burrow JA, Hopkin RJ, Leslie ND, Tinkle BT, Grabowski GA. 2007. Enzyme reconstitution/replacement therapy for lysosomal storage diseases. *Curr Opin Pediatr* **19:** 628–635.

Connock M, Burls A, Frew E, Fry-Smith A, Juarez-Garcia A, McCabe C, Wailoo A, Abrams K, Cooper N, Sutton A, et al. 2006. The clinical effectiveness and cost-effectiveness of enzyme replacement therapy for Gaucher's disease: A systematic review. *Health Technol Assess* **10:** iii-iv, ix-136.

Desnick RJ. 2004. Enzyme replacement and enhancement therapies for lysosomal diseases. *J Inherit Metab Dis* **27:** 385–410.

Desnick RJ, Dean KJ, Grabowski G, Bishop DF, Sweeley CC. 1979. Enzyme therapy in Fabry disease: Differential in vivo plasma clearance and metabolic effectiveness of plasma and splenic α-galactosidase A isozymes. *Proc Natl Acad Sci* **76:** 5326–5330.

Eng CM, Guffon N, Wilcox WR, Germain DP, Lee P, Waldek S, Caplan L, Linthorst GE, Desnick RJ; International Collaborative Fabry Disease Study Group. 2001.

Safety and efficacy of recombinant human α-galactosidase A replacement therapy in Fabry's disease. *New Engl J Med* **345:** 9–16.

Ficicioglu C. 2008. Review of miglustat for clinical management in Gaucher disease type 1. *Ther Clin Risk Manag* **4:** 425–431.

Furbish FS, Blair HE, Shiloach J, Pentchev PG, Brady RO. 1977. Enzyme replacement therapy in Gaucher's disease: Large-scale purification of glucocerebrosidase suitable for human administration. *Proc Natl Acad Sci* **74:** 3560–3563.

Grabowski GA, Barton NW, Pastores G, Dambrosia JM, Banerjee TK, McKee MA, Parker C, Schiffmann R, Hill SC, Brady R. 1995. Enzyme therapy in type I Gaucher disease: Comparative efficacy of mannose-terminated glucocerebrosidase from natural and recombinant sources. *Ann Intern Med* **122:** 33–39.

Hemsley KM, Hopwood JJ. 2011. Emerging therapies for neurodegenerative lysosomal stoarage disorders—From concept to reality. *J Inherit Metab Dis* **34:** 1003–1012.

Henley WE, Anderson LJ, Wyatt KM, Nikolaou V, Anderson R, Logan S. 2014. The NCS-LSD cohort study: A description of the methods and analyses used to assess the long-term effectiveness of enzyme replacement therapy and substrate reduction therapy in patients with lysosomal storage disorders. *J Inherit Metab Dis* **37:** 939–944.

Hollak CE, Wijburg FA. 2014. Therapy of lysosomal storage disorders: Success and challenges. *J Inherit Metab Dis* **37:** 587–598.

Kishnani D, Beckemeyer AA, Mendelsohn NJ. 2012. The new era of Pompe disease: Advances in detection, understanding of the phenotypic spectrum, pathophysiology, and management. *Am J Med Genet C Semin Med Genet* **160C:** 1–7.

Kishnani PS, Corzo D, Nicolino M, Byrne B, Mandel H, Hwu WL, Leslie N, Levine J, Spencer C, McDonald M, et al. 2007. Recombinant human acid α-glucosidase: Major clinical benefits in infantile-onset Pompe disease. *Neurology* **68:** 99–109.

Laforêt P, Laloui K, Granger B, Hamroun D, Taouagh N, Hogrel JY, Orlikowski D, Bouhour F, Lacour A, Salort-Campana E, et al. 2013. The French Pompe Registry. Baseline characteristics of 126 patients with Pompe disease. *Rev Neurol* **168:** 595–602.

Matern D, Oglesbee D, Tortorelli S. 2013. Newborn screening for lysosomal storage disorders and other neuronopathic conditions. *Dev Disabil Res Rev* **17:** 247–253.

Mehta A, Beck M, Linhart A, Sunder-Plassmann G, Widmer U. 2006. History of lysosomal storage disorders: An overview. In *Fabry disease: Perspectives from 5 years of FOS* (ed. Mehta A, Beck M, Sunder-Plassmann G), pp. 1–18. Oxford Pharmagenesis, Oxford University Press, Oxford.

Mehta A, Ricci R, Widmer U, Dehout F, Garcia de Lorenzo A, Kampmann C, Linhart A, Sunder-Plassmann G, Ries M, Beck M. 2004. Fabry disease defined: Baseline clinical manifestations of 366 patients in the Fabry outcome survey. *Eur J Clin Invest* **34:** 236–242.

Neufeld EF. 2011. From serendipity to therapy. *Ann Rev Biochem* **80:** 1–15.

Peters FP, Vermeulen A, Kho TL. 2001. Anderson-Fabry's disease: α galactosidase deficiency. *Lancet* **357:** 138–140.

Wyatt K, Henley W, Anderson L, Anderson R, Nikolaou V, Stein K, Klinger L, Hughes D, Waldek S, Lachmann R, et al. 2012. The effectiveness and cost-effectiveness of enzyme and substrate therapies: A longitudinal cohort study of people with lysosomal storage disorders. *Health Tech Assess* **16:** 1–543.

Zeidman LA. 2012. Johannes C. Pompe, MD, hero of neuroscience: The man behind the syndrome. *Muscle Nerve* **46:** 134–138.

Further Reading

Barranger JA, Cabrera-Salazar MA, eds. 2007. *Lysosomal storage disorders.* Springer, New York.

Crowley J. 2010. *Chasing miracles: The Crowley family journey of strength, hope and joy.* Harper Collins, New York.

Elstein D, Altarescu G, Beck M, eds. 2010. *Fabry disease.* Springer, New York.

Futerman AH, Zimran A, eds. 2006. *Gaucher disease.* CRC Press, Boca Raton, FL.

WWW Resources

www.bmrn.com Biomarin, History.

www.fabrydisease.org National Fabry Disease Foundation.

www.gaucherdisease.org National Gaucher Foundation.

http://www.genzyme.com/Patients/Educational-Info/The-Cost-of-Enzyme-Replacement-Therapy.aspx Genzyme (A Sanofi Company), The Cost of Therapy.

www.nursingworld.org American Nurses Association, An Overview of Enzyme Replacement Therapy for Lysosomal Storage Disorders.

www.unitedpompe.org United Pompe Foundation.

Chapter 7: Gene Therapy
Using Viruses to Deliver Normal Genes

References

Acland GM, Aguirre GD, Ray J, Zhang Q, Aleman TS, Cideciyan AV, Pearce-Kelling SE, Anand V, Zeng Y, Maguire AM, et al. 2001. Gene therapy restores vision in a canine model of childhood blindness. *Nat Genet* **28:** 92–95.

Bainbridge JW, Smith AJ, Barker SS, Robbie S, Henderson R, Balaggan K, Viswanathan A, Holder GE, Stockman A, Tyler N, et al. 2008. Effect of gene therapy on visual function in Leber's congenital amaurosis. *New Engl J Med* **358**: 2231–2239.

Boztug K, Schmidt M, Schwarzer A, Banerjee PP, Díez IA, Dewey RA, Böhm M, Nowrouzi A, Ball CR, Glimm H, et al. 2010. Stem-cell gene therapy for the Wiskott-Aldrich syndrome. *New Engl J Med* **363:** 1918–1927.

Cavazzana-Calvo M, Hacein-Bey S, de Saint Baile G, Gross F, Yvon E, Nusbaum P, Selz F, Hue C, Certain S, Casanova JL, et al. 2000. Gene therapy of human severe combined immunodeficiency (SCID)-X1 disease. *Science* **288:** 669–672.

Cideciyan AV, Hauswirth WW, Aleman TS, Kaushal S, Schwartz SB, Boye SL, Windsor EA, Conlon TJ, Sumaroka A, Pang JJ, et al. 2009. Human *RPE65* gene therapy for Leber congenital amaurosis: Persistence of early visual improvements and safety at 1 year. *Hum Gene Ther* **20:** 999–1004.

Cideciyan AV, Jacobson SG, Beltran WA, Sumaroka A, Swider M, Iwabe S, Roman AJ, Olivares MB, Schwartz SB, Komáromy AM, et al. 2013. Human retinal gene therapy for Leber congenital amaurosis shows advancing retinal degeneration despite enduring visual improvement. *Proc Natl Acad Sci* **110:** E517–E525.

Culver KW. 1994. *Gene therapy: A handbook for physicians.* MaryAnn Liebert, Inc., New York.

Edelstein ML, Abedi MR, Wixon J. 2007. Gene therapy clinical trials worldwide to 2007—An update. *J Gene Med* **9:** 833–842.

Friedmann T, ed. 1999. *The development of gene therapy.* Cold Spring Harbor Laboratory Press, Cold Spring Harbor, NY.

Haldane JBS. 1928. *Possible worlds,* pp. 90–93. Harper Brothers, New York.

Jacobson SG, Cideciyan AV, Ratnakaram R, Heon E, Schwartz SB, Roman AJ, Peden MC, Aleman TS, Boye SL, Sumaroka A, et al. 2012. Gene therapy for Leber congenital amaurosis caused by *RPE65* mutations: Safety and efficacy in 15 children and adults followed up to 3 years. *Arch Opthalmol* **130:** 9–24.

Jaeger W. 1988. The foundation of experimental ophthalmology by Theodor Leber. *Doc Ophthalmol* **68:** 71–77.

Leone P, Shera D, McPhee SW, Francis JS, Kolodny EH, Bilaniuk LT, Wang DJ, Assadi M, Goldfarb O, Goldman HW, et al. 2012. Long-term follow-up after gene therapy for Canavan disease. *Sci Transl Med* **4:** p165ra163.

Maguire AM, Simonelli F, Pierce EA, Pugh EN Jr, Mingozzi F, Bennicelli J, Banfi S, Marshall KA, Testa F, Surace EM, et al. 2008. Safety and efficacy of gene transfer for Leber's congenital amaurosis. *New Engl J Med* **358:** 2240–2248.

Miller WP, Rothman SM, Nascene D, Kivisto T, DeFor TE, Ziegler RS, Eisengart J, Leiser K, Raymond G, Lund TC, et al. 2011. Outcomes after allogeneic hematopoietic cell transplantation for childhood cerebral adrenoleukodystrophy: The largest single-institution cohort report. *Blood* **118:** 1971–1978.

Mount JD, Herzog RW, Tillson DM, Goodman SA, Robinson N, McCleland ML, Bellinger D, Nichols TC, Arruda VR, Lothrop CD Jr, et al. 2002. Sustained phenotypic correction of hemophilia B dogs with a Factor IX null mutation by liver-directed gene therapy. *Blood* **99:** 2670–2676.

Nathwani AC, Tuddenham EG, Rangarajan S, Rosales C, McIntosh J, Linch DC, Chowdary P, Riddell A, Pie AJ, Harrington C, et al. 2011. Adenovirus-associated virus vector–mediated gene transfer in hemophilia B. *New Engl J Med* **365:** 2357–2365.

Stent G. 1969. *The coming of the Golden Age.* The Natural History Press, Garden City, NY.

Stolberg SG. 1999. "The biotech death of Jesse Geisinger." *The New York Times,* Nov. 28, 1999.

Testa F, Maguire AM, Rossi S, Pierce EA, Melillo P, Marshall K, Banfi S, Surace EM, Sun J, Acerra C, et al. 2013. Three-year follow-up after unilateral subretinal delivery of adeno-associated virus in patients with Leber congenital amaurosis type 2. *Ophthalmology* **120:** 1283–1291.

Yla-Herttuala S. 2011. Gene therapy moves forward in 2010. *Mol Ther* **19:** 219–220.

Further Reading

Laurence J, Franklin M, eds. 2014. *Translating gene therapy to the clinic: Techniques and approaches.* Academic, New York.

Lewis R. 2013. *The forever fix.* St. Martin's Press, New York.

Perin EC, Miller LC, Taylor D, Willerson JT, eds. 2015. *Stem cells and gene therapy for cardiovascular disease.* Academic, San Diego.

Walters L, Palmer JG. 1996. *The ethics of human gene therapy.* Oxford University Press, Oxford.

WWW Resources

www.asgct.org American Society for Gene and Cell Therapy.

www.bluebirdbio.com bluebird bio.

www.bsgct.org British Society for Gene and Cell Therapy, A history of gene therapy.

www.esgct.eu European Society of Gene and Cell Therapy.

www.ghr.nlm.nih.gov National Library of Medicine, Genetics Home Reference, What is Gene Therapy?

www.history.nih.gov/exhibits/genetics/sect4.htm National Institutes of Health, Gene Therapy—A Revolution in Progress: Human Genetics.

www.nytimes.com *The New York Times,* Government Halts 27 Gene Therapy Trials.

Chapter 8: Overcoming Mutations

References

Bagni C, Tassone F, Neri G, Hagerman R. 2012. Fragile X syndrome: Causes, diagnosis, mechanisms, and therapeutics. *J Clin Invest* **122:** 4314–4322.

Bowen JM, Connolly HM. 2014. Of Marfan's syndrome, mice, and medications. *New Engl J Med* **371:** 2127–2128.

Brooke BS, Habashi JP, Judge DP, Patel N, Loeys B, Dietz HC III. 2008. Angiotensin II blockade and aortic root dilatation in Marfan syndrome. *N Engl J Med* **358:** 2787–2795.

Cameron DE, Alejo DE, Patel ND, Nwakanma LU, Weiss ES, Vricella LA, Dietz HC, Spevak PJ, Williams JA, Bethea BT, et al. 2009. Aortic root replacement in 372 Marfan patients: Evolution of repair over 30 years. *Ann Thorac Surg* **87:** 1344–1349.

Cuisset JM, Estournet B; French Ministry of Health. 2012. Recommendations for the diagnosis and management of typical childhood spinal muscular atrophy. *Rev Neurol* **168:** 902–909.

Dietz HC. 2007. Marfan syndrome: From molecules to medicines. *Am J Hum Genet* **81:** 662–667.

Farrar MA, Vucic S, Johnston HM, du Sart D, Kiernan MC. 2013. Pathophysiological insights derived by natural history and motor function of spinal muscular atrophy. *J Pediatr* **162:** 155–159.

Gott VL. 1998. Antoine Marfan and his syndrome: One hundred years later. *Md Med J* **47:** 247–252.

Gottesfeld JM, Rusche JR, Pandolfo M. 2013. Increasing frataxin gene expression with histone deacetylase inhibitors as a therapeutic approach for Friedreich's ataxia. *J Neurochem* **1** (Suppl)**:** 147–154.

Gregoretti C, Ottonello G, Chiarini Testa MB, Mastella C, Ravà L, Bignamini E, Veljkovic A, Cutrera R. 2013. Survival of patients with spinal muscular atrophy type 1. *Pediatrics* **131:** e1509–e1514.

Habashi JP, Judge DP, Holm TM, Cohn RD, Loeys BL, Cooper TK, Myers L, Klein EC, Liu G, Calvi C, et al. 2006. Losartan, an AT1 antagonist, prevents aortic aneurysm in a mouse model of Marfan syndrome. *Science* **312:** 117–121.

Hecht FA, Sutherland G. 1984. Detection of the fragile X chromosome and other fragile sites. *Clin Genet* **26:** 301–303.

Kolata G. 2013. "Learning to defuse the aorta." *The New York Times*, Dec 8, 2013, pD1.

Krueger DD, Bear MF. 2011. Toward fulfilling the promise of molecular medicine in fragile X syndrome. *Ann Rev Med* **62:** 411–429.

Lacro RV, Dietz HC, Sleeper LA, Yetman AT, Bradley TJ, Colan SD, Pearson GD, Selamet Tierney ES, Levine JC, Atz AM, et al. 2014. Atenolol versus losartan in children and young adults with Marfan's syndrome. *New Engl J Med* **371:** 2061–2071.

Lorson M, Lorson LC. 2012. SMN-inducing compounds for the treatment of spinal muscular atrophy. *Future Med Chem* **4:** 2067–2084.

Lubs HA, Stevenson RE, Schwartz CE. 2012. Fragile X and X-linked intellectual disability: Four decades of discovery. *Am J Hum Genet* **90:** 579–590.

Lynch DR, Pandolfo M, Schulz JB, Perlman S, Delatycki MB, Payne RM, Shaddy R, Fischbeck KH, Farmer J, Kantor P, et al. 2013. Common data elements for clinical research in Friedreich's atraxia. *Mov Disord* **28:** 190–195.

Martínez-Hernández R, Bernal S, Also-Rallo E, Alías L, Barceló MJ, Hereu M, Esquerda JE, Tizzano EF. 2013. Synpatic defects in type 1 spinal muscular atrophy in human development. *J Pathol* **229:** 49–61.

McKusick VA. 1956. *Heritable Disorders of Connective Tissue*. Mosby, St Louis.

McKusick VA. 1991. The defect in Marfan syndrome. *Nature* **352:** 279–281.

Pandolfo M. 2012. Freidriech's ataxia: New pathways. *J Child Neurol* **27:** 1204–1211.

Pandolfo M, Hausmann L. 2013. Deferiprone for the treatment of Friedreich's ataxia. *J Neurochem* **1** (Suppl)**:** 142–146.

Saito M, Kurokawa M, Oda M, Oshima M, Tsutsui K, Kosaka K, Nakao K, Ogawa M, Manabe R, Suda N, et al. 2011. ADAMTSL6β protein reduces fibrillin-1 microfibril disorder in a Marfan syndrome mouse model through the promotion of fibrillin-1 assembly. *J Biol Chem* **286:** 38602–38613.

Schulz JB, Pandolfo M. 2013. 150 years of Friedreich ataxia: From its discovery to therapy. *J Neurochem* **1** (Suppl): 1–3.

Thomas NH, Dubowitz V. 1994. The natural history of type 1 (severe) spinal muscular atrophy. *Neuromuscular Disord* **4**: 497.

Further Reading

Carvajal IF. 2011. *Understanding Fragile X syndrome: A guide for families and professionals.* Jessica Kingsley, New York.

Laws J, Brrokfield G. 2012. *Marfan syndrome: Causes, tests, and treatment options.* Create Space Independent Publishing Platform, Seattle.

Parker JN, Parker PM, eds. 2002. *The official parents sourcebook on Friedreich's ataxia: A revised and updated directory for the internet age.* ICON Health Publications, San Diego.

Wilmott R. 2013. The natural history of spinal muscular atrophy. *J Pediatr* **162**: 4–11.

WWW Resources

www.curefa.org The Friedreich's Ataxia Research Alliance (FARA).

www.curesma.org Cure SMA.

www.faparents.org Friedreich's Ataxia Parents' Group.

www.fara.org Friedreich's Ataxia Research Association.

www.fragilex.org National Fragile X Foundation.

www.fraxa.org Fragile X Research Foundation.

www.marfan.org The Marfan Foundation.

www.marfanworld.org International Federation of Marfan Syndrome Organizations.

www.isispharm.com Isis Pharmaceuticals, Inc. Press Release (November 12, 2013), Isis Pharmaceuticals earns $1.5 million from the advancement of the phase 2 study of ISIS-SMN Rx in infants with spinal muscular atrophy.

www.smafoundation.org/ Spinal Muscular Atrophy Foundation.

Chapter 9: Butterfly Children

Rebuilding the Skin

References

Berk DR, Jazayeri L, Marinkovich MP, Sundram UN, Bruckner AL. 2013. Diagnosing epidermolysis bullosa type and subtype in infancy using immunofluorescence microscopy: The Stanford experience. *Pediatr Dermatol* **30**: 226–233.

Bonneman CG, Wang CH, Quijano-Roy S, Deconinck N, Bertini E, Ferreiro A, Muntoni F, Sewry C, Béroud C, Mathews KD, et al. 2014. Diagnostic approaches to the congenital muscular dystrophies. *Neuromuscular Disord* **24**: 289–311.

Clement EM, Feng L, Mein R, Sewry CA, Robb SA, Manzur AY, Mercuri E, Godfrey C, Cullup T, Abbs S, et al. 2012. Relative frequency of congenital muscular dystrophy subtypes: Analysis of the UK diagnostic service 2001–2008. *Neuromuscular Disord* **22:** 522–527.

Fine JD, Eady RA, Bauer EA, Bauer JW, Bruckner-Tuderman L, Heagerty A, Hintner H, Hovnanian A, Jonkman MF, Leigh I, et al. 2008. The classification of inherited epidermolysis bullosa (EB): Report of the third international consensus meeting on diagnosis and classification of epidermolysis bullosa. *J Am Acad Dermatol* **58:** 931–950.

Gilbreath HR, Castro D, Iannaccone ST. 2014. Congenital myopathies and muscular dystrophies. *Neurol Clin* **32:** 689–703.

Gorzelany JA, de Souza MP. 2013. Protein replacement therapies for rare diseases: A breeze for regulatory approval? *Sci Transl Med* **5:** 178fs10.

Hovnanian A. 2013. Systemic protein therapy for recessive dystrophic epidermolysis bullosa: How far are we from clinical translation? *J Invest Dermatol* **133:** 1719–1721.

Kim HJ, Choi YC, Park HJ, Lee YM, Kim HD, Lee JS, Kang HC. 2014. Congenital muscular dystrophy type 1A with residual merosin expression. *Korean J Pediatr* **57:** 149–152.

Kirscher J. 2013. Congenital muscular dystrophies. *Handb Clin Neurol* **113:** 1377–1385.

Lin AN, Carter DM (eds.). 1992. *Epidermolysis bullosa: Basic and clinical aspects.* Springer-Verlag, New York.

Ramshaw JA, Werkmeister JA, Dumsday GJ. 2014. Bioengineered collagens: Emerging directions for biomedical materials. *Bioengineered* **5:** 227–233.

Remington J, Wang X, Hou Y, Zhou H, Burnett J, Muirhead T, Uitto J, Keene DR, Woodley DT, Chen M. 2009. Injection of recombinant human type VII collagen corrects the disease phenotype in a murine model of dystrophic epidermolysis bullosa. *Mol Ther* **17:** 26–33.

Sybert VP. 2010. Genetic counseling in epidermolysis bullosa. *Dermatol Clin* **28:** 239–243.

Van Ry PM, Minogue P, Hodges BL, Burkin DJ. 2014. Lamin-111 improves muscular repair in a mouse model of merosin-deficient congenital muscular dystrophy. *Hum Mol Genet* **23:** 383–396.

Vanden Oever MJ, Tolar J. 2014. Advances in understanding and treating dystrophic epidermolysis bullosa. *F1000Prime Rep* **6:** 6–35.

Vogel JH, Nguyen H, Giovannini R, Ignowski J, Garger S, Salgotra A, Tom J. 2012. A new large-scale manufacturing platform for complex biopharmaceuticals. *Biotechnol Bioeng* **109:** 3049–3058.

Wagner J, Ishida-Yamamoto A, McGrath JA, Hordinsky M, Keene DR, Woodley DT, Chen M, Riddle MJ, Osborn MJ, Lund T, et al. 2010. Bone marrow transplantation for recessive dystrophic epidermolysis bullosa. *New Engl J Med* **363:** 629–639.

Woodley D, Wang X, Amir M, Hwang B, Remington J, Hou Y, Uitto J, Keene D, Chen M. 2013. Intravenously injected recombinant human type VII collagen homes to skin wounds and restores skin integrity of dystrophic epidermolysis bullosa. *J Invest Dermatol* **133:** 1910–1913.

Woodley DT, Keene DR, Atha T, Huang Y, Lipman K, Li W, Chen M. 2004. Injection of a human type VII collagen restores collagen function in dystrophic epidermolysis bullosa. *Nat Med* **10:** 693–695.

Further Reading

Perry D. 2005. *Living with X-linked hypohidrotic ectodermal dysplasia.* Ectodermal Dysplasia Society, Cheltenham, UK.

WWW Resources

www.curecmd.org Cure CMD.

www.debra.org The Dystrophic Epidermolysis Bullosa Research Association of America.

www.niams.nih.gov/ National Institutes of Health, National Institute of Arthritis and MusculoSkeletal and Skin Diseases, What is Epidermolysis Bullosa?

Chapter 10: Ligands
Turning Genes On

References

Bluschke G, Nüsken KD, Schneider H. 2010. Prevalence and prevention of severe complications of hypohidrotic ectodermal dysplasia in infancy. *Early Hum Dev* **86:** 397–399.

Burger K, Schneider AT, Wohlfart S, Kiesewetter F, Huttner K, Johnson R, Schneider H. 2014. Genotype-phenotype correlation in boys with X-linked hypohidrotic ectodermal dysplasia. *Am J Med Genet A* **164A:** 2424–2432.

Casal M, Lewis JR, Mauldin EA, Tardivel A, Ingold K, Favre M, Paradies F, Demotz S, Gaide O, Schneider P. 2007. Significant correction of disease after postnatal administration of recombinant ectodysplasin A in canine X-linked ectodermal dysplasia. *Am J Hum Genet* **81:** 1050–1056.

Casal ML, Jezyk PF, Greek JM, Goldschmidt MH, Patterson DF. 1997. X-linked ectodermal dysplasia in the dog. *J Hered* **88:** 513–517.

Clarke A. 1987. Hypohidrotic ectodermal dysplasia. *J Med Genet* **24:** 659–663.

Fete M, Hermann J, Behrens J, Huttner KM. 2014. X-linked hypohidrotic ectodermal dysplasia (XLHED): Clinical and diagnostic insights from an international patient registry. *Am J Med Genet A* **164A:** 2437–2442.

Gaide O, Schneider P. 2003. Permanent correction of an inherited ectodermal dysplasia with recombinant EDA. *Nat Med* **9:** 614–618.

Huttner K. 2014. Future developments in XLHED treatment approaches. *Am J Med Genet A* **164A:** 2433–2436.

Jones KB, Goodwin AF, Landan M, Seidel K, Tran DK, Hogue J, Chavez M, Fete M, Yu W, Hussein T, et al. 2013. Characterization of X-linked hypohidrotic ectodermal dysplasia (XL-HED) hair and sweat gland phenotyping using phototrichogram analysis and live confocal imaging. *Am J Med Genet A* **161A:** 1585–1593.

Kere J, Srivastava AK, Montonen O, Zonana J, Thomas N, Ferguson B, Munoz F, Morgan D, Clarke A, Baybayan P, et al. 1996. X-linked anhidrotic (hypohidrotic) ecotodermal dysplasia is caused by mutation in a novel transmembrane protein. *Nat Genet* **13:** 409–416.

Koch PJ, Dinella J, Fete M, Siegfried EC, Koster MI. 2014. Modeling AEC—New approaches to study rare genetic disorders. *Am J Med Genet A* **164A:** 2443–2447.

Mauldin EA, Gaide O, Schneider P, Casal ML. 2009. Neonatal therapy with recombinant ectodysplasin prevents respiratory disease in dogs with X-linked ectodermal dysplasia. *Am J Med Genet A* **149A:** 2045–2049.

Mustonen N, Ilmonen M, Pummila M, Kangas AT, Laurikkala J, Jaatinen R, Pispa J, Gaide O, Schneider P, Thesleff I, et al. 2004. Ectodysplasin A1 promotes placodal cell fate during early morphogenesis of ectodermal appendages. *Development* **131:** 4907–4919.

Nguyen-Nielsen M, Skovbo S, Svaneby D, Pedersen L, Fryzek J. 2013. The prevalence of X-linked hypohidrotic ectodermal dysplasia (XLHED) in Denmark, 1995–2010. *Eur J Med Genet* **56:** 236–242.

Rough BJ. 2010. *Carrier: Untangling the danger in my DNA.* Counterpoint Press, Berkeley, CA.

Rough BJ. 2013. Genetic drift. Three phone calls: A carrier's journey to motherhood. *Am J Med Genet A* **161:** 2119–2121.

Schneider P, Street SL, Gaide O, Hertig S, Tardivel A, Tschopp J, Runkel L, Alevizopoulos K, Ferguson BM, Zonana J. 2001. Mutations leading to X-linked hypohidrotic ectodermal dysplasia affect three major functional domains in the tumor necrosis factor family member ectodysplasin-A. *J Biol Chem* **276:** 18819–18827.

Zonona J. 1993. Hypohidrotic (anhidrotic) ectodermal dysplasia: Molecular genetic research and its clinical applications. *Semin Derm Sep* **12:** 241–246.

Further Reading

Fine J, Hinter H, eds. 2009. *Life with epidermolysis bullosa (EB): Etiology, diagnosis, multidisciplinary care and therapy.* Springer, New York.

WWW Resources

www.dermnetnz.org/hair-nails-sweat/ectodermal-dysplasia.html DermNet NZ, Ectodermal dysplasia.

www.ectodermaldysplasia.org Ectodermal Dysplasia Society.

www.Edimerpharma.com See website for an extensive list of abstracts and posters concerning recent research on this disorder.

www.nfed.org National Foundation for Ectodermal Dysplasias.

Chapter 11: Mending Broken Proteins

References

Accurso FJ, Rowe SM, Clancy JP, Boyle MP, Dunitz JM, Durie PR, Sagel SD, Hornick DB, Konstan MW, Donaldson SH, et al. 2010. Effect of VX-770 in persons with cystic fibrosis and the G551D-CFTR mutation. *New Eng J Med* **363:** 1991–2003.

Anthony K, Feng L, Arechavala-Gomeza V, Guglieri M, Straub V, Bushby K, Cirak S, Morgan J, Muntoni F. 2012. Exon skipping quantification by quantitative reverse-transcription polymerase chain reaction in Duchenne muscular dystrophy patients treated with antisense-oligomer eteplirsen. *Hum Gene Ther Methods* **23:** 336–345.

Caldwell A, Grove DE, Houck SA, Cyr DM. 2011. Increased folding and channel activity of a rare cystic fibrosis mutant with CFTR modulators. *Am J Physiol Lung Cell Mol Physiol* **301:** L346–L352.

Chaudhuri TK, Paul S. 2006. Protein-misfolding diseases and chaperone-based therapeutic approaches. *FEBS J* **273:** 1331–1349.

Cirak S, Arechavala-Gomeza V, Guglieri M, Feng L, Torelli S, Anthony K, Abbs S, Garralda ME, Bourke J, Wells DJ, et al. 2011. Exon skipping and dystrophin restoration in patients with Duchenne muscular dystrophy after systemic phosphorodiamidate morpholino oligomer threatment: An open-label, phase 2, dose-escalation study. *Lancet* **378:** 595–605.

Clunes MT, Boucher RC. 2008. Front-runners for pharmacotherapeutic correction of the airway ion transport defect in cystic fibrosis. *Curr Opin Pharmacol* **8:** 292–299.

Desnick RJ, Ioannou YA, Eng CM. 2001. α-Galactosidase A deficiency: Fabry disease. In *The metabolic and molecular bases of inherited disease* (ed. Scriver CR, Sly WS), pp. 3733–3774. McGraw-Hill, New York.

Germain DE, Giugliani R, Hughes DA, Mehta A, Nicholls K, Barisoni L, Jennette CJ, Bragat A, Castelli J, Sitaraman S, et al. 2012. Safety and pharmacodynamic effects of a pharmacological chaperone on α-galactosidase A activity and globotriaosylceramide clearance in Fabry disease: Report from two phase 2 clinical studies. *Orphanet J Rare Dis* **7:** 91.

Giugliani R, Waldek S, Germain DP, Nicholls K, Bichet DG, Simosky JK, Bragat AC, Castelli JP, Benjamin ER, Boudes PF. 2013. A phase 2 study of migalastat hydrochloride in females with Fabry disease: Selection of population, safety and pharmacodynamic effects. *Mol Genet* **109:** 86–92.

Globe Newswire (September 20, 2013). GSK and Prosensa announce primary endpoint not met in Phase III study of drisapersen in patients with Duchenne muscular dystrophy. www.gsk-clinicalstudyregister.com.

Groopman J. 2009. "Annals of medicine: Open channels" *The New Yorker*, May 4, 2009, pp. 30–34.

Gupta S, Ries M, Kotsopoulos S, Schiffmann R. 2005. The relationship of vascular glycolipid storage to clinical manifestations of Fabry disease: A cross-sectional study of a large cohort of clinically affected heterozygous women. *Medicine* **84:** 261–268.

Kerem E. 2006. Mutation specific therapy in CF. *Paediatr Respir Rev* **7** (Suppl 1): S166–S169.

Lu QL, Yokota T, Takeda S, Garcia L, Muntoni F, Partridge T. 2011. The status of exon skipping as a therapeutic approach to Duchenne muscular dystrophy. *Mol Ther* **19:** 9–15.

Mendell JR, Rodino-Klapac LR, Sahenk Z, Roush K, Bird L, Lowes LP, Alfano L, Gomez AM, Lewis S, Kota J, et al. 2013. Eteplirsen for the treatment of Duchenne muscular dystrophy. *Ann Neurol* **74:** 637–647.

Molinski SV, Gonska T, Huan LJ, Baskin B, Janahi IA, Ray PN, Bear CE. 2014. Genetic, cell biological, and clinical interrogation of the *CFTR* mutation c.3700A>G (p.Ile1234Val) informs strategies for future medical intervention. *Genet Med* **16:** 625–632.

Pollack A. 2014. "For cystic fibrosis nonprofit, a windfall in hope and cash." *The New York Times*, Nov. 19, 2014, pA1.

Ramsey BW, Davies J, McElvaney NG, Tullis E, Bell SC, Dřevínek P, Griese M, McKone EF, Wainwright CE, Konstan MW, et al. 2011. A CFTR potentiator in patients with cystic fibrosis and the *G551D* mutation. *New Engl J Med* **365:** 1663–1672.

Rowe SM, Varga K, Rab A, Bebok Z, Byram K, Li Y, Sorscher EJ, Clancy JP. 2007. Restoration of W1282X CFTR activity by enhanced expression. *Am J Respir Cell Mol Biol* **37:** 347–356.

Sermet-Gaudelus A. 2013. Ivacaftor treatment in patients with cystic fibrosis and the *G551D-CFTR* mutation. *Eur Respir Rev* **22:** 66–71.

Shelley ED, Shelley WB, Kurczynski TW. 1995. Painful fingers, heat intolerance, and telangiectases of the ear: Easily ignored childhood signs of Fabry disease. *Pediatr Dermatol* **12:** 215–219.

Siva K, Covello G, Denti MA. 2014. Exon-skipping antisense oligonucleotides to correct missplicing in neurogenetic diseases. *Nucleic Acid Ther* **24:** 69–86.

Suzuki Y. 2013. Chaperone therapy update: Fabry disease, GM1-gangliosidosis and Gaucher disease. *Brain Dev* **35:** 515–523.

Welsh M. 2010. Targeting the basic defect in cystic fibrosis. *New Eng J Med* **363:** 2056–2058.

Young-Gqamana B, Brignol N, Chang HH, Khanna R, Soska R, Fuller M, Sitaraman SA, Germain DP, Giugliani R, Hughes DA, et al. 2013. Migalastat HCL reduces globotriaosylsphingosine (lyso-Gb3) in Fabry transgenic mice and in the plasma of Fabry patients. *PLoS ONE* **8:** e57631.

Further Reading

Emery A. 2008. *Muscular dystrophy: The facts*, 3rd ed. Oxford University Press, Oxford.

Makarow M, Braakman I, eds. 2010. *Chaperones*. Springer, New York.

Thomson AH, Harris A. 2008. *Cystic fibrosis,* 4th ed. Oxford University Press, Oxford.

WWW Resources

www.amicusrx.com Amicus Therapeutics.

www.cff.org Cystic Fibrosis Foundation.

www.mda.org Muscular Dystrophy Association.

www.musculardystrophy.uk.org/ Muscular Dystrophy Campaign, What is exon skipping and how does it work?

www.sarepta.com Sarepta Therapeutics.

www.vrtx.com Vertex Pharmaceuticals.

Chapter 12: What Is Next

Emerging Therapies

References

ACOG. 2013. Ob-Gyns recommend chromosomal microarray analysis for genetic evaluation of fetal anomalies. www.acog.org/About/News_Room_Releases/2013.

Ahmed SS, Li H, Cao C, Sikoglu EM, Denninger AR, Su Q, Eaton S, Liso Navarro AA, Xie J, Szucs S, et al. 2013. A single intravenous rAAV injection as late as P20 achieves efficacious and sustained CNS gene therapy in Canavan mice. *Mol Ther* **21:** 2136–2147.

Aiuti A, Cattaneo F, Galimberti S, Nenninghoff U, Cassani B, Callegaro L, Scaramuzza S, Andolfi G, Mirolo M, Brigida I, et al. 2009. Gene therapy for immunodeficiency due to adenosine deaminase deficiency. *N Engl J Med* **360:** 447–458.

Bainbridge JW, Mehat MS, Sandaram V, Robbie SJ, Barker SE, Ripamonti C, Georgiadis A, Mowat FM, Beattie SW, Gardner PJ, et al. 2015. Long-term effect of gene therapy on Leber's congenital amaurosis. *N Engl J Med* **372:** in press.

Baltimore D, Berg P, Botchan M, Carroll D, Charo RA, Church G, Corn JE, Daley GQ, Doudna JA, Fenner M, et al. 2015. A prudent path forward for genomic engineering and germline gene modification. *Science* **348:** 36–38.

Bamshad MJ, Ng SB, Bigham AW, Tabor HK, Emond MJ, Nickerson DA, Shendure J. 2011. Exome sequencing as a tool for Mendelian disease gene discovery. *Nat Rev Genet* **12:** 745–755.

Bianchi DW. 2012. From prenatal genomic diagnosis to fetal personalized medicine: Progress and challenges. *Nature Med* **18:** 1–11.

Bianchi DW, Parker RL, Wentworth J, Madankumar R, Saffer C, Das AF, Craig JA, Chudova DI, Devers PL, Jones KW, et al. 2014. DNA sequencing versus standard prenatal aneuploidy screening. *New Engl J Med* **370:** 799–808.

Bjornsson HT, Fallin MD, Feinberg AP. 2004. An integrated epigenetic and genetic approach to common human diseases. *Trends Genet* **20:** 350–358.

Bryant LM, Christopher DM, Giles AR, Hinderer C, Rodriguez JL, Smith JB, Traxler EA, Tycko J, Wojno AP, Wilson JM. 2013. Lessons learned from the clinical development and market authorization of Glybera. *Hum Gene Ther Clin Dev* **24:** 55–64.

Dalkara D, Byrne LC, Klimczak RR, Visel M, Yin L, Merigan WH, Flannery JG, Schaffer DV. 2013. In vivo–directed evolution of new adeno-associated virus for therapeutic outer retinal gene delivery from the vitreous. *Sci Transl Med* **5:** 189ra76.

Das I, Park J-M, Shin J, Jeon SK, Lorenzi H, Luder DJ, Worley P, Reeves RH. 2013. Hedgehog agonist therapy corrects structural and cognitive defects in a Down syndrome mouse model. *Sci Transl Med* **5:** 201ra120.

de Ligt J, Willemsen MH, van Bon BW, Kleefstra T, Yntema HG, Kroes T, Vulto-van Silfhout AT, Koolen DA, de Vries P, et al. 2012. Diagnostic exome sequencing in persons with severe intellectual disability. *N Engl J Med* **367:** 1921–1929.

Flannick J, Thorleifsson G, Beer NL, Jacobs SB, Grarup N, Burtt NP, Mahajan A, Fuchsberger C, Atzmon G, Benediktsson R, et al. 2014. Loss-of-function mutations in *SLC30A8* protect against type 2 diabetes. *Nat Genet* **45:** 357–363.

Fox IJ, Daley GQ, Goldman SA, Huard J, Kamp TJ, Trucco M. 2014. Use of differentiated pluripotent stem cells as replacement therapy for treating disease. *Science* **345:** 889.

Grati FR, Malvestiti F, Ferreira JC, Bajaj K, Gaetani E, Agrati C, Grimi B, Dulcetti F, Ruggeri AM, De Toffol S, et al. 2014. Fetoplacental mosaicism: Potential implications for false-positive and false-negative noninvasive prenatal screening results. *Genet Med* **16:** 620–624.

Gurdon JB. 1962. The developmental capacity of nuclei taken from intestinal epithelial cells of feeding tadpoles. *J Embryol Exp Morph* **10:** 622–640.

Haurwitz RE, Jinek M, Wiedenheft B, Zhou K, Doudna JA. 2010. Sequence- and structure-specific RNA processing in a CRISPR endonuclease. *Science* **329:** 1355–1358.

Haydar TF, Reeves RH. 2012. Trisomy 21 and early brain development. *Trends Neurosci* **35:** 81–91.

Huh D, Matthews BD, Mammoto A, Montoya-Zavala M, Hsin HY, Ingber DE. 2010. Reconstituting organ-level lung functions on a chip. *Science* **328:** 1662–1668.

Inoue H, Nagata N, Kurokawa H, Yamanaka S. 2014. iPS cells: A game changer for future medicine. *EMBO J* **33:** 409–417.

Jiang F, Ren J, Chen F, Zhou Y, Xie J, Dan S, Su Y, Xie J, Yin B, Su W, et al. 2012. Noninvasive Fetal Trisomy (NIFTY) test: An advanced noninvasive prenatal diagnosis methodology for fetal autosomal and sex chromosomal aneuploidies. *BMC Med Genomics* **5:** 57.

Jinik M, Chylinski K, Fonfara I, Hauer M, Doudna JA, Charpentier E. 2012. A programmable dual RNA-guided DNA endonuclease in adaptive bacterial immunity. *Science* **337:** 816–821.

Kay MA. 2011. State-of-the-art gene based therapies: The road ahead. *Nat Rev Genet* **12:** 316–328.

Kohn DB, Candotti F. 2009. Gene therapy fulfilling its promise. *N Engl J Med* **360:** 518–521.

Lawrence J. 2013. Interview: From Down's syndrome to basic epigenetics and back again. *Epigenomics* **5:** 611–614.

Lieber DS, Calvo SE, Shanahan K, Slate NG, Liu S, Hershman SG, Gold NB, Chapman BA, Thorburn DR, Berry GT, et al. 2013. Targeted exome sequencing of suspected mitochondrial disorders. *Neurology* **80:** 1762–1770.

Lu QL, Yokota T, Takeda S, Garcia L, Muntoni F, Partridge T. 2011. The status of exon skipping as a therapeutic approach to Duchenne muscular dystrophy. *Mol Ther* **19:** 9–15.

Lupski J, Reid JG, Gonzaga-Jauregui C, Rio Deiros D, Chen DC, Nazareth L, Bainbridge M, Dinh H, Jing C, Wheeler DA, et al. 2010. Whole-genome sequencing in a patient with Charcot–Marie–Tooth Neuropathy. *New Engl J Med* **310:** 1181–1191.

Morris J, Ableman E. 2009. Trends in Down's syndrome live births and antenatal diagnoses in England and Wales from 1989 to 2008: Analysis of data from the National Down Syndrome Cytogenetic Registry. *Br Med J* **339:** b3794.

Mouawia H, Saker A, Jais JP, Benachi A, Bussières L, Lacour B, Bonnefont JP, Frydman R, Simpson JL, Paterlini-Brechot P. 2012. Circulating trophoblastic cells provide genetic diagnosis in 63 fetuses at risk for cystic fibrosis or spinal muscular atrophy. *Reprod Biomed Online* **25:** 508–520.

Mozersky J, Mennuti M. 2013. Cell-free fetal DNA testing: Who is driving implementation? *Gen Med* **15:** 433–434.

Need AC, Shashi V, Hitomi Y, Schoch K, Shianna KV, McDonald MT, Meisler MH, Goldstein DB. 2012. Clinical application of exome sequencing in undiagnosed genetic conditions. *J Med Genet* **49:** 353–361.

Obokata H, Sasai Y, Niwa H, Kadota M, Andrabi M, Takata N, Tokoro M, Terashita Y, Yonemura S, Vacanti CA, et al. 2014. Bidirectional developmental potential in reprogrammed cells with acquired pluripotency. *Nature* **505:** 676–679.

Obokata H, Wakayama T, Sasai Y, Kojima K, Vacanti MP, Niwa H, Yamato M, Vacanti CA. 2014. Stimulus-triggered fate conversion of somatic cells into pluripotency. *Nature* **505:** 641–647.

Obokata H, Wakayama T, Sasai Y, Kojima K, Vacanti MP, Niwa H, Yamato M, Vacanti CA. 2014. Retraction: Stimulus-triggered fate conversion of somatic cells into pluripotency. *Nature* **511:** 112.

Saunders CJ, Miller NA, Soden SE, Dinwiddie DL, Noll A, Alnadi NA, Andraws N, Patterson ML, Krivohlavek LA, Fellis J, et al. 2012. Rapid whole-genome sequencing for genetic disease diagnosis in neonatal intensive care units. *Sci Trans Med* **4:** 1–14.

Sheridan C. 2011. Gene therapy finds its niche. *Nat Biotechnol* **29:** 121–128.

Smith A. 2014. Potency unchained. *Nature* **505:** 623–624.

Soldner F, Jaenisch R. 2012. iSPC disease modeling. *Science* **338:** 1155–1156.

Sparks AB, Wang ET, Struble CA, Barrett W, Stokowski R, McBride C, Zahn J, Lee K, Shen N, Doshi J, et al. 2012. Selective analysis of cell-free DNA in maternal blood for evaluation of fetal trisomy. *Prenatal Diag* **32:** 3–9.

Takahashi K, Yamanaka S. 2006. Induction of pluripotent stem cells from mouse embryonic and adult fibroblast clultures by defined factors. *Cell* **126:** 633–637.

Talkowski ME, Ordulu Z, Pilalamarri V, Benson CB, Blumenthal I, Connolly S, Hanscom C, Hussain N, Pereira S, Picker J, et al. 2012. Clinical diagnosis by whole-genome sequencing of a prenatal sample. *N Engl J Med* **367:** 2226–2232.

Tebas P, Stein D, Tang WW, Frank I, Wang SQ, Lee G, Spratt SK, Surosky RT, Giedlin MA, Nichol G, et al. 2014. Gene editing of *CCR5* in autologous CD4 T cells of persons infected with HIV. *New Engl J Med* **370:** 901–910.

Umbarger MA, Kennedy CJ, Saunders P, Breton B, Chennagiri N, Emhoff J, Greger V, Hallam S, Maganzini D, Micale C, et al. 2014. Next generation carrier screening. *Genet Med* **16:** 132–140.

Wade N. 2015. "Scientists seek ban on method of making gene-edited babies." *The New York Times*, March 20, 2015, p. A1.

Wapner RJ, Martin CL, Levy B, Ballif BC, Eng CM, Zachary JM, Savage M, Platt LD, Saltzman D, Grobman WA, et al. 2012. Chromosomal microarray versus karyotyping for prenatal diagnosis. *New Engl J Med* **367:** 2175–2184.

You Y, Sun Y, Li X, Li Y, Wei X, Chen F, Ge H, Lan Z, Zhu Q, Tang Y, et al. 2014. Integration of targeted sequencing and NIPT into clinical practice in a Chinese family with maple syrup urine disease. *Genet Med* **16:** 594–600.

Zhang Y, Sontheimer EJ. 2014. Cascading into focus. *Science* **345:** 1452–1453.

Further Reading

Hadjivasiliou A. 2014. *EvaluatePharma Orphan Drug Report 2014*. www.evaluate.com.

Mali P, Yang L, Esvelt KM, Aach J, Guell M, DiCarlo JE, Norville JE, Church GM. 2013. RNA-guided human genome engineering via Cas9. *Science* **339:** 823–826.

PhRMA. 2013. *Medicines in development: Rare diseases: A report on orphan drugs in the pipeline.* www.pharma.org.

WWW Resources

www.acog.org/About-ACOG/News-Room/News-Releases/2013/Ob-Gyns-Recommend-Chromosomal-Microarray-Analysis ACOG. 2013. Ob-Gyns recommend chromosomal microarray analysis for genetic evaluation of fetal anomalies.

www.23andme.com 23andMe.

www.bio.org Biotechnology Industry Organization.

www.canavanfoundation.org Canavan Foundation.

www.clinicaltrials.gov National Institutes of Health, Clinical trials.

www.ndscenter.org/ National Down Syndrome Congress.

www.sparktx.com Spark Therapeutics.

www.ultragenyx.com Ultragenyx Pharmaceutical.

www.voyagertherapeutics.com Voyager Therapeutics.

Chapter 13: We Are All Orphans
Lessons for Common Diseases

References

Aschard H, Vilhjálmsson BJ, Greliche N, Morange PE, Trégouët DA, Kraft P. 2014. Maximizing the power of principal-component analysis of correlated phenotypes in genome-wide association studies. *Am J Hum Genet* **94:** 662–676.

Brennan TA, Wilson JM. 2014. The special case of gene therapy pricing. *Nat Biotechnol* **32:** 874–876.

Brice A. 2005. Genetics of Parkinson's disease: LRRK2 on the rise. *Brain* **128:** 2760–2762.

Carlsson A, Lindqvist M, Magnusson T. 1957. 3-4-Dihydroxyphenylalanine and 5-hydroxytryptophan as reserpine antagonists. *Nature* **180:** 1200.

Celestino-Soper PB, Shaw CA, Sanders SJ, Li J, Murtha MT, Ercan-Sencicek AG, Davis L, Thomson S, Gambin T, Chinault AC, et al. 2011. Use of array CGH to detect exonic copy number variants throughout the genome in autism families detects a novel deletion in TMLHE. *Hum Mol Genet* **20:** 4360–4370.

Cotzias G, Papavasiliou PS, Gellene R. 1968. Experimental therapy of parkinsonism with L-Dopa. *Neurology* **18:** 276–277.

De Rubeis S, He X, Goldberg AP, Poultney CS, Samocha K, Cicek AE, Kou Y, Liu L, Fromer M, Walker S, et al. 2014. Synaptic, transcriptional and chromatin genes disrupted in autism. *Nature* **515:** 209–215.

Druker BJ. 2002. Perspectives on the development of a molecularly targeted agent. *Cancer Cell* **1:** 31–36.

Fernandez HH, Vanagunas A, Odin P, Espay AJ, Hauser RA, Standaert DG, Chatamra K, Banesh J, Pritchett Y, Hass SL, Lenz RA. 2013. Levodopa–carbidopa intestinal gel in advanced Parkinson's disease open-label study: Interim results. *Parkinsonism Relat Disord* **19:** 339–345.

Hauser PS, Ryan RO. 2013. Impact of apolipoprotein E on Alzheimer's disease. *Curr Alzheimer Res* **10:** 809–817.

High K, Skinner MW. 2011. Cell phones and landlines: The impact of gene therapy on the host and availability of treatment for hemophilia. *Mol Ther* **19:** 1749–1750.

Iossifov I, O'Roak BJ, Sanders SJ, Ronemus M, Krumm N, Levy D, Stessman HA, Witherspoon KT, Vives L, Patterson KE, et al. 2014. The contribution of de novo coding mutations to autism spectrum disorders. *Nature* **515:** 216–221.

Kainthla R, Kim KB, Falchook GS. 2013. Dabrafenib for treatment of BRAF-mutant melanoma. *Pharmacogenomics Pers Med* **7:** 21–29.

Kanner L. 1943. Autistic disturbances of affective contact. *Nerv Child* **2:** 217–250.

Letter to the Editor. 2014. Michael J Fox Foundation *LRRK2* Consortium: Geographical differences in returning genetic research to study participants. *Genet Med* **16:** 644–645.

Lubs H, Chiurazzi P, Arena J, Schwartz C, Tranebjaerg L, Neri G. 1999. XLMR genes: Update 1998. *Am J Med Genet* **83:** 237–247.

Mefford HC, Batshaw ML, Hoffman EP. 2012. Genomics, intellectual disability, and autism. *N Engl J Med* **366:** 733–743.

Might M, Wilsey M. 2014. The shifting mode in clinical diagnosis: How next-generation sequencing and families are altering the way rare diseases are discovered, studied, and treated. *Genet Med* **16:** 736–737.

Miles JH. 2011. Autism spectrum disorders—A genetics review. *Genet Med* **13:** 278–294.

Mnookin S. 2014. "One of a kind." *The New Yorker,* July 21, 2014.

PhRMA. 2013. *Rare diseases: A report on orphan drugs in the pipeline.* www.phrma.org.

Pollack A. 2015. "Sales of new hepatitis C drug soar to $10.3 billion, straining budgets." *The New York Times,* February 4, 2015, p. B2.

Rowley J. 2013. Genetics: A story of swapped ends. *Science* **340:** 1412–1413.

Saint Pierre A, Genin E. 2014. How important are rare variants in common disease? *Brief Funct Genomics* **13:** 353–361.

Sánchez Longo LP, Cruz de León C, Rodríguez del Valle J, Hernández Ortiz T. 1971. The history of the discovery of L-dopa as a treatment in parkinsonism. *Bol Asoc Med P R* **63:** 36–40.

Sekiyama K, Takamatsu Y, Waragai M, Hashimoto M. 2014. Role of genomics in translational research for Parkinson's disease. *Biochem Biophys Res Commun* **452:** 226–235.

Tawbi H. 2014. Selective BRAF inhibitors make inroads in mutated metastatic melanoma. *J Commun Support Oncol* **1292:** 46–47.

Tüysüz B, Bayrakli F, DiLuna ML, Bilguvar K, Bayri Y, Yalcinkaya C, Bursali A, Ozdamar E, Korkmaz B, Mason CE, et al. 2008. Novel NTRK1 mutations cause hereditary sensory and autonomic neuropathy type IV: Demonstration of a founder mutation in the Turkish population. *Neurogenetics* **9:** 119–125.

Van der Sijde MR, Ng A, Fu J. 2014. Systems genetics: From GWAS to disease pathways. *Biochim Biophys Acta* **1842:** 1903–1909.

Volkmar FR, Cohen DJ. 1988. Neurobiologic aspects of autism. *New Engl J Med* **318:** 1390–1392.

Wellman-Labadie O, Zhou Y. 2010. The US Orphan Drug Act: Rare disease research stimulator or commercial opportunity? *Health Policy* **95:** 216–228.

Wood J, Sames L, Moore A, Ekins S. 2013. Multifaceted roles of ultra-rare and rare disease patients/parents in drug discovery. *Drug Discov Today* **18:** 1043–1052.

Further Reading

Anand G. 2009. *The cure: How a father raised $100 million—and bucked the medical establishment—in a quest to save his children.* Harper Collins, New York.

Coleman M, ed. 2005. *The neurology of autism.* Oxford University Press, Oxford.

Eurordis (Rare Diseases Europe). 2009. *The voice of 12,000 patients: Experiences and expectations of rare disease patients on diagnosis and care in Europe.* Eurordis, Paris.

Goetz T. 2010. "Sergey Brin's search for a Parkinson's cure." *Wired,* June 22, 2010. www.wired.com.

Graboys TB. 2008. *Life in the balance: A physician's memoir of life, love, and loss with Parkinson's disease and dementia.* Union Square Press, New York.

Hollander E, Anagnostou E, eds. 2007. *Clinical manual for the treatment of autism.* American Psychiatric Association, Arlington, VA.

Silverman C. 2012. *Understanding autism: Parents, doctors, and the history of a disorder.* Princeton University Press, Princeton, NJ.

WWW Resources

https://clinicaltrials.gov This site lists more than 170,000 studies in an easily searchable database.

www.accessdata.fda.gov/scripts/opdlisting/oopd/index.cfm FDA, Search Orphan Drug Designations and Approvals.

www.autism-society.org Autism Society of America.

www.autismspeaks.org Autism Speaks.

www.forbes.com *Forbes*, The World's Most Expensive Drugs.

www.gracewilsey.org Grace Wilsey Foundation.

www.michaeljfox.org The Michael J. Fox Foundation for Parkinson's Research.

www.parkinson.org National Parkinson Foundation.

www.pdf.org Parkinson Disease Foundation.

www.pharmac.health.nz/assets/high-cost-medicines-discussion-document-2014-04.pdf Pharmac, The High Cost of Medicines for Rare Diseases.

www.rsrt.org Rett Syndrome Research Trust.

www.sfari.org Simons Foundation Autism Research Initiative.

www.SimonsVIPConnect.org Simons Variation in Individuals Project—a platform to partner families and researchers to help understand autism and developmental delay.

Index

C

G

H

I

N

Q

R

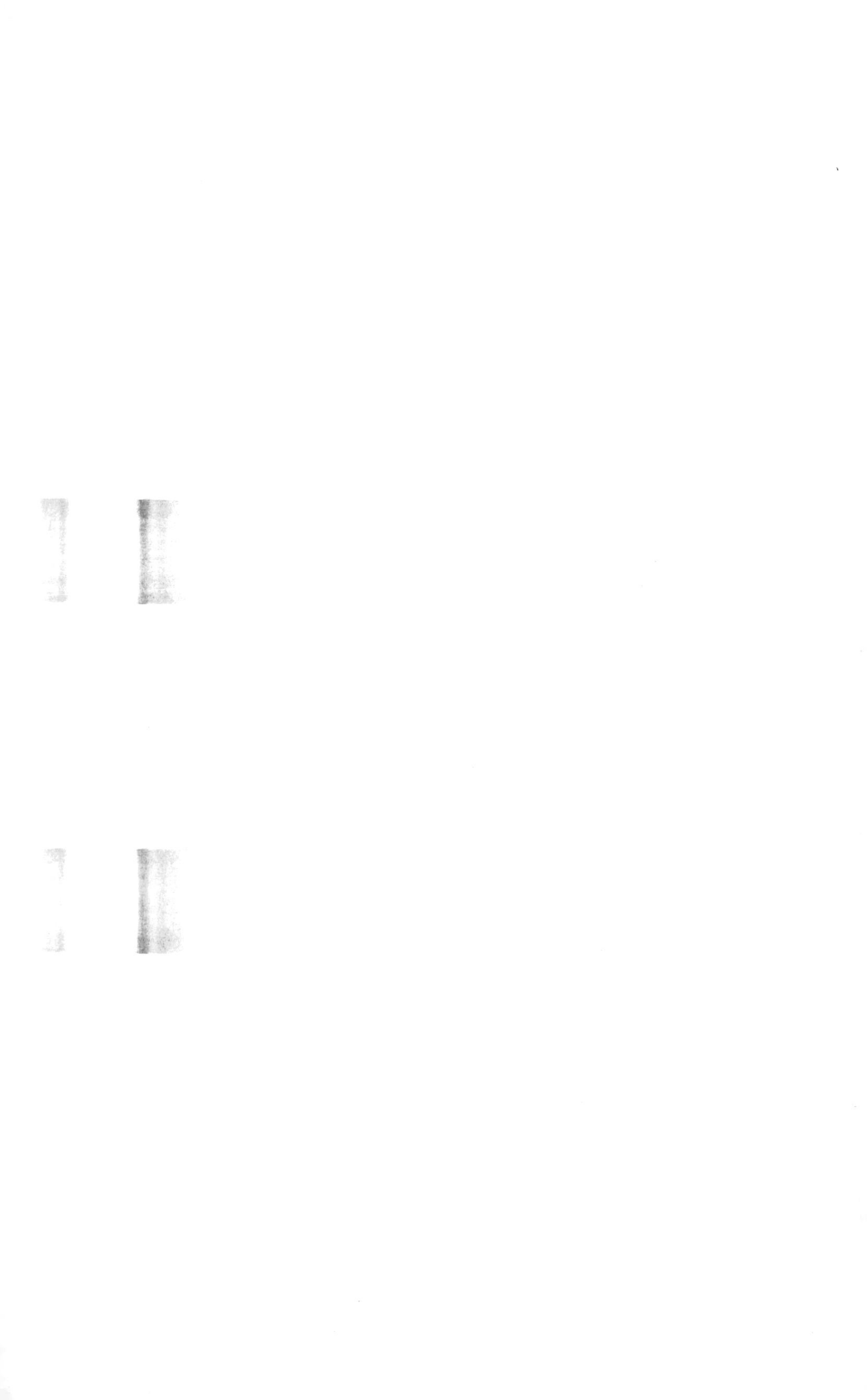